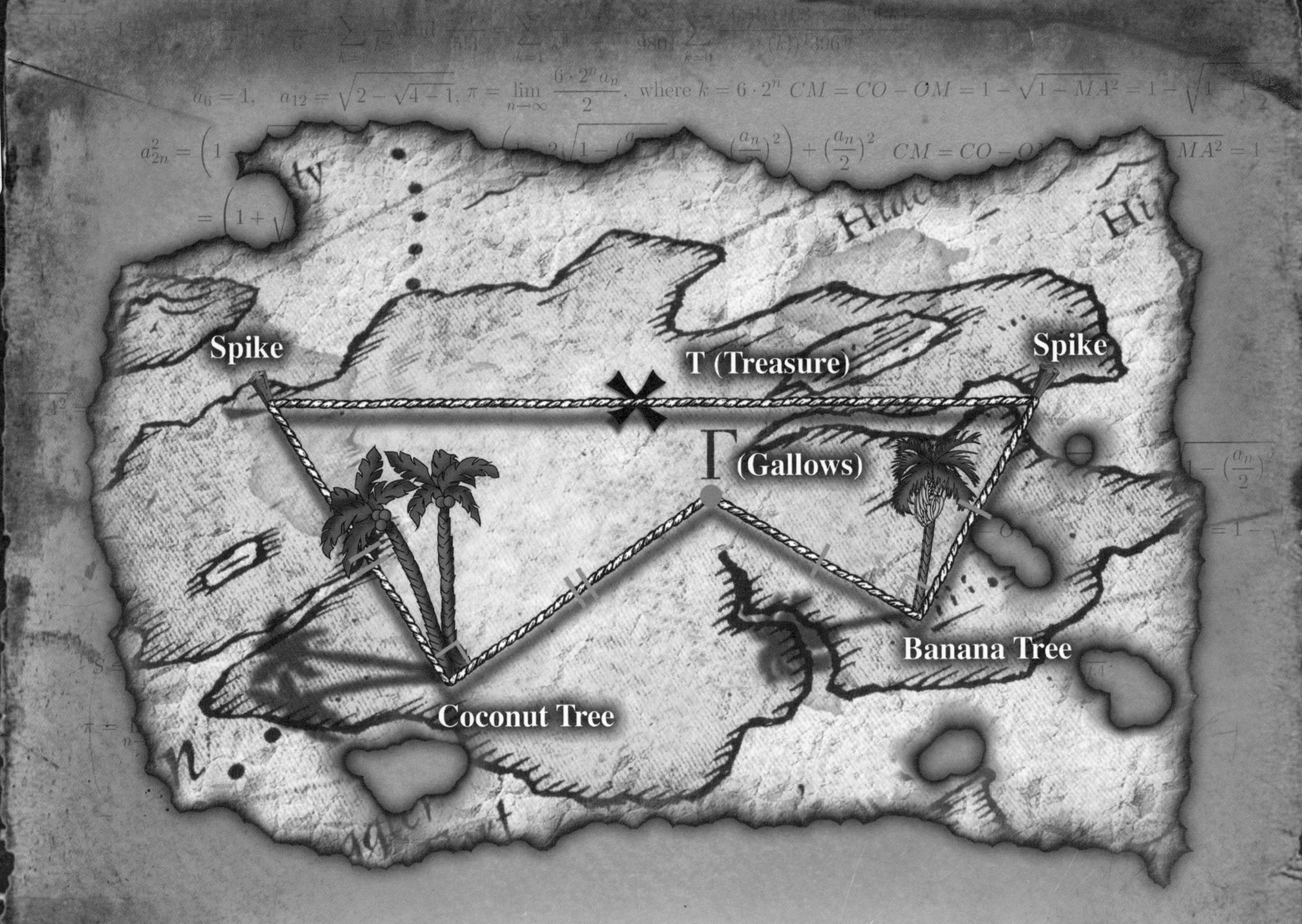
Spike
T (Treasure)
Spike
Γ (Gallows)
Banana Tree
Coconut Tree

EUCLIDEAN AND TRANSFORMATIONAL GEOMETRY

A Deductive Inquiry

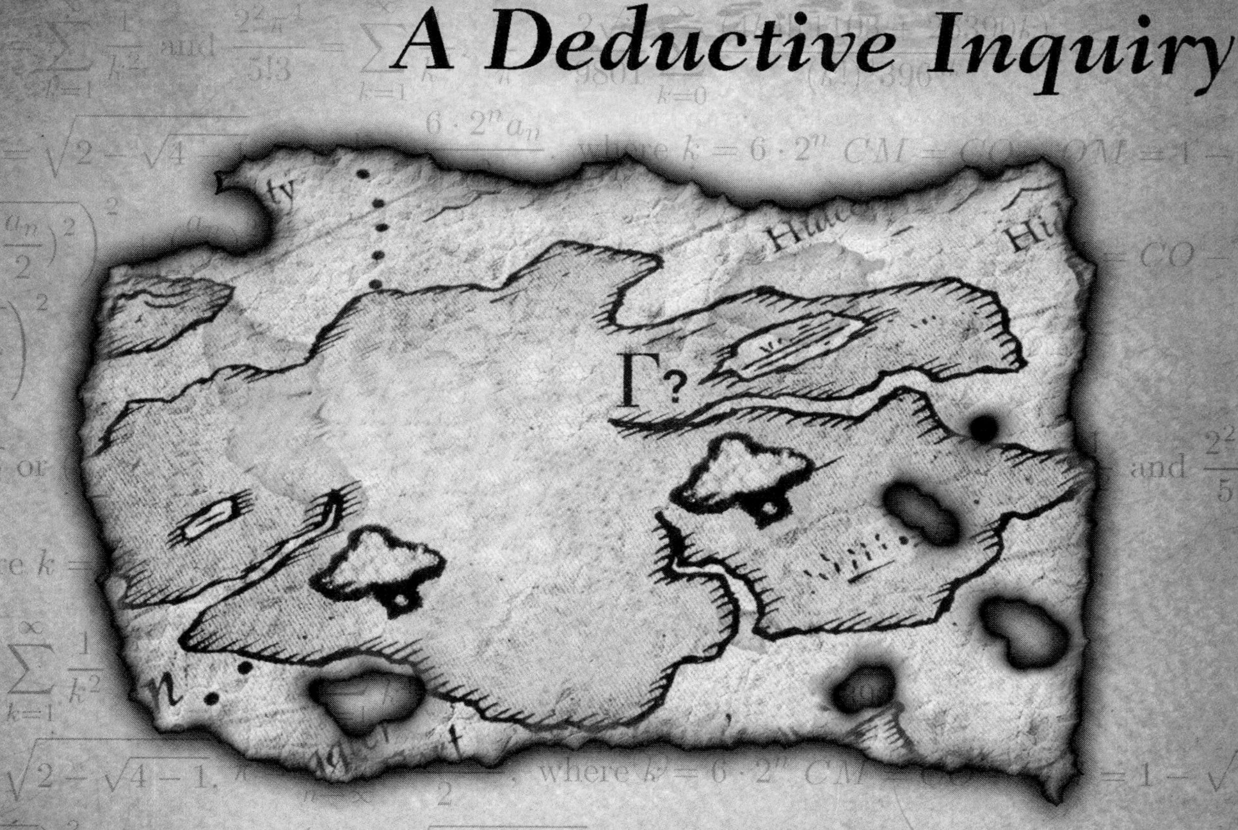

Shlomo Libeskind

University of Oregon

JONES AND BARTLETT PUBLISHERS

Sudbury, Massachusetts

BOSTON TORONTO LONDON SINGAPORE

World Headquarters
Jones and Bartlett Publishers
40 Tall Pine Drive
Sudbury, MA 01776
978-443-5000
info@jbpub.com
www.jbpub.com

Jones and Bartlett Publishers Canada
6339 Ormindale Way
Mississauga, Ontario L5V 1J2
CANADA

Jones and Bartlett Publishers International
Barb House, Barb Mews
London W6 7PA
UK

Jones and Bartlett's books and products are available through most bookstores and online booksellers. To contact Jones and Bartlett Publishers directly, call 800-832-0034, fax 978-443-8000, or visit our website, www.jbpub.com.

Substantial discounts on bulk quantities of Jones and Bartlett's publications are available to corporations, professional associations, and other qualified organizations. For details and specific discount information, contact the special sales department at Jones and Bartlett via the above contact information or send an email to specialsales@jbpub.com.

Production Credits

Chief Executive Officer: Clayton Jones
Chief Operating Officer: Don W. Jones, Jr.
President, Higher Education and Professional Publishing: Robert W. Holland, Jr.
V.P., Sales and Marketing: William J. Kane
V.P., Design and Production: Anne Spencer
Acquisitions Editor: Timothy Anderson
Production Director: Amy Rose
Production Assistant: Sarah Bayle
Marketing Manager: Andrea DeFronzo
Manufacturing Buyer: Therese Connell
Editorial Assistant: Laura Pagluica
Editorial Assistant: Melissa Elmore
Composition: Northeast Compositors, Inc.
Text Design: Anne Spencer
Cover Design: Kristin E. Ohlin
Cover Image: Leather Background: © Najin/ShutterStock, Inc., Extra Texture: © Stephen Mulcahey/ShutterStock, Inc., Map: © Scott Rothstein/ShutterStock, Inc.
Printing and Binding: Courier Kendallville
Cover Printing: Courier Kendallville

Library of Congress Cataloging-in-Publication Data

Libeskind, Shlomo.
Euclidean and transformational geometry / Shlomo Libeskind. — 1st ed.
p. cm.
ISBN-13: 978-0-7637-4366-6
ISBN-10: 0-7637-4366-6
1. Geometry. 2. Transformations (Mathematics) I. Title.
QA453.L53 2008
516—dc22

2007031433

6048

Printed in the United States of America
11 10 09 08 07 10 9 8 7 6 5 4 3 2 1

praxis password rlszpor

Rtrrrn11

Dedication

אברם ליבסקינד ז"ל

To my uncle Janek Zarzycki, of blessed memory, whose resourcefulness, bravery, intelligence, and vigilance kept me and others in my family safe during the Holocaust.

~~Read 1.5? 4.2–4.3~~

~~thrm 3.6 3.7~~

~~4.6~~

~~Problem set 4.1 2,3,5~~

~~wed 250 words ; 2 diagrams~~

~~Golden rectangles / ratio~~

- 40% multiple choice questions
- 1 or more construction (explain the construction)
- Use theorem list for proofs
- proofs + [illegible]
- numerical problems

- go over all the homework
Exams + quizes

p. 83 cor 2.2 Exam 2

thrm 2.4
2.6

p. 99 #3
102 #14
thrm 3.3

p. 127 (129 #) 11, 17, 19

pythagorean proof - any proof

- construct $\sqrt{3}$ $\sqrt{5}$ $\sqrt{7}$

- prove distance formula

- thrm 3.6, 3.7, 4.3 ← top of p. 143

- prove tangent / sec

thrm 4.6

4.1 #2, 3, 5

geometric mean def p. 182

construct G. M. of 2 + 5 (p. 182)

thrm 4.13

construct golden rectangle (p. 200)

- octagon radius 1 → approximate π

show is 1-to-1 functions

Def. of isometry p. 240

Contents

Preface

My goal with this text is to kindle in college geometry students a passion for problem solving and a love for mathematics. The book evolved out of my notes for a two-term sequence of courses in college geometry I have taught for many years. Because the sequence is required for prospective high school mathematics teachers, the classes are usually heavily enrolled with these students, but I have found that other math majors also derive great benefit from taking the courses. In the courses and in the text, the strategies for approaching proofs and solving problems guide students toward successfully solving unfamiliar problems and toward doing proofs on their own. Some of the questions students will ask themselves and often find answers to in the text include:

- How does one know where to begin and how to proceed?
- Which approach is more promising, and why?
- Are different solutions possible, and how do they compare?

Many of my students, at first frustrated with the challenge of non-routine, proof-oriented problems, in the end become passionate problem solvers. They especially appreciate realizing that proofs and constructions do not come "out of the blue," and that the thinking processes leading to them are explored in the teaching and learning from this text. Learning these problem-solving strategies and ways of thinking about math can set students up to transfer such a deductive reasoning approach to other areas of their learning and eventually their teaching.

In order to improve the teaching of mathematics—and particularly the teaching of geometry—in high schools, we must graduate teachers who know the subject on a level higher than that which they will teach, who enjoy and are successful in solving non-routine problems, and who love the subject. I trust that teaching from this text will help set us on the path to accomplishing these goals.

Experience at most universities shows that the majority of students remember very little geometry from high school. Similarly, most do not have experience with constructing proofs or with solving more challenging problems. Because of these gaps in student experience, the text does not assume any previous knowledge of geometry. Topics covered in high school geometry are included but are covered on a higher level and in greater depth, with many related challenging problems.

Transformational geometry is often included in the high school curriculum, and therefore a thorough introduction to the subject is of great importance to future teachers. Moreover, transformational geometry is useful in solving otherwise difficult-to-tackle problems in Euclidean geometry, in understanding functions and their graphs, and as a fascinating introduction to more advanced mathematical topics such as transformation groups and abstract algebra. For these reasons, this text focuses both on Euclidean and transformational geometry. To get to more interesting and less intuitively obvious geometric results early on, the text presents the rudiments of an axiomatic approach to Euclidean geometry in the appendix. Some instructors may want to start with the appendix; others may comfortably start with Chapter 0 and urge students to consult the appendix when needed. However, a significant part of Chapter 1 is dedicated to neutral geometry and to the role of the parallel postulate.

The text introduces many challenging problems that over the years have fascinated students. For example, in Chapter 0 and on the inside front cover, I introduce the Treasure Island Problem, which has captivated students—including myself—at all levels, from high school through graduate school. When I was sixteen and in high school, I read Gamow's book *One Two Three ... Infinity: Facts and Speculations of Science*, and was fascinated by the problem. I found Gamow's solution strange because it involved complex numbers, so I solved it using an elementary geometrical approach. Then, as an undergraduate, I learned about complex numbers and liked the solution via that approach. Much later, when I taught transformational geometry, I solved the problem using transformations.

In the text, this problem is first investigated experimentally, then deductively through several approaches in different chapters, and finally through two transformational approaches, the last of which is the "best" because it shows how the problem could have been discovered and explores how to generalize it and create similar problems of this kind.

In teaching from my notes for this text, I meet with students for one of the four weekly classes in a computer laboratory. There I work with them on explorations using Geometer's Sketchpad (GSP), a wonderful tool with which to explore geometry. While the entire book can be taught without GSP, there will be references to GSP next to problems where use of this software would be particularly useful for constructions. The website described later in this introduction will contain ideas on using GSP with the text.

Additional unique features of this text:

1. **In-Depth Discussion of Constructions** Nearly every chapter includes constructions that will captivate and inspire students at all levels. Most constructions are discussed in three stages:
   - **a.** Investigation—discovery of how to construct the required figures is explored so that students can do new constructions on their own.
   - **b.** Description of the construction steps, and the actual construction.
   - **c.** Proof of the construction.
2. **Trigonometry** Optional Section 4.6 reviews trigonometry and explores its use in proving geometrical theorems, in particular the amazing "Morley's Miracle." This section explores a variety of proofs and derivations of trigonometric formulas and works at convincing students that they can derive all of trigonometry on their own rather than memorizing the various trigonometric identities.
3. **Introduction to Transformation Groups** Transformation groups are explored in Problem Set 6.4 and in Problem Set 7.2 using complex numbers representation. In fact, in the latter, all isometries and size transformations are described via complex functions.
4. **Thorough Introduction to Complex Numbers** Chapter 7 introduces more recent results in Euclidean geometry such as the celebrated Nine-Point Circle Theorem. Section 7.2 gives a thorough introduction to complex numbers and especially applications of complex numbers to proving geometrical results. Through a property of multiplication of complex numbers, the trigonometric additional formulas are derived again, showing students the interconnection of mathematical topics.

Additional Pedagogy

In addition to extensive problem sets in "Now Solve This," the text asks students to tackle questions that are directly related to the material just covered. These occur frequently so that instructors can assign them at the end of lecture even if they haven't reached the end of a section. This approach ensures that the reader can tackle related problems without having to wait until the end of the section.

Notation

Notation in mathematics is not always standardized. At first—as is common in high school geometry—we distinguish between an angle and its measure. Later, to simplify notation and for easy reference, we use Greek letters for both the angles and their measure; it is always clear from the context which is which. We distinguish between segment AB, ray AB, and line AB by either writing the words "segment," "ray," or "line" preceding AB or by using the notation $\overline{AB}$ for segment AB, AB for the length of AB, $\overrightarrow{AB}$ for the ray AB, or $\overleftrightarrow{AB}$ for the line AB.

Icons

Icons placed in the margins of the text signal readers about key elements of the text:

- The Thinking Person icon indicates that the reader should pause and think before reading further.
- The Construction icon indicates that the problem involves a construction.
- The square icon □ indicates the end of a proof.

Theorems, Corollaries, and Axioms

Theorems, corollaries, and axioms are highlighted and numbered for easy reference, allowing readers to quickly find them. In addition, many theorems and corollaries are followed by restatements, which are intended to reinforce their meaning.

Definitions and Properties

Key terms appear in bold. Lists of critical properties appear as boxed features, emphasizing their importance and enabling readers to identify them with ease.

Examples and Problems

The text includes numerous solved examples that illustrate the concepts and theorems. Important questions are labeled as problems; they are answered in the text and often when solved are stated as theorems.

Historical Notes

Mathematics has a long and rich history that can serve to inspire students. The historical notes bring to life some of the fascinating characters that populate this history—from Pythagoras to U.S. President James Garfield, from Jacob Steiner to Emperor Napoleon I.

End-of-Section Problem Sets

At the end of most sections in the book are problem sets containing problems at various levels of difficulty. A green dot indicates that a hint or answer to the problem can be found at the end of the book. Starred problems are particularly challenging.

Student and Instructor Website

Along with the text, a special website for students and for instructors is available: http://www.jbpub.com/catalog/9780763743666/. It will include additional problems with solutions and additional topics, and will provide the following:

- Solutions to most of the problems in the text.
- Additional topics such as suggestions for using Geometer's Sketchpad (GSP), 3–D geometry, inversion and introduction to non-Euclidean geometry. (These additional topics will also be available on the website.)

Acknowledgments

First and foremost I am grateful to my wife Debbie for her continuous support and marvelous editing.

I am deeply indebted to the wonderful team at Jones and Bartlett Publishers. In particular I want to thank Tim Anderson, Amy Rose, and Sarah Bayle. I owe Sarah a special debt of gratitude for her dedication to perfecting the manuscript. Special thanks also to Greg Urbaniak for refining the art.

In addition, I truly appreciate the constructive suggestions offered by these reviewers:

James W. Wilson, University of Georgia
Greisy Winicki-Landman, California State Polytechnic University, Pomona
Jeanette Palmiter, Portland State University

Finally, I would like to thank the following people for their invaluable help:

Marius Andronie
Teal Guidici
Dragos Neacsu
Tammy Nezol
Michelle Sabato
Kathleen Wilkinson

Prologue 0

It may well be doubted whether, in all the range of science, there is any field so fascinating to the explorer—so rich in hidden treasures—so fruitful in delightful surprises—as Pure Mathematics.

—Lewis Carroll (1832–1898), best known as author of *Alice's Adventures in Wonderland* and *Through The Looking Glass*

Introduction: Surprising Results

Frequently a course in high school geometry starts with a long list of definitions and proofs of theorems that are intuitively obvious. It rarely includes theorems and proofs that would make a lasting impression on the student.

Pause for a moment and recall some geometrical theorems. Are there any on your list that contain an unexpected or surprising result? Which are especially beautiful? Are any of them especially important and, if so, why?

This short section introduces some problems and theorems from Euclidean geometry whose solutions or statements you will very likely find surprising. In several cases, you will be asked to conjecture a solution or a theorem through a sequence of suggested experiments. Later in this text, you will learn how to prove some of these conjectures and statements. Several of the proofs are particularly beautiful owing to their unexpected simplicity and applicability to new problems.

For the work you will be asked to do in this introductory chapter (and in other chapters), you will need a compass, a ruler, and some blank sheets of paper. A geometry utility software such as GSP (Geometer's Sketchpad) is especially helpful in investigating geometrical properties and making conjectures. Throughout the text we will, when appropriate, suggest optional activities using the software. It should be noted, however, that the text is independent of GSP and can be read and studied without the software.

0.1 The Treasure Island Problem

Among his great-grandfather's papers, Marco found a parchment describing the location of a pirate treasure buried on a deserted island. The island contained a coconut tree, a banana tree, and a gallows (Γ) where traitors were hung. A reproduction of the map appears in Figure 0.1. It was accompanied by the following directions:

> *Walk from the gallows to the coconut tree, counting the number of steps. At the coconut tree, turn 90° to the right. Walk the same distance and put a spike in the ground. Return to the gallows and walk to the banana tree, counting your steps. At the banana tree, turn 90° to the left, walk the same number of steps, and put another spike in the ground. The treasure is halfway between the spikes.*

Marco found the island and the two trees but could find no trace of the gallows or the spikes, as both had probably rotted. In desperation, he began to dig at random but soon gave up because the island was too large. Your quest is to devise a plan to find the exact location of the treasure.

If you have spent enough time pondering a solution but could not find one, try the following: On a piece of paper, fix the positions of the two trees. Choose an arbitrary position for the gallows and mark it Γ_1. Follow the directions to find the corresponding spikes and the midpoint between them, T_1. Next choose another position for the gallows Γ_2 and follow the directions to find the corresponding location of the treasure, T_2. Repeat the procedure for at least two more positions of the gallows (GSP is especially convenient here). What do you notice about T_1, T_2, T_3, and T_4? Now try to conjecture how to locate the treasure.

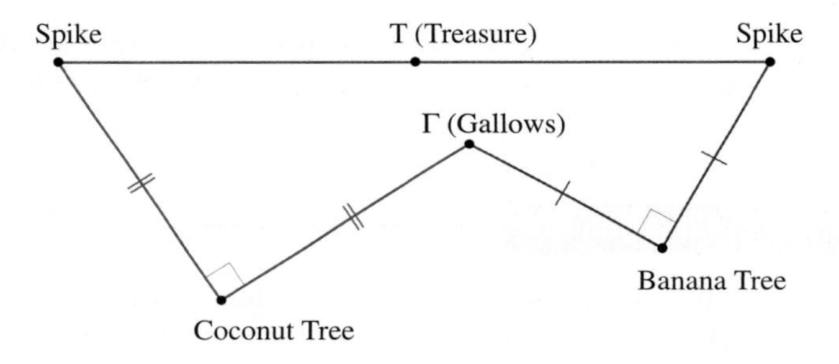

Figure 0.1

0.2 The Nine-Point Circle

The nineteenth century experienced a renewed interest in classical Euclidean geometry. Probably the most spectacular discovery was the **Nine-Point Circle**, which was investigated simultaneously by the French mathematicians Charles Jules Brianchion (1785–1864) and Jean-Victor Poncelet (1788–1867), who published their work jointly in 1821. The theorem is, however, commonly attributed to the German mathematician and high school teacher Karl Wilhem Feurbach (1800–1834), who independently discovered the theorem and published it with some related results in 1822.

With any triangle ABC, nine particular points can be associated with it as shown in Figure 0.2. The first three points—M_1, M_2, and M_3—are the midpoints of the three sides of the triangle. The next three points—N_1, N_2, and N_3—are the midpoints of the segments joining the vertices A, B, and C with the point H; H is the point of intersection of the three altitudes of the triangle (we will prove in Chapter 1 that the three altitudes of any triangle intersect in a single point). The final three points—F_1, F_2, and F_3—are the points of intersection of each altitude with each corresponding side (these points are known as the "feet" of the altitudes).

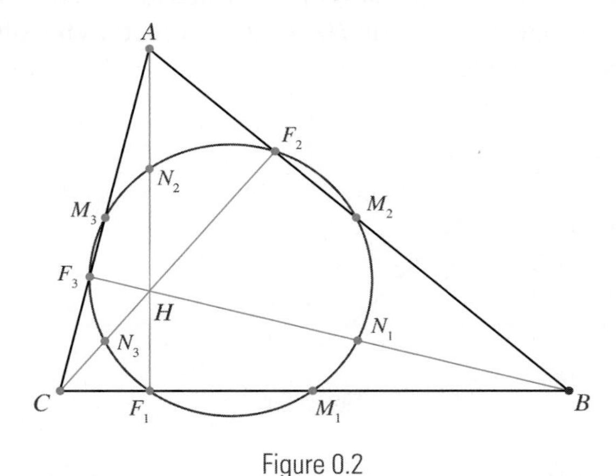

Figure 0.2

The theorem states that all nine points lie on one circle called the Nine-Point Circle. The Nine-Point Circle Theorem and some related results will be proved in Chapter 7.

0.3 Morley's Theorem

In 1899, Frank Morley, professor of mathematics at Haverford College and later at Johns Hopkins University, discovered an unusual property of the trisectors of the three angles of any triangle: If the angle trisectors are drawn for each angle of any triangle, then the adjacent trisectors of the angles meet at vertices of an equilateral triangle. In Figure 0.3, the adjacent trisectors of angles A and B meet at D, the adjacent trisectors of angles A and C meet at E, and the adjacent trisectors of angles C and B meet at F. Morley's Theorem states that triangle FDE is equilateral (i.e., its three sides are all the same length). Since 1899, many different proofs of this theorem have been published, including several since 2000. A proof of Morley's Theorem will be presented in Chapter 4, and additional proofs in Chapter 7.

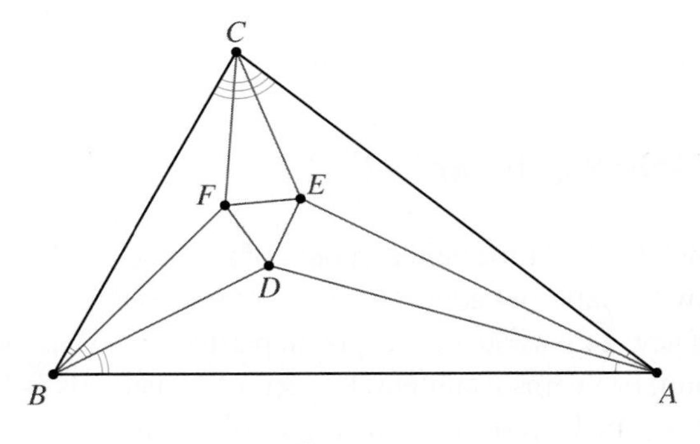

Figure 0.3

0.4 The Hiker's Path

A hiker H in Figure 0.4 needs to get first to the river r and then to her tent T. Find the point X on the bank of the river so that the hiker's total trip $HX + XT$ is as short as possible. (This problem will be investigated in Chapters 1 and 5.)

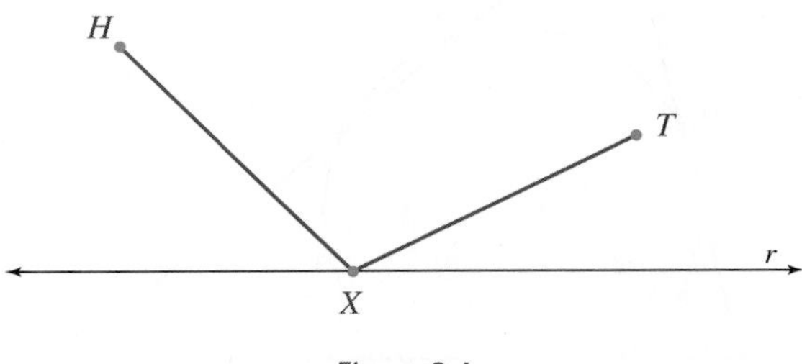

Figure 0.4

0.5 The Shortest Highway

A highway connecting two cities A and B as in Figure 0.5 needs to be built so that part of the highway is on a bridge perpendicular to the parallel banks b_1 and b_2 of a river. Where should the bridge be built so that the path $AXYB$ is as short as possible? (This problem will be investigated in Chapter 5.)

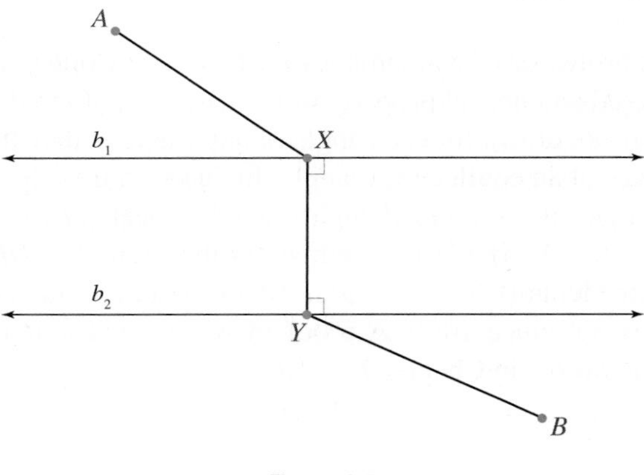

Figure 0.5

0.6 Steiner's Minimum Distance Problem

One of the greatest geometers of all time, and certainly of the nineteenth century, was Jacob Steiner (1796–1863). Born in Switzerland but educated in Germany, Steiner discovered and proved new theorems and introduced new geometrical concepts. In particular, he was interested in the solutions of maximum and minimum problems using purely geometric methods—that is, without using calculus or algebra. Among others, he proved the theorem illustrated in Figure 0.6.

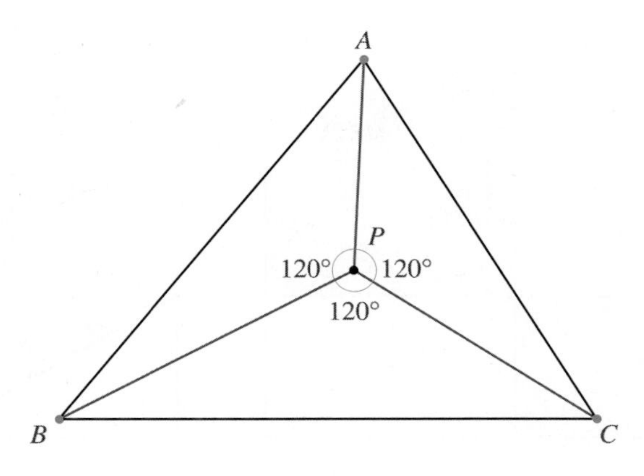

Figure 0.6

If A, B, and C are three points forming a triangle such that each of the angles of $\triangle ABC$ is less than 120°, then the point for which the sum of the distances from P to the vertices of the triangle (that is, $PA + PB + PC$) is at its minimum has the property that each of the angles at P is 120°. Steiner also dealt with the case when one of the angles is 120° or greater than 120°. In addition, he generalized the problem for n points.

Many maximum and minimum problems can be solved more efficiently using purely geometrical methods rather than calculus. Such problems will be investigated using transformational geometry in Chapter 5.

0.7 The Pythagorean Theorem

One of the best-known and most useful theorems in geometry and perhaps all of mathematics is the Pythagorean Theorem, which was discovered in the sixth century B.C.E. (Before the Common Era). There are numerous known proofs of the theorem (*The Pythagorean Proposition* by E. Loomis contains hundreds of them). Do you recall any proof of the Pythagorean Theorem? Can you prove it on your own?

The following is one way to state the theorem: If squares are constructed on the sides of any right triangle (a triangle with a 90° angle), then the area of the largest square equals the sum of the area of the other squares. In Figure 0.7, if the areas of the squares are A, B, and C, then $A + B = C$ or, equivalently, $a^2 + b^2 = c^2$, where a and b are the lengths of the legs of the triangle and c is the length of the hypotenuse (the side opposite the right angle).

Figure 0.7 and Figure 0.8 can be used to justify the Pythagorean Theorem. Can you see how?

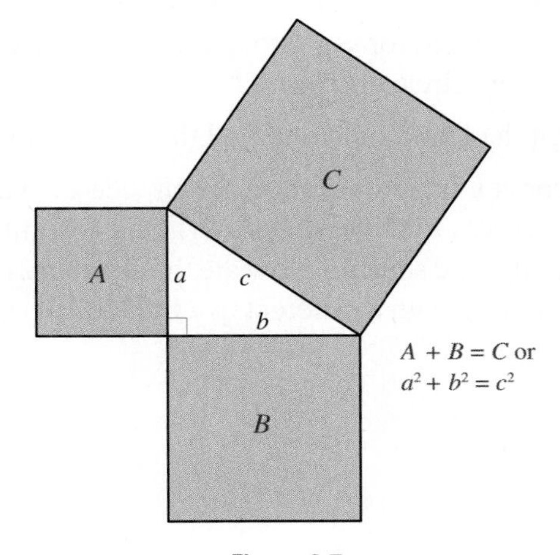

Figure 0.7

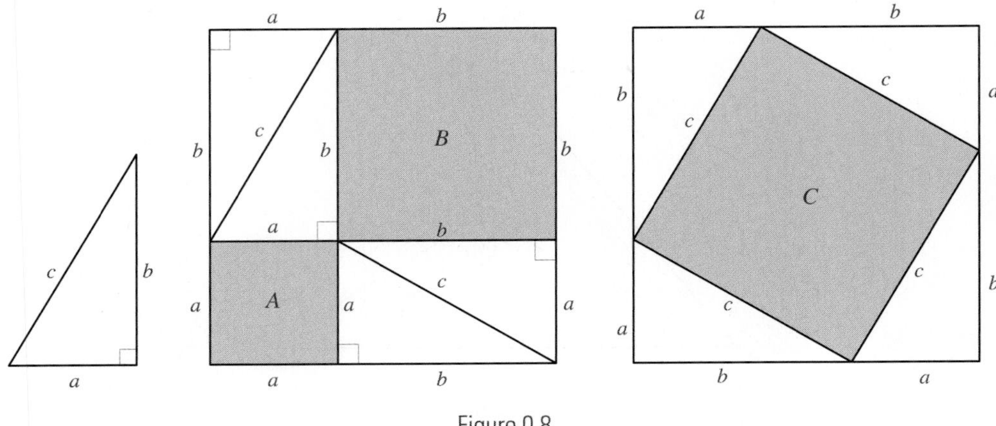

Figure 0.8

In Chapters 3 and 4, we will give several proofs of the Pythagorean Theorem. In Chapter 4 we will also explore what other figures can be constructed on the sides of a right triangle so that the area of the figure constructed on the hypotenuse is equal to the sum of the areas of the figures constructed on the other two sides. We will also generalize the theorem for triangles that are not necessarily right triangles.

Problem Set 0

In each of the following problems, you may use any tools to perform the experiments (a geometry drawing utility such as GSP, the Geometer's Sketchpad, is especially convenient but not necessary).

- **1.** **a.** Conjecture the solution to the Treasure Island Problem by choosing at least four different positions for the gallows as described in the text.
  - **b.** Does your conjecture hold for some Γ positioned below the line connecting the coconut and banana tree? (Make an appropriate construction.)
  - **c.** Place Γ at one of the trees and find the corresponding treasure.
  - **d.** Place Γ on the line connecting the trees halfway between them and find the corresponding treasure.
  - **e.** Based on your experiments in (a) through (d), what seems to be the simplest way to find the treasure?
- **2.** **a.** Draw a circle on transparent (or see-through) paper. How would you find the center of the circle by folding the circle onto itself?
  - **b.** Draw an arc of a circle. How could you find the center now?
- **3.** Let $ABCD$ be any convex quadrilateral. On each side of the quadrilateral, construct a square as shown in Figure 0.9. Find the centers C_1, C_2, C_3, and C_4 of the squares, where C_1 and C_3 are centers of opposite squares. How are the segments C_1C_3 and C_2C_4 related? Repeat the experiment starting with a different quadrilateral.

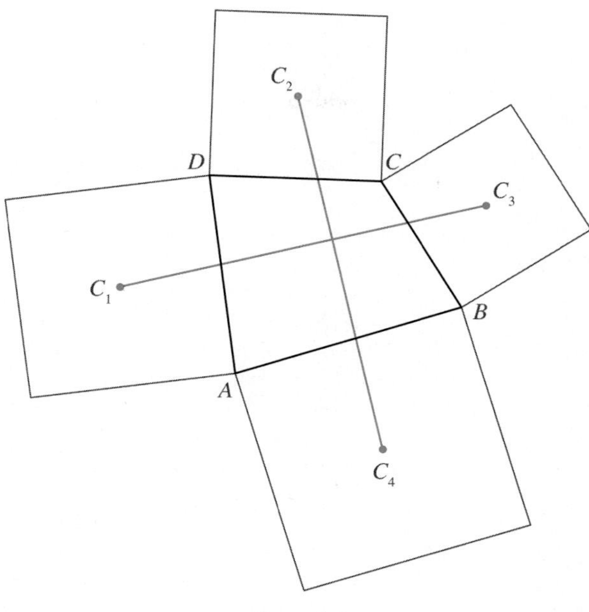

Figure 0.9

4. Choose an arbitrary triangle *ABC* and construct the corresponding Nine-Point Circle. (You may need the result from Problem 2.)

5. Check Morley's Theorem experimentally for some triangle *ABC*.

•**6.** Use Figure 0.8 to prove the Pythagorean Theorem.

1 Congruence, Constructions, and the Parallel Postulate

At the age of eleven, I began Euclid, with my brother as my tutor. This was one of the great events of my life, as dazzling as first love. I have not imagined there was anything so delicious in the world . . . From that moment until . . . I was thirty-eight, mathematics was my chief interest and my chief source of happiness.

—Bertrand Russell (1872–1970)
The Autobiography of Bertrand Russell
(London: G. Allen and Unwin, 1968)

Historical Note: Bertrand Russell

Lord Bertrand Arthur William Russell was a British philosopher, logician, and mathematician who made important contributions to the foundations of mathematics. In 1910, he became a lecturer at Cambridge University but was dismissed and later jailed for making pacifist speeches during World War I. He abandoned pacifism during World War II in the face of the Nazi atrocities in Europe and the Nazi threat to Great Britain. After the war he reverted back to pacifism, becoming a leader in the anti-nuclear and anti–Vietnam War movements. Later Russell taught at several U.S. universities, including Harvard University and the University of Chicago. He won the Nobel Prize for Literature in 1950. Russell died in 1970 at the age of 98.

Introduction

In this chapter we lay the foundation for Euclidean and non-Euclidean geometry. About 250 B.C.E., Euclid systematically collected and organized the geometrical knowledge of his time in a treatise composed of 13 books, called *The Elements*. (This treatise also included number theory and covered topics in algebra.) Euclid started out with a list of statements called axioms or postulates that he assumed to be true and then showed that geometric statements followed logically from his assumptions. However, he did not realize that some geometric terms could not be defined. Some terms must be left undefined to avoid "circular" definitions. (For example, in one dictionary a *line* is defined as "the path traced by a moving point" and then a *path* is defined as "a line obtained by a moving point.") Euclid "defined" a *point* as that which has no part and a *line* as breadthless length, when neither *part* nor *breadth* nor *length* had been defined.

In his proofs Euclid used unstated assumptions and relied on diagrams to make what seemed to him to be obvious conclusions. Despite these shortcomings, Euclid's achievements were monumental. In presenting a vast amount of mathematics by starting with a few basic assumptions and then logically deducing other mathematical statements, Euclid laid the foundation for a deductive approach to mathematics. Only toward the end of the nineteenth century was a rigorous foundation for Euclidean geometry established. In 1899, David Hilbert in his book *Foundations of Geometry* established a set of axioms along with undefined terms for Euclidean geometry and succeeded in proving Euclid's theorems relying solely on logic. Hilbert's success had its roots in a revolution that had taken place in geometry some years earlier. In 1829, the Russian mathematician Nikolai Lobachevsky—and independently two years later the Hungarian mathematician Janos Bolyai—established a new geometry referred to as non-Euclidean geometry. At the same time, the great German mathematician Karl Friedrich Gauss was very likely aware of the new results, but did not publish them as he was worried about the controversy that they might arouse.

Non-Euclidean geometry is based on an axiom that denies Euclid's Fifth Postulate. (Euclid used the term *postulate* for geometric assumptions and the term *axiom* for general mathematical non-geometrical assumptions. For example, one of Euclid's axioms is "The part is smaller than the whole" and one of his postulates is "A straight line segment can be drawn joining any two points.") Euclid's Fifth Postulate or the Parallel Postulate is equivalent to the following:

Euclidean Parallel Postulate Through a given point P not on a line ℓ, there is exactly one line parallel to ℓ. (See Figure 1.1.)

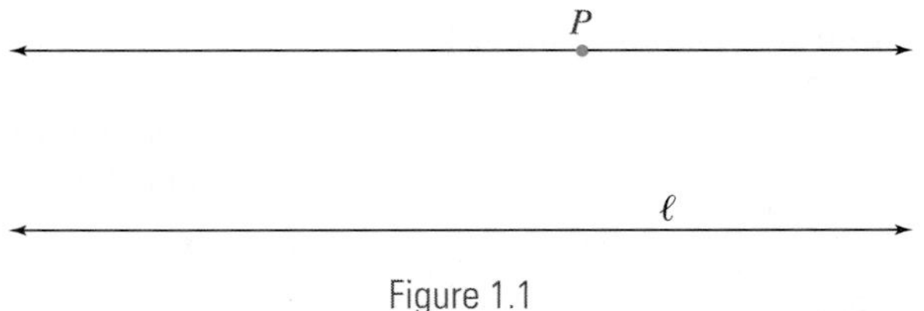

Figure 1.1

For generations, mathematicians believed it was possible to prove the Parallel Postulate using Euclid's other postulates or axioms, and many tried to do so. Only Lobachevsky, Bolyai, and Gauss were bold enough to substitute Euclid's Parallel Postulate by another one that denied Euclid's postulate. It is referred to as the Hyperbolic Parallel Postulate.

Hyperbolic Parallel Postulate Given a line ℓ and a point P not on ℓ, there exist at least two lines through P that are parallel to ℓ.

Euclid's approach to geometry was directed by physical reality. Thus points and lines were idealized mathematical terms for what we perceive as a point and a straight line. Postulates were "self-evident" truths. To the originators of hyperbolic geometry, points and lines were undefined terms satisfying certain axioms; as such they did not necessarily represent any idealized physical objects. (It is possible to find familiar objects to represent points, lines, and planes that satisfy the Hyperbolic Parallel Postulate.)

In Sections 1.2 and 1.3, we develop neutral geometry (also called absolute geometry) in which the Parallel Postulate is not used. Consequently, theorems in neutral geometry are true in Euclidean geometry as well as hyperbolic geometry. We adopt a set of axioms that constitute a modification of the axioms introduced by the American mathematician George David Birkoff

(1884–1944). The axioms are *consistent* and *independent*. A set of axioms is said to be consistent if no contradictions can be derived from the set; it is said to be independent if no axiom of the set is implied by the other axioms in the set.

To get quickly to interesting results, we placed the axioms and related material in the Appendix.

Historical Note: Euclid (Third Century B.C.E.)

The place and exact year of Euclid's birth are not known. Historians believe he was the first mathematics professor at the University of Alexandria. Euclid's *The Elements* was the first treatise on mathematics as a deductive system and is considered to be one of the greatest achievements of humankind. It has been more widely studied than any other book except for the Bible. More than 1000 editions of *The Elements* have appeared since its first printing in 1482. However, no Greek copy from Euclid's time has been found; instead, *The Elements* reached the West from an Arabic translation. The teaching of geometry has been dominated by *The Elements* for more than 20 centuries, and Euclid's work has had a profound influence on scientific thinking.

1.1 Angles and Their Measurement

Pairs of Angles

Certain pairs of angles occur often enough that it is convenient to give them special names.

Two angles are **adjacent** if they lie in the same plane, they share a common side, and the interiors of the angles have no point in common. $\angle CAD$ and $\angle DAB$ in Figure 1.2 are adjacent angles. The common side of two adjacent angles of equal measure is called an **angle bisector**.

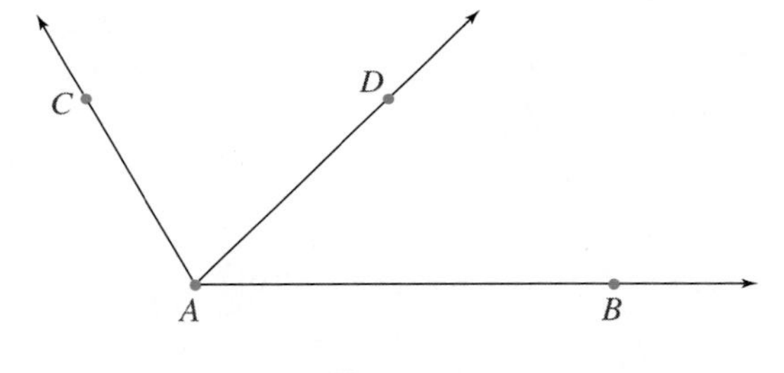

Figure 1.2

Two angles form a **linear pair** if they are adjacent and the noncommon sides are opposite rays (see Figure 1.3a). Two angles are **vertical** if their sides form two pairs of opposite rays (see Figure 1.3b).

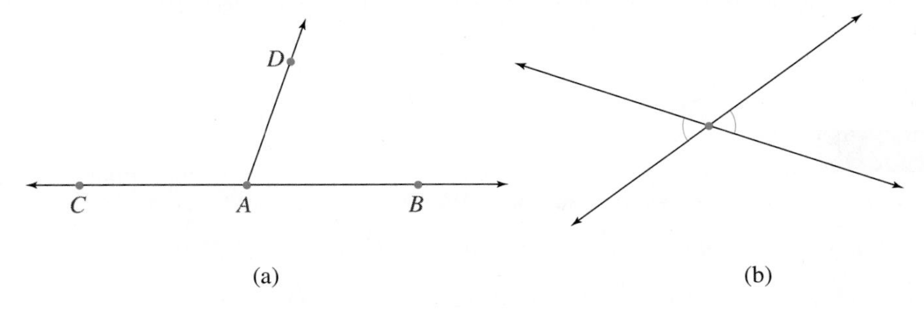

Figure 1.3

It seems obvious and can be readily deduced that the sum of the measures of two angles in a linear pair is 180°. We often encounter pairs of angles whose measures add up to 180° or 90°.

Such angles have a special name. Two angles are **supplementary** if the sum of their measures is 180°; each angle is called the supplement of the other. Two angles are **complementary** if the sum of their measures is 90°; each angle is called the complement of the other.

Notice that two angles in a linear pair are supplementary but not every two supplementary angles form a linear pair. In fact, it follows that if two angles are supplementary and adjacent, then they form a linear pair. An angle measuring less than 90° is termed **acute**, and one measuring more than 90° is termed **obtuse**.

It is common to call a 90° angle a **right angle**. An angle measuring 180° is a **straight angle**. **Congruent angles** are angles that have the same measure (we denote the measure of an angle by m). Thus, if $m(\angle A) = m(\angle B)$, $\angle A$ and $\angle B$ are **congruent**. We use the symbol $\cong$ for congruent. Hence $\angle A \cong \angle B$. Notice that the equality symbol (=) is reserved for real numbers or for sets that have the same elements (the degree measure is a real number). Because two angles that have the same measure are not necessarily the same set of points, we do not use the equality symbol between congruent angles. Similarly, we do not use the equality symbol between congruent segments.

Notation for Angles and Their Measures

It is often cumbersome to use the notation $m(\angle A)$ or $m(\angle BAC)$ for the measure of an angle. For this reason, we commonly use small Latin or Greek alphabet letters for measures of angles.

In Figure 1.4 we designated the measures of the three angles of $\triangle ABC$ by α, β, and γ (notice the correspondence to A, B, and C). Thus $m(\angle A) = \alpha$, $m(\angle B) = \beta$, and $m(\angle C) = \gamma$. Because α is already the measure of $\angle A$, we do not write $m(\alpha)$ or $m(\angle\alpha)$.

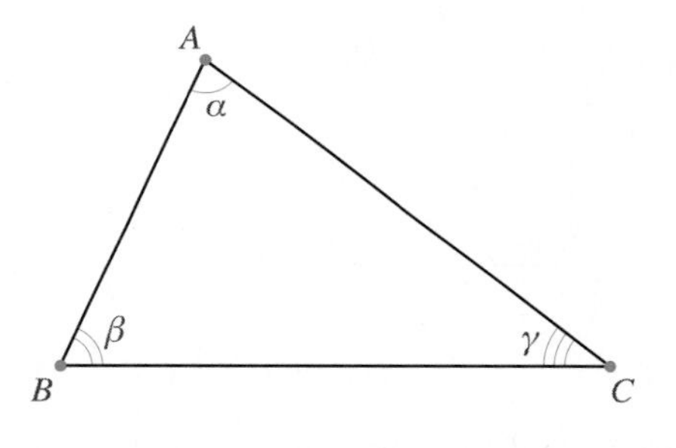

Figure 1.4

The following theorems follow immediately from the preceding definitions. We prove the second theorem.

Theorem 1.1

1. *Supplements of congruent angles are congruent.*
2. *Complements of congruent angles are congruent.*
3. *The sum of the measures of two angles in a linear pair is* 180°.

Theorem 1.2

Vertical angles are congruent.

Proof

Designating the measures of the angles in Figure 1.5 by α, β, γ, and δ, we need to show that $\alpha = \beta$ and $\gamma = \delta$. Notice that $\alpha + \gamma = 180$ and $\beta + \gamma = 180$, which implies that $\alpha = \beta$. (Alternatively, we could say that α is a supplement of γ and that β is a supplement of γ and, therefore, by Theorem 1.1, $\alpha = \beta$.) □

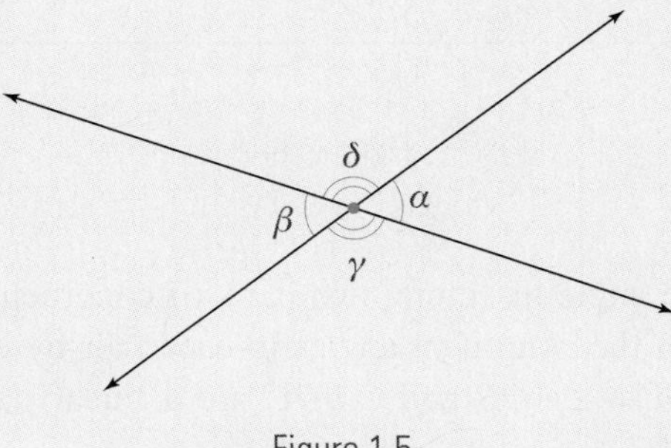

Figure 1.5

When two lines intersect, they form four angles as shown in Figure 1.5. If one of the angles formed is a right angle, the lines are perpendicular. Previous theorems imply that in this case all four angles are right angles. The fact that lines are perpendicular or equivalently that an angle formed is a right angle is designated in drawings by the symbol ⌐, as shown in Figure 1.6. We also write $\mathbf{a} \perp \mathbf{b}$ to designate that lines a and b are perpendicular.

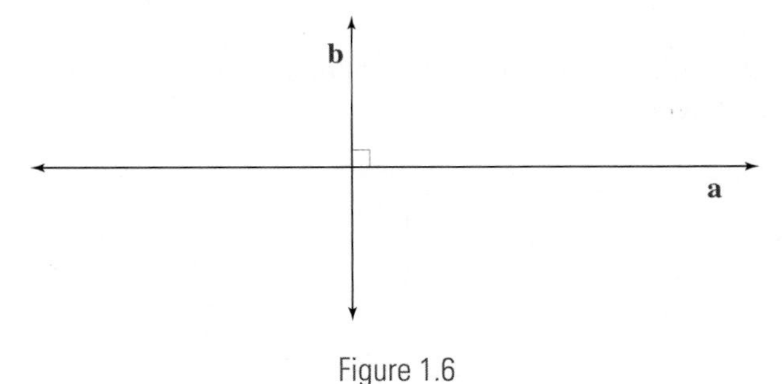

Figure 1.6

When a line intersects a segment at its midpoint and is perpendicular to the segment, it is called the **perpendicular bisector** of the segment. The line ℓ in Figure 1.7 is the perpendicular bisector of $\overline{AB}$. Notice that M is the midpoint of $\overline{AB}$ and that the congruent segments $\overline{AM}$ and $\overline{MB}$ are marked by the same symbol.

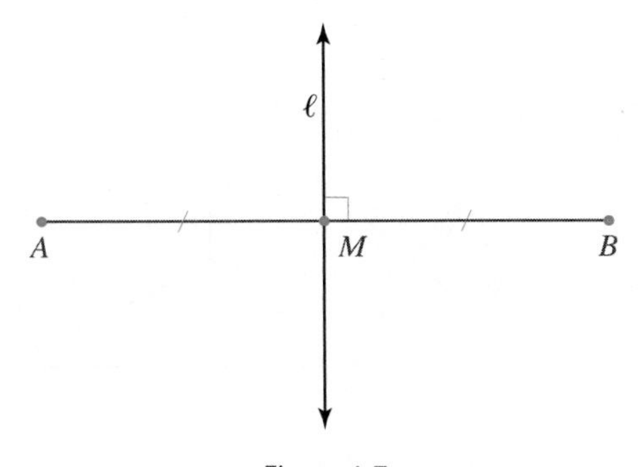

Figure 1.7

Example 1.1 In Figure 1.8, $\angle AOB$ and $\angle BOC$ form a linear pair. The rays $\overrightarrow{OD}$ and $\overrightarrow{OE}$ are their angle bisectors, respectively. Prove that $\overleftrightarrow{OE} \perp \overleftrightarrow{OD}$.

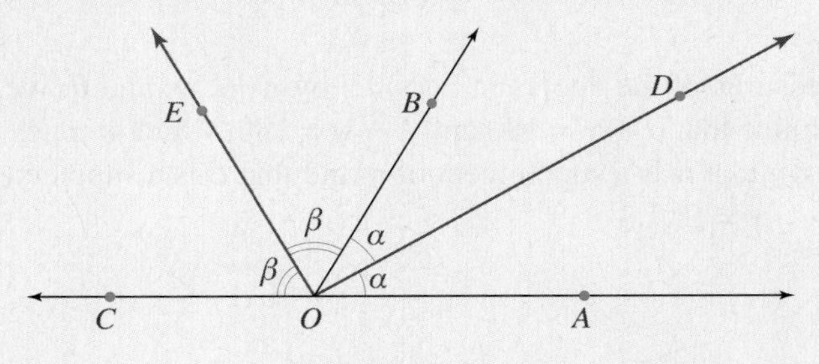

Figure 1.8

Solution

Because $\overrightarrow{OD}$ and $\overrightarrow{OE}$ are angle bisectors, two pairs of congruent angles are formed. We designate the measures of the congruent angles in each pair by α and β, respectively, as shown in Figure 1.8. Because $\angle AOB$ and $\angle BOC$ are a linear pair, we have $2\alpha + 2\beta = 180°$. Hence $\alpha + \beta = 90°$.

Because $m(\angle EOD) = \alpha + \beta$, it follows that $m(\angle EOD) = 90°$ and therefore $\overleftrightarrow{OE} \perp \overleftrightarrow{OD}$.

Problem Set 1.1

1. $\overline{OA} \perp \overline{OC}$ and $\overline{OB} \perp \overline{OD}$. Which of the non-right angles formed in the diagram are congruent? Justify your answer.

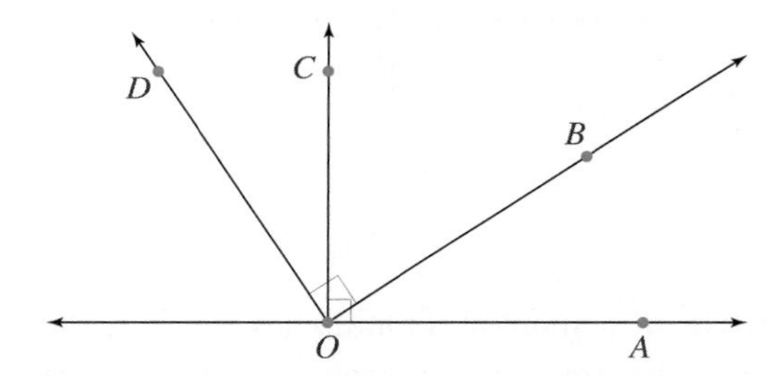

2. Examine the following argument showing that the sum of the measures of the angles in any triangle is 180º. Is the argument valid? Justify your answer.

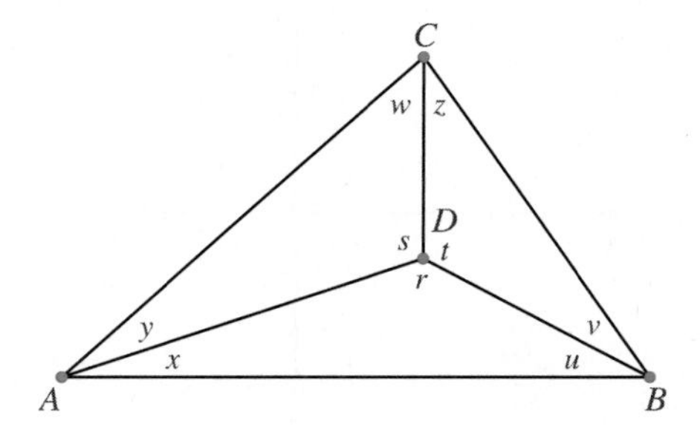

In $\triangle ABC$ we choose an arbitrary point D in the interior of the triangle and connect it to the three vertices of the triangle. We then mark the measures of the nine angles as shown in the figure and let the sum of the measures of the three angles in a triangle be k. We show that $k = 180$.

Adding up all the measures of the angles in the three triangles $\triangle ABC$, $\triangle ACD$, and $\triangle CBD$, we get $3k$. Hence $(x + u + r) + (v + z + t) + (w + y + s) = 3k$. By the commutative and associative properties of addition, we get $(x + u + v + z + w + y) + (r + t + s) = 3k$. Notice that the quantity in the first parentheses represents the sum of the measures of the angles in $\triangle ABC$ and for that reason equals k. This and the fact that $r + t + s = 360$ imply that $k + 360 = 3k$ or $k = 180$.

3. Jaimee announced that she has her own somewhat different proof of Example 1.1. Examine Jaimee's proof and compare it with the one given in the text. Answer the following questions:

a. What are the strengths and weaknesses of Jaimee's exposition? In the case of shortcomings suggest improvements.

b. Which proof do you like better? Why?

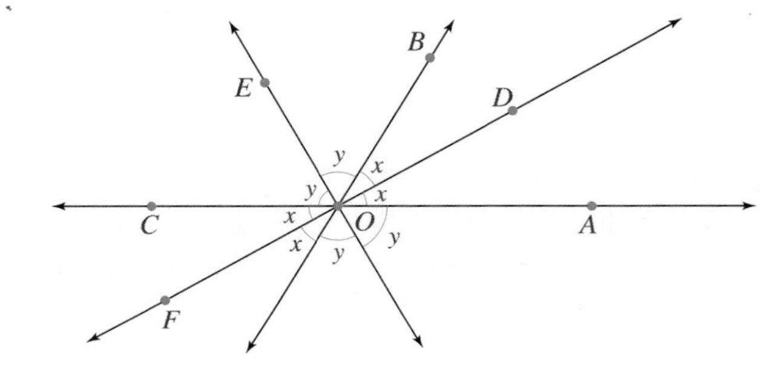

Jaimee:

I know that if a ray divides a straight angle into two congruent angles, each must be 90°. Thus, to show that $EO \perp OD$, I need to prove only that $\angle FOE = \angle EOD$. I extend the rays in Figure 1.8 and use the fact that the bisector of an angle divides it into two congruent angles and the fact that vertical angles are congruent. I mark the angles by x or by y as shown in my figure above. $\angle FOD$ is a straight angle because I extended the ray OD. I see that ray OE divides $\angle FOD$ into two angles, each equal to $x + y$. Hence each angle is 90°.

1.2 Congruence of Triangles

Informally, when two figures have the same size and shape, they are congruent. In Section 1.1, we defined congruent segments as segments that have the same length, and congruent angles as angles that have the same measure. When two figures are congruent, it is always possible to fit one figure onto the other so that matching sides and angles are congruent. This is the basis for defining congruent triangles.

Definition of Congruent Triangles Triangles are congruent if there is a one-to-one correspondence between their vertices so that corresponding sides are congruent and corresponding angles are congruent.

In Figure 1.9, $\triangle ABC$ and $\triangle DEF$ are congruent. Corresponding sides and angles are marked as shown. Notice that if we fit one triangle onto the other, vertex A will correspond to vertex D, B to E, and C to F. (You should be able to superimpose one triangle onto the other by tracing $\triangle DEF$ on a sheet of paper. In some cases you will need to flip the sheet and trace the triangle on the other side of the paper before superimposing it on a congruent triangle.) The fact that the triangles are congruent is written $\triangle ABC \cong \triangle DEF$. Whenever the symbol $\cong$ is used, corresponding vertices must be written in the same position. Thus, without reference to a figure, the congruence $\triangle ABC \cong \triangle DEF$ implies $\angle A \cong \angle D$, $\angle B \cong \angle E$, $\angle C \cong \angle F$, $\overline{AB} \cong \overline{DE}$, $\overline{BC} \cong \overline{EF}$, and $\overline{AC} \cong \overline{DF}$.

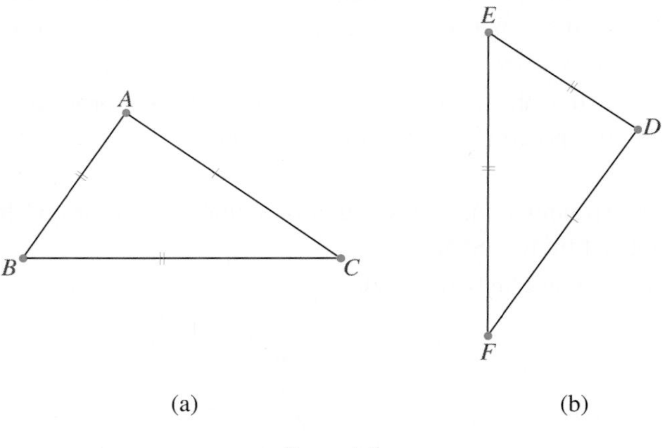

Figure 1.9

If the three sides and three angles of one triangle are congruent to the corresponding sides and angles of another triangle, by definition the two triangles are congruent. It turns out that congruence of fewer corresponding parts is sufficient to determine that two triangles are congruent.

Figure 1.10 shows $\triangle ABC$ and $\triangle DEF$ in which $\overline{AB} \cong \overline{DE}$, $\angle A \cong \angle D$, and $\overline{AC} \cong \overline{DF}$.

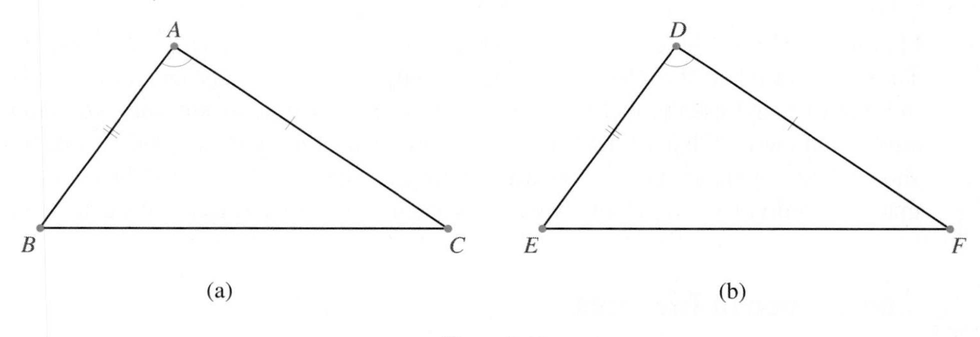

Figure 1.10

These conditions assure that the triangles are congruent, which can be seen as follows. Because $\angle A \cong \angle D$, we could superimpose $\angle D$ on top of $\angle A$ so that ray $\overrightarrow{DE}$ falls on ray $\overrightarrow{AB}$ and $\overrightarrow{DF}$ falls on $\overrightarrow{AC}$. Because $\overline{AB} \cong \overline{DE}$ and $\overline{AC} \cong \overline{DF}$, the vertex E will fall on B and the vertex F on C. Thus the vertices D, E, and F will fall on the corresponding vertices A, B, and C and, therefore, the triangles are congruent.

Notice, however, that this argument is an intuitive justification and not a formal proof. We do not have a rigorous definition of what it means to superimpose one figure on top of another. In Chapter 5, during our study of transformations, we will be able to give a precise definition of what it means to *move* a figure and hence a definition of congruent figures. Meanwhile we state the congruence condition discussed above as an axiom.

Axiom 1.1 **The Side, Angle, Side (SAS) Congruence Condition** *If two sides and the angle included between these sides are congruent to two sides and the included angle of the second triangle, then the triangles are congruent.*

We can use the SAS condition to prove other congruence conditions, but first we need a theorem concerning the angles of an **isosceles triangle**. A triangle is isosceles if at least two of its sides are congruent. If all the sides of a triangle are congruent, the triangle is **equilateral**. A triangle with no two sides congruent is called **scalene**. A triangle with all acute angles is called an **acute** triangle. A triangle with an obtuse angle is called an **obtuse** triangle.

Theorem 1.3

The Isosceles Triangle Theorem

If two sides of a triangle are congruent, then the angles opposite these sides are congruent.

Theorem 1.3 is frequently stated as follows: *The base angles of an isosceles triangle are congruent.*

Figure 1.11a gives a quick pictorial representation of this theorem. The marks on the sides indicate that the sides are congruent and the marks on the base angles indicate the conclusion that the angles are congruent.

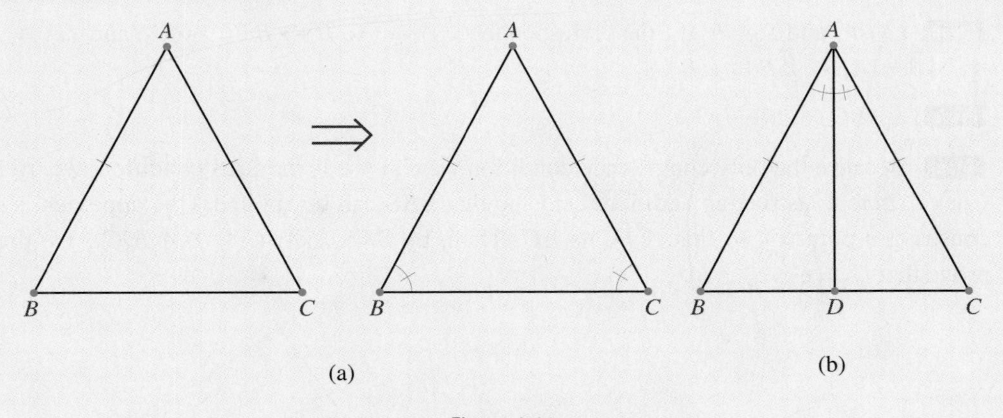

Figure 1.11

We formally state the hypothesis (what is given) and the conclusion (what we need to prove) of the theorem:

Given In $\triangle BAC$ (see Figure 1.11a), $\overline{AB} \cong \overline{AC}$.

Prove $\angle B \cong \angle C$.

Plan One approach is to "divide" $\triangle ABC$ into two triangles that seem to be congruent and in which $\angle B$ and $\angle C$ will be corresponding angles. Anticipating the use of the SAS condition, we bisect $\angle A$ in Figure 1.11b. If we could show that the two triangles are congruent, it would follow that the corresponding angles $\angle B$ and $\angle C$ are congruent.

Proof 1 Bisect $\angle A$. The angle bisector intersects $\overline{BC}$ at D as shown in Figure 1.11b.

We have now $\triangle ABD \cong \triangle ACD$ by SAS ($\overline{AB} \cong \overline{AC}$, $\angle BAD \cong \angle CAD$, and $\overline{AD} \cong \overline{AD}$). Because the corresponding parts of the triangle are congruent, $\angle ABD \cong \angle ACD$. □

Proof 2 $\triangle BAC \cong \triangle CAB$ by SAS because $\overline{BA} \cong \overline{CA}$, $\angle BAC \cong \angle CAB$, and $\overline{AC} \cong \overline{AB}$. Because the corresponding angles in these triangles are congruent, we have $\angle B \cong \angle C$. □

Remark You may feel uneasy about the second proof of Theorem 1.3, perhaps because you are used to seeing two separate triangles when you are proving that triangles are congruent. Note that the definition of congruence of triangles does not preclude the possibility that if $\overline{BA} \cong \overline{CA}$,

$\triangle BAC \cong \triangle CAB$ as indicated in the second proof. Notice that if $\triangle BAC$ is not isosceles, then $\triangle BAC \not\cong \triangle CAB$.

Theorem 1.3 implies the following:

Corollary 1.1 *All the angles of an equilateral triangle are congruent; that is, an equilateral triangle is equiangular.*

We are now ready to prove two other congruence conditions for triangles.

Theorem 1.4

The Angle, Side, Angle (ASA) Condition

Given a one-to-one correspondence between the vertices of two triangles, if two angles and the included side of one triangle are congruent to the corresponding parts of the second triangle, the two triangles are congruent.

Given $\triangle ABC$ and $\triangle A_1B_1C_1$, the correspondence $A \leftrightarrow A_1$, $B \leftrightarrow B_1$, $C \leftrightarrow C_1$, and $\angle A \cong \angle A_1$, $\overline{AB} \cong \overline{A_1B_1}$, $\angle B \cong \angle B_1$.

Prove $\triangle ABC \cong \triangle A_1B_1C_1$.

Plan Because the only congruence condition we can use is the SAS condition, we try in Figure 1.12b to construct an additional side so that SAS can be applied. One approach is to construct a point C_2 so that $\overline{A_1C_2} \cong \overline{AC}$. Then, by SAS, $\triangle ABC \cong \triangle A_1B_1C_2$. We then prove that $C_2 = C_1$.

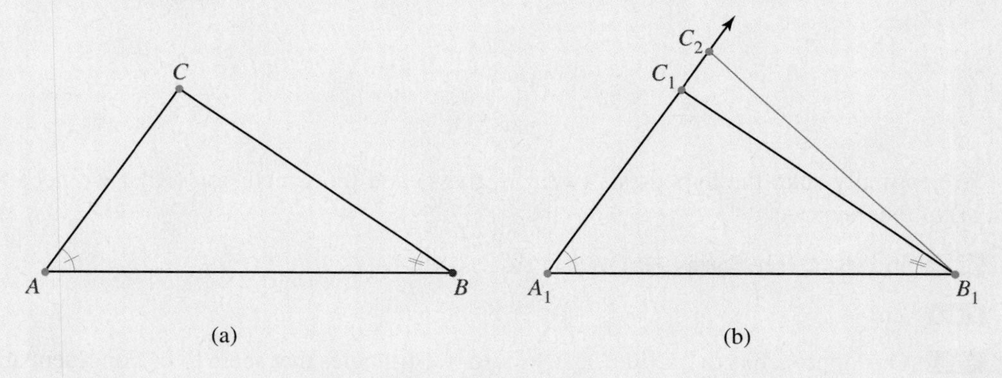

Figure 1.12

Proof

Construct on $\overrightarrow{A_1C_1}$ in Figure 1.12b a point C_2 such that $A_1C_2 = AC$. By SAS (regardless whether $C_2 \neq C_1$ or $C_2 = C_1$), we have

1. $\triangle ABC \cong \triangle A_1B_1C_2$

   To show that $C_2 = C_1$, notice that (1) implies

2. $\angle A_1B_1C_2 \cong \angle ABC$

   In addition, the hypothesis tells us that

3. $\angle ABC \cong \angle A_1B_1C_1$

Statements 2 and 3, along with the transitive property of congruence, imply that $\angle A_1B_1C_1 \cong \angle A_1B_1C_2$. We show now that this congruence of angles implies $C_2 = C_1$. Indeed, if $C_1 \neq C_2$, then either C_1 is between A_1 and C_2 or C_2 is between A_1 and C_1. In the first case, C_1 would be in the interior of $\angle A_1B_1C_2$, which would imply that $m(\angle A_1B_1C_1) < m(\angle A_1B_1C_2)$ and would contradict the fact that the angles are congruent. Similarly, we can show that C_2 between A_1 and C_1 contradicts the equality of the angles. Thus $C_2 = C_1$, and from (1) we obtain the fact that $\triangle ABC \cong \triangle A_1B_1C_1$. □

Now Solve This 1.1

Prove the following statements, which follow from the ASA condition:

1. **Converse of Theorem 1.3**: *If two angles of a triangle are congruent, then the sides opposite these angles are congruent.*
2. *An equiangular triangle is equilateral (all sides congruent).*

Medians, Altitudes, and Additional Properties of Isosceles Triangles

The segment connecting the vertex of a triangle with the midpoint of the opposite side is called a **median**. If M is the midpoint of $\overline{BC}$ in Figure 1.13a, then $\overline{AM}$ is a median. The segment from a vertex of a triangle perpendicular to the line containing the opposite side is an **altitude** of the triangle. In Figure 1.13a, $\overline{AH} \perp \overline{BC}$ and thus $\overline{AH}$ is the altitude to side $\overline{BC}$. In Figure 1.13b, $\overline{AH}$ is perpendicular to the line containing $\overline{BC}$ and, therefore, is also an altitude to $\overline{BC}$. Every triangle has three medians and three altitudes, one for each of its sides.

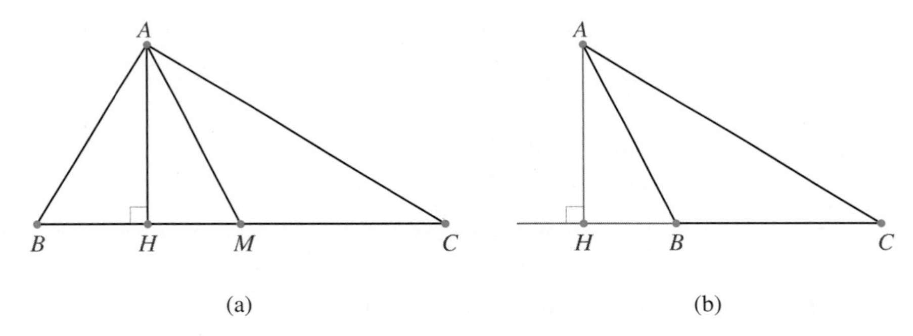

Figure 1.13

Now Solve This 1.2

To investigate the relationship among a median, altitude, angle bisector, and perpendicular bisector of a base in an isosceles triangle, fold or draw an isosceles triangle ABC as shown in Figure 1.14a, where $AB = AC$ and $\overline{BC}$ is the base. Fold the paper so that C falls on top of B and then unfold it. The unfolded triangle and the crease are shown in Figure 1.14b.

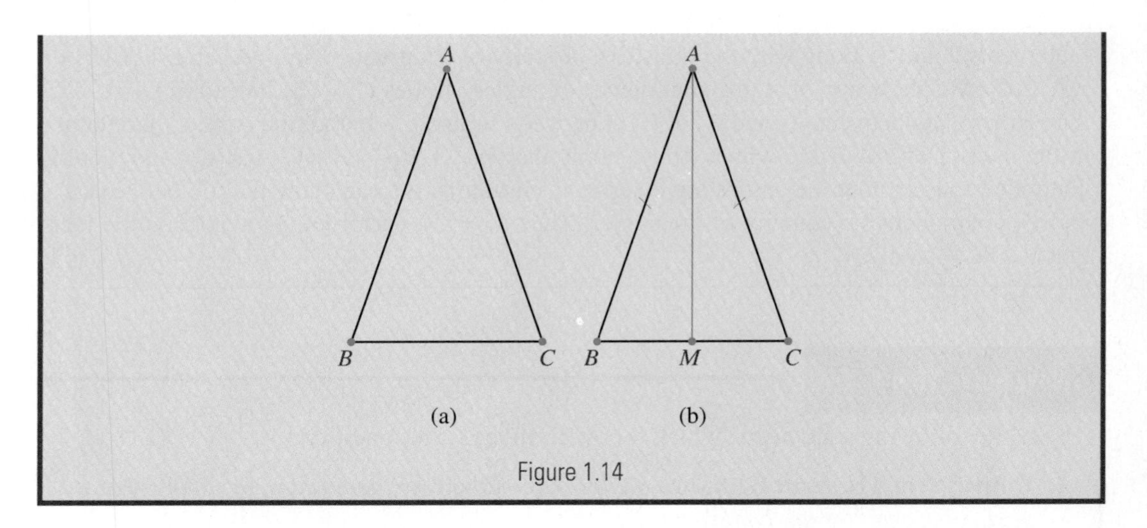

Figure 1.14

While performing the paper-folding activity, you likely noticed that $\triangle ACM$ was folded onto $\triangle ABM$ (see Figure 1.14b). Thus it seems that $\angle CAM \cong \angle BAM$, $\overline{BM} \cong \overline{MC}$, and $\angle AMB \cong \angle AMC$. The last congruence of angles means that $\overline{AM} \perp \overline{BC}$. These observations tell us that in an isosceles triangle, the angle bisector of the angle opposite the base is also the perpendicular bisector to the base and the altitude to the base. We state these observations in the following theorem:

Theorem 1.5

The median to the base of an isosceles triangle is the perpendicular bisector as well as the angle bisector of the angle opposite the base.

Proof

Assume that in Figure 1.14b, $\overrightarrow{AM}$ is the angle bisector of $\angle A$. It is sufficient to prove that $\triangle ABM \cong \triangle ACM$. The congruency of these triangles follows from SAS. □

Remark In the proof of Theorem 1.5 and from now on we assume that an angle has a unique angle bisector. This fact can be proved (see Moise, p. 109).

Theorem 1.5 can be viewed as follows: If vertex A in Figure 1.14b is equidistant from the endpoints of the segment $\overline{BC}$ (that is, $AC = AB$), then A lies on the perpendicular bisector of $\overline{BC}$. It seems that the converse, which we state in the next theorem, is also true.

Theorem 1.6

Every point on the perpendicular bisector of a segment is equidistant from the endpoints of the segment.

Restatement

Given m is the perpendicular bisector of $\overline{AB}$ (see Figure 1.15) and P is any point on m.

Prove $\overline{AP} \cong \overline{BP}$.

Proof

There are two cases. If P is on $\overline{AB}$, then because it is on the perpendicular bisector of $\overline{AB}$, it must be the midpoint of $\overline{AB}$ and hence equidistant from A and B. If P is not on $\overline{AB}$ as in Figure 1.15, we have $\triangle AMP \cong \triangle BMP$ by SAS because $\overline{AM} \cong \overline{BM}$, $\overline{MP} \cong \overline{MP}$, and the included angles at M are congruent as each is a right angle. From the congruence of the triangles, we have $\overline{AP} \cong \overline{BP}$. □

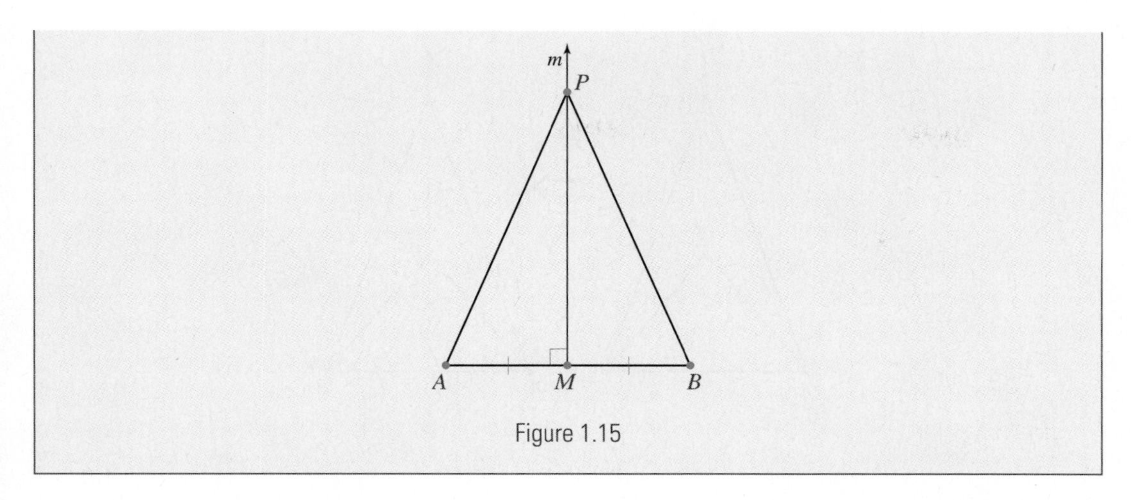

Figure 1.15

Theorem 1.6 tells us that every point on the perpendicular bisector of a segment is equidistant from the endpoints of the segment. Are there any other such points? That is, are there points not on the perpendicular bisector that are equidistant from the endpoints of the segment? By Theorem 1.5, if P is equidistant from A and B, it must be on the perpendicular bisector of $\overline{AB}$. Consequently, the perpendicular bisector of a segment is the set of all points equidistant from the endpoints of the segment. In geometry, a set of points satisfying a certain property is often called a **locus**. Thus we have the following corollary to Theorem 1.6:

Corollary 1.2 *A point is equidistant from the endpoints of a segment if and only if it is on the perpendicular bisector of the segment. Equivalently, the locus of all points equidistant from the endpoints of a segment is the perpendicular bisector of the segment.*

Figure 1.16 shows two points P and Q that are equidistant from A and B. From Corollary 1.2 we know that P and Q are on the perpendicular bisector of $\overline{AB}$. Because a line is determined by two points, any two points on the perpendicular bisector of a segment determine the perpendicular bisector. Hence $\overleftrightarrow{PQ}$ is the perpendicular bisector of $\overline{AB}$. This is stated in the following corollary to Theorems 1.5 and 1.6:

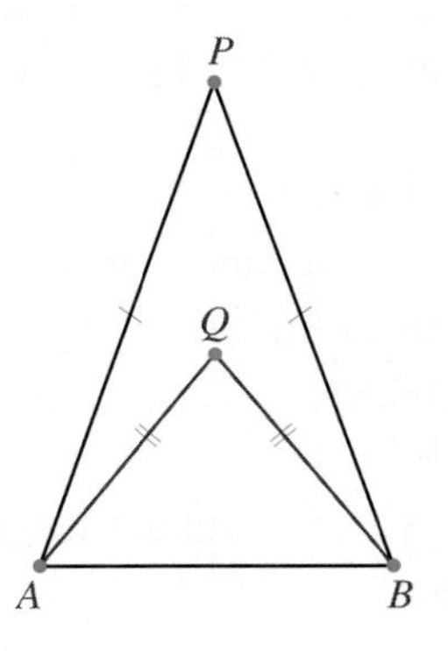

Figure 1.16

Corollary 1.3 *If each of two points is equidistant from the endpoints of a segment, then the line through these points is the perpendicular bisector of the segment.*

Corollary 1.3 can be represented pictorially as shown in Figure 1.17.

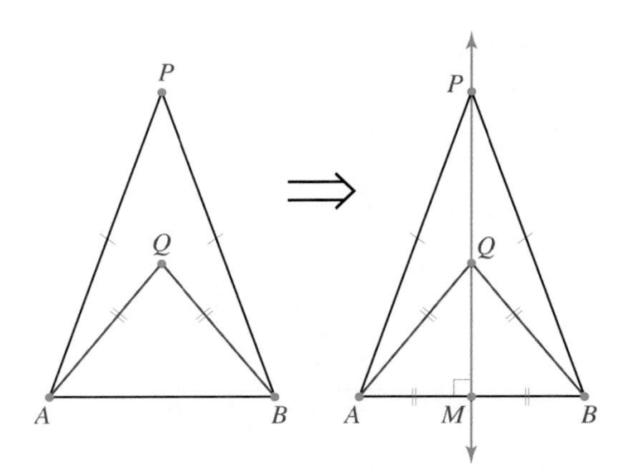

Figure 1.17

Now Solve This 1.3

1. Use any tools to construct an acute scalene triangle and the three perpendicular bisectors, three angle bisectors, three medians, and three altitudes. What do you notice?
2. Repeat Problem 1 for an obtuse triangle.

Euclidean Constructions

Euclid and other Greek mathematicians required that construction of geometric figures be done using only a compass and a straight edge (an unmarked ruler). The Greek compass was a collapsing compass, which loses its radius once it is moved. As a consequence, once a circle was constructed one could not move the compass to construct another circle with a new center but the same radius. Thus the Greek compass could not be used to mark off distances. Although today it is common practice to use a noncollapsible compass, any construction using a modern compass can also be accomplished using a collapsing compass (for a proof, see the webpage http://homepage.mac.com/teast/collapse.html).

The following rules apply to Euclidean constructions (constructions with a compass and straight edge in a plane):

1. Given two points, a unique straight line can be drawn containing the points as well as the unique segment connecting the points. (This is accomplished by aligning the straight edge across the points.)
2. It is possible to extend any part of a line.
3. A circle can be drawn given its center and radius.
4. Any finite number of points can be chosen on a given line, segment, or circle.
5. Points of intersection of two lines, two circles, or a line and a circle can be used to construct segments, lines, or circles.
6. No other instruments (such as a marked ruler, triangle, or protractor) or procedures can be used to perform constructions.

In reality, compass and ruler constructions are subject to error. For example, a geometrical line is an ideal line with zero width. However, a drawing of a line, no matter how sharp the pencil and how good the ruler and compass are, has a non-zero width.

From now on when we ask for a **construction** of a geometrical object, we will mean a compass and straight edge construction that follows the rules given previously. For convenience, we will frequently substitute "straight edge" with the one word: **ruler**. To construct a line or a line segment, the ruler must be aligned across two fixed points. Students sometimes ignore this requirement, leading to invalid constructions.

Example 1.2 **Invalid Construction of a Tangent to a Circle.** Given a circle with center O and point P as shown in Figure 1.18, construct through P a tangent to the circle (a line that touches the circle in a single point).

Invalid Procedure Rotate the ruler about point P counterclockwise until it just touches the circle as shown in Figure 1.18.

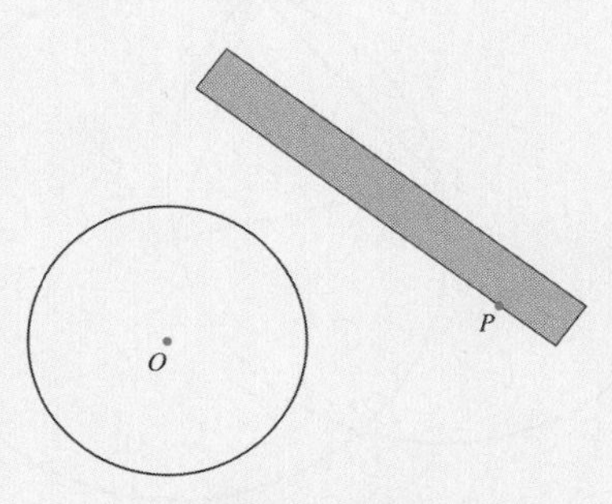

Figure 1.18

This approach, which is sometimes called "eyeballing," involves looking at the figure and guessing. (A valid construction will be explored in Chapter 2.)

Example 1.3 **Invalid Construction of a Perpendicular to a Line.** Given a line ℓ and a point P as in Figure 1.19, construct the perpendicular to the line through the point.

Invalid Procedure Use a drafting triangle to align one of the legs of the triangle with the line as shown in Figure 1.19. Then move the triangle along the leg until the other leg passes through the given point.

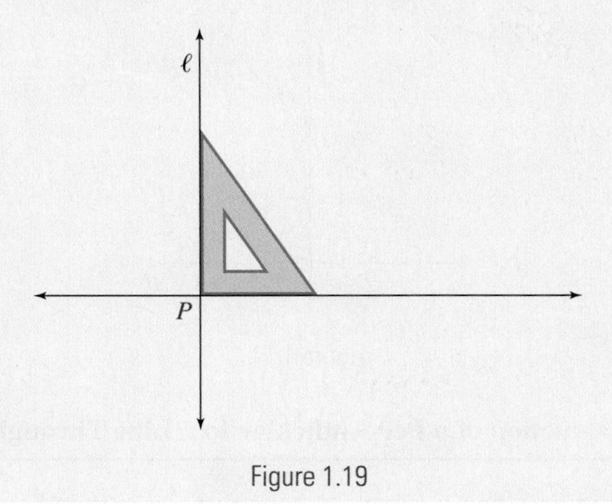

Figure 1.19

In what follows, we assume that the reader can construct a circle (the locus of all points at a given distance from a given point—the center) given its center and radius and can construct a segment congruent to a given segment. Let's proceed to our first construction.

Construction 1.1: Constructing an Equilateral Triangle

To construct an equilateral triangle on a given side AB as in Figure 1.20, we want to find point C whose distance from point A as well as from point B is AB. We know how to construct all the points whose distance from A is AB: It is the circle whose center is A and whose radius is AB. Our desired point C lies somewhere on this circle. To determine its location, we find all the points whose distance from B is AB. Such points are on the circle with center B and radius AB. Because vertex C lies on both circles, it can only be where the two circles intersect—that is, at point C or C' as shown in Figure 1.20. Thus $\triangle ABC$ or $\triangle ABC'$ is a required equilateral triangle. □

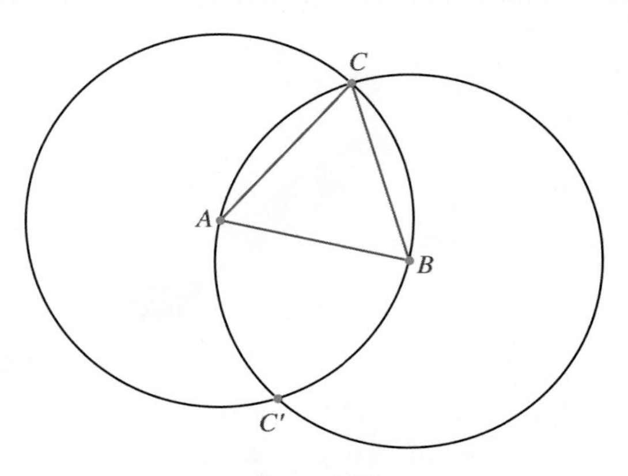

Figure 1.20

Construction 1.2: Construction of a Perpendicular Bisector of a Segment

Corollary 1.3 tells us that if we find any two points equidistant from the endpoints of a segment, the line through these points is the perpendicular bisector of the segment. In Figure 1.21, two such points P and Q have been constructed using a compass. $\overleftrightarrow{PQ}$ is the perpendicular bisector of $\overline{AB}$. □

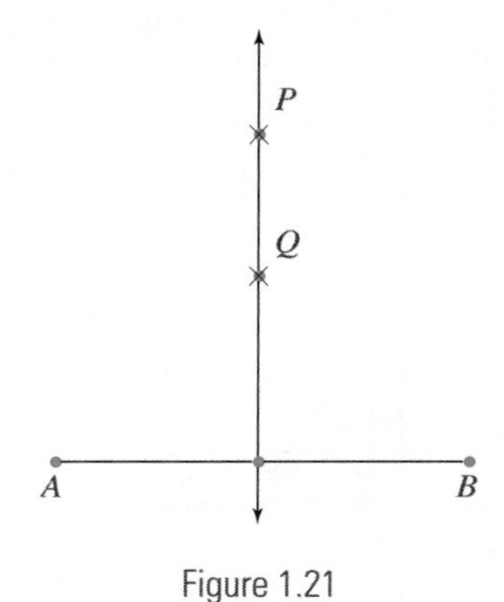

Figure 1.21

Construction 1.3: Construction of a Perpendicular to a Line Through a Point on the Line

In Figure 1.22a, M is a point on ℓ and we want to construct the perpendicular to ℓ through M. We wish to use the previous construction; to that end, we construct an arbitrary segment $\overline{AB}$ so that M is its midpoint. This is done by drawing a circle (or semicircle) centered at M as pictured in Figure 1.22b. The circle intersects ℓ at points A and B, both of which are equidistant from M. We now need only to find another point P equidistant from A and B. The line $\overleftrightarrow{PM}$ is the perpendicular bisector of $\overline{AB}$ and, in particular, is perpendicular to ℓ through M as required. □

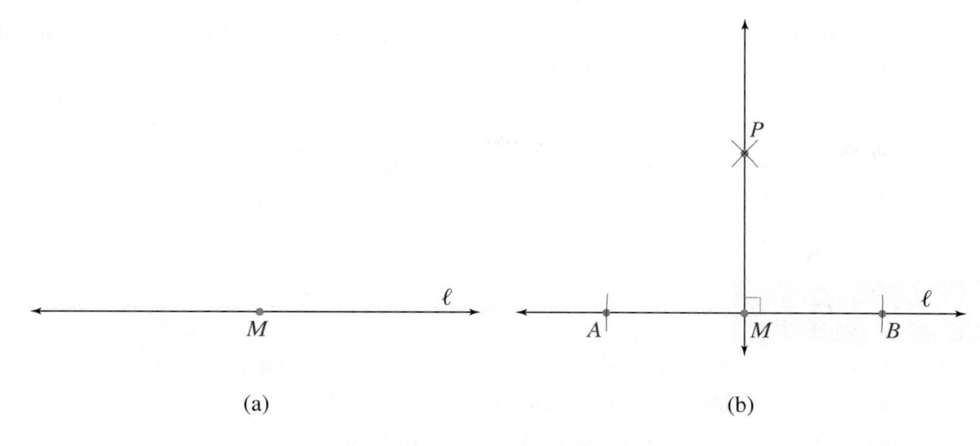

Figure 1.22

Properties of a Kite

A quadrilateral that has two pairs of congruent adjacent sides is called a **kite**. It can also be described as the quadrilateral created when two isosceles triangles share a common base (the base becomes then a diagonal of the kite). In Figure 1.23a, *ABCD* is a convex kite; in Figure 1.23b, *ABCD* is a non-convex kite.

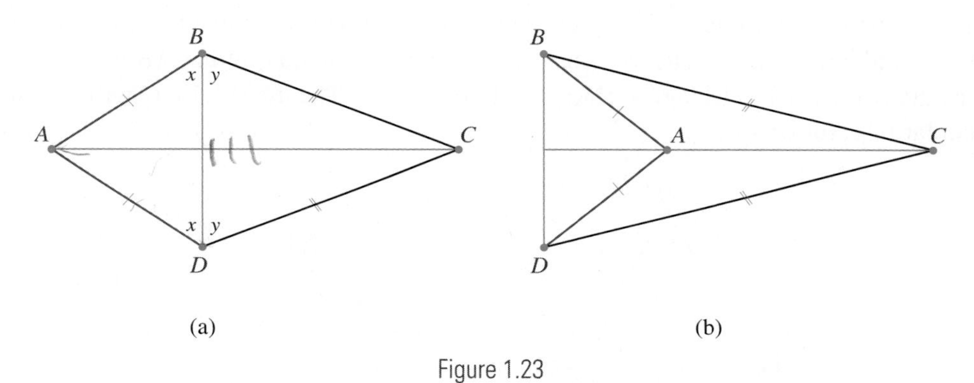

Figure 1.23

If we fold each of the kites in Figure 1.23 along the diagonal $\overline{AC}$ or the line containing $\overline{AC}$, we observe that B falls on D. Hence the diagonals of a kite are perpendicular to each other, and $\overline{AC}$ bisects both $\overline{BD}$ and the angles at A and C. We state these properties in the following theorem:

Theorem 1.7

The diagonal of a kite connecting the vertices where the congruent sides intersect bisects the angles at these vertices and is the perpendicular bisector of the other diagonal.

Proof

In Figure 1.23, the vertices A and C are equidistant from the endpoints of $\overline{BD}$. Hence $\overleftrightarrow{AC}$ is the perpendicular bisector of $\overline{BD}$ (by Corollary 1.3). Because the perpendicular bisector of the base of an isosceles triangle is also an angle bisector of the angle opposite the base, $\overleftrightarrow{AC}$ bisects each of the angles at A and C. In Figure 1.23b, $\overleftrightarrow{AC}$ bisects $\angle BAD$, which implies that $\angle BAC \cong \angle DAC$. □

A quadrilateral in which all the sides are congruent is called a **rhombus**. Because a rhombus is a kite, we have the following corollary to Theorem 1.7:

Corollary 1.4 *The diagonals of a rhombus are perpendicular bisectors of each other, and each bisects a pair of opposite angles.*

Now Solve This 1.4

1. Prove the converse of the statement in Corollary 1.4: A quadrilateral in which the diagonals are perpendicular bisectors of each other is a rhombus.
2. State and prove a statement similar to that given in Problem 1 for a kite.
3. Classify each of the following quadrilaterals using angle bisectors. Your statement should start like the one in Problem 1, using (i) a rhombus (ii) a kite. Justify your answers.

Basic Constructions Using the Properties of a Kite

The properties of a kite or a rhombus can be used to perform some basic constructions. To **bisect a given segment** $\overline{AB}$, we construct a kite or a rhombus so that $\overline{AB}$ is the common base of the two isosceles triangles, shown as in Figure 1.24a. The diagonal $\overline{DC}$ will be the perpendicular bisector of $\overline{AB}$.

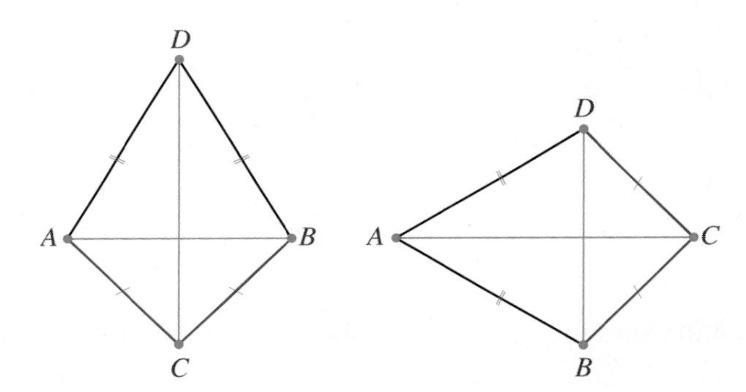

Figure 1.24

Construction 1.4: Bisect a Given Angle

To bisect a given angle like $\angle A$ in Figure 1.24b, we create a kite (or a rhombus) in which A is the vertex where the two congruent sides of a kite intersect. To do so, we construct any two isosceles triangles with a common base. The diagonal $\overline{AC}$ will bisect the angles at A and C. Other constructions utilizing the properties of a kite will be investigated in the problem set. □

Remark Notice that the constructions taking advantage of the properties of a kite involve basically the same steps as the corresponding ones done earlier using the properties of a perpendicular bisector and an isosceles triangle. Because a kite (or a rhombus) is a concrete and appealing object, some students find it easier to recall how to perform these constructions by referring to the properties of a kite (or a rhombus).

The properties of a kite can be used to prove the side, side, side congruency condition stated in the following theorem:

Theorem 1.8

Side, Side, Side (SSS) Congruency Condition

Given a one-to-one correspondence among the vertices of two triangles, if the three sides of one triangle are congruent to the corresponding sides of the second triangle, then the triangles are congruent.

Restatement

Given $\triangle ABC$, $\triangle A_1B_1C_1$, and the correspondence $A \leftrightarrow A_1$, $B \leftrightarrow B_1$, $C \leftrightarrow C_1$ and $\overline{AB} \cong \overline{A_1B_1}$, $\overline{AC} \cong \overline{A_1C_1}$, $\overline{BC} \cong \overline{B_1C_1}$.

Prove $\triangle ABC \cong \triangle A_1B_1C_1$.

Plan The idea is to place a flipped-over copy $\triangle A_1B_1C_1$ so that the copy and $\triangle ABC$ share a side (side $\overline{AC}$ in Figure 1.25). In this way a kite is formed. We can then use what we know about kites to show that the triangles are congruent.

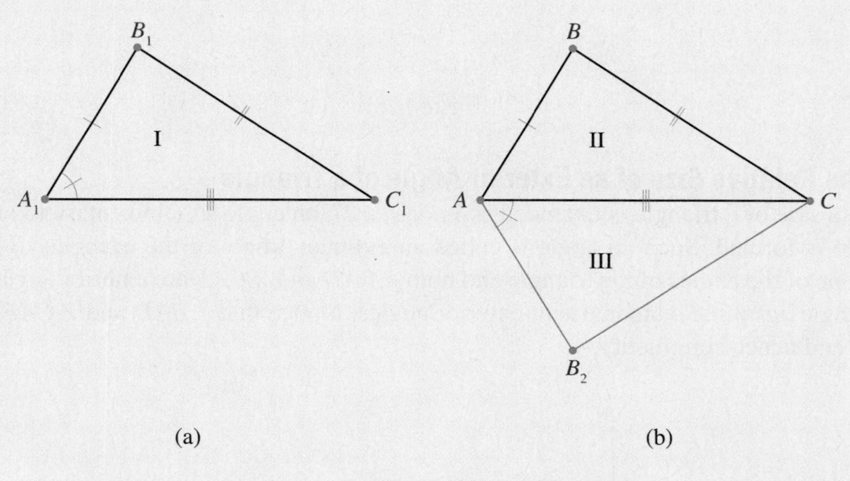

Figure 1.25

Proof

In Figure 1.25b, construct triangle III congruent to triangle I using the SAS condition. This can be accomplished by constructing $\angle CAB_2$ congruent to $\angle C_1A_1B_1$ and $AB_2 = A_1B_1$. It follows that $ABCB_2$ is a kite. [$AB_2 = A_1B_1$ by construction, $A_1B_1 = AB$ (given), and, therefore, $AB_2 = AB$. Also by **CPCT** (corresponding parts of congruent triangles), $B_2C = B_1C_1$ and $B_1C_1 = BC$ (given), which implies $B_2C = BC$.] From the properties of a kite, $\overline{CA}$ bisects $\angle BCB_2$ and, consequently, triangles III and II are congruent by ASA. From the congruence $\triangle I \cong \triangle III$ and $\triangle III \cong \triangle II$, it follows (by transitivity) that $\triangle I \cong \triangle II$ i.e. that $\triangle A_1B_1C_1 \cong \triangle ABC$. □

An approach similar to the one used in the proof of the SSS theorem can be used to determine a useful condition for two right triangles to be congruent. A triangle in which one of the angles is a 90° angle (that is, a **right angle**) is called a **right triangle**. The side opposite the right angle is called the hypotenuse and the other two sides are called **legs**. (We will soon prove that a triangle cannot have two right angles.)

Theorem 1.9

Hypotenuse–Leg (H-L) Congruence Condition

If the hypotenuse and a leg of one triangle are congruent to the hypotenuse and a leg of another right triangle, then the triangles are congruent.

The proof of this theorem is suggested in Figure 1.26 and will be explored in the problem set at the end of this section.

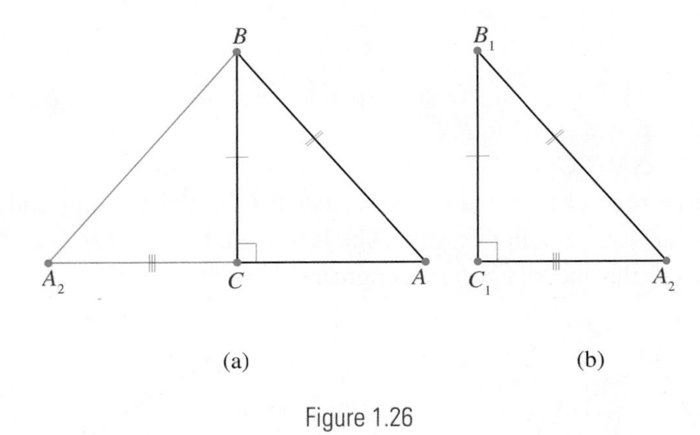

Figure 1.26

The Relative Size of an Exterior Angle of a Triangle

When a side of a triangle is extended as in Figure 1.27, an angle supplementary to an angle of the triangle is formed. Such an angle is called an **exterior angle** of the triangle. In Figure 1.27, $\angle BAC$ is one of the angles of the triangle and both $\angle BAD$ and $\angle CAE$ are exterior angles. In general, a triangle has three noncongruent exterior angles. Notice that $\angle BAD$ and $\angle CAE$ are vertical angles and hence congruent.

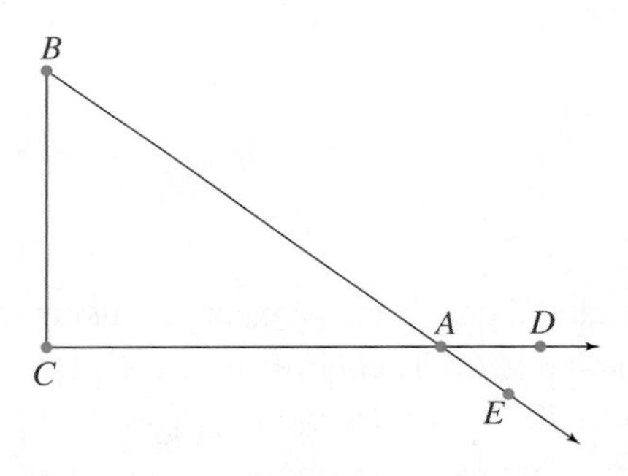

Figure 1.27

Any angle of the triangle that is not adjacent to an exterior angle is called a **remote interior** angle of the exterior angle. Thus $\angle B$ and $\angle C$ are remote interior angles of $\angle BAD$. Theorem 1.10 compares the size of an exterior angle and its remote interior angle. (We say that one angle is greater than another if the measure of that angle is greater than the measure of the other angle.) This major theorem will be used to prove several other theorems.

Theorem 1.10

The Exterior Angle Theorem

An exterior angle of a triangle is greater than either of the remote interior angles.

Restatement

Given $\triangle ABC$ in which $\overline{BA}$ has been extended as in Figure 1.28.

Prove $m(\angle CAD) > m(\angle ACB)$.

Plan To prove that $\angle CAD$ is greater than $\angle C$, we try to "fit" the latter angle inside the exterior angle. This can be done by creating congruent triangles.

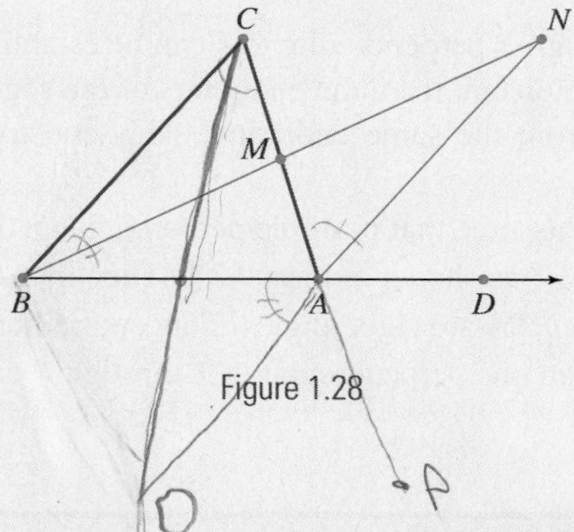

Figure 1.28

Proof

Let M be the midpoint of $\overline{AC}$. Extend $\overline{BM}$ such that $\overline{BM} \cong \overline{MN}$. Now $\triangle MCB \cong \triangle MAN$ by SAS (why?) and, therefore, $\angle CAN \cong \angle C$. Because N is in the interior of $\angle CAD$, $\angle CAN$ is smaller than $\angle CAD$. Because $\angle CAN \cong \angle C$, it follows that $\angle C$ is smaller than $\angle CAD$. To prove that $\angle CAD$ is greater than $\angle B$, extend $\overline{CA}$ to create the ray $\overrightarrow{CA}$. Then an exterior angle congruent to $\angle CAD$ is formed. Now "fit" $\angle B$ inside that angle by a proper construction. The details are left as an exercise. □

Notice that we have not justified the intuitively obvious fact that N is in the interior of $\angle CAD$. This can be proved using the Plane Separation Axiom of the Appendix. We leave the details to the interested reader.

Corollary 1.5 follows immediately from the Exterior Angle Theorem:

Corollary 1.5 *Through a point not on a line, there is a unique (one and only one) perpendicular to the line.*

Restatement

Given A line ℓ as in Figure 1.29a and P as a point not on ℓ.

Prove There exists one and only one line through P perpendicular to ℓ.

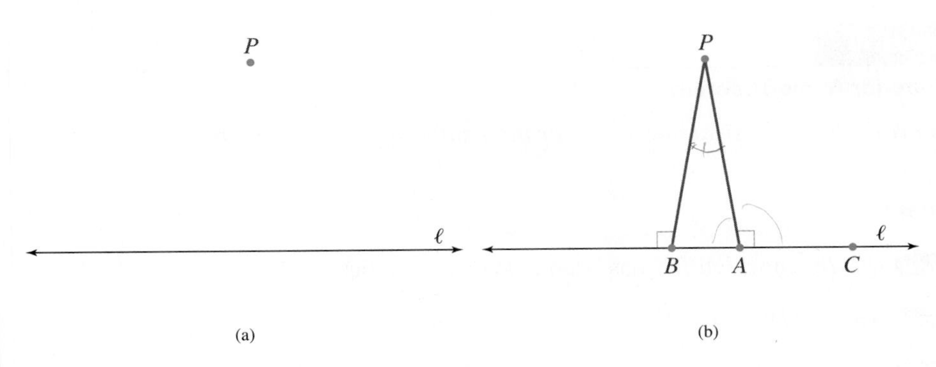

Figure 1.29

Proof

The existence of a line through P perpendicular to ℓ can be established by constructing a segment on ℓ such that P is equidistant from the endpoints of the segment and then constructing another point equidistant from the same endpoints. This construction is explored in Now Solve This 1.5.

To prove the uniqueness, assume that multiple perpendiculars through P exist. Let $\overline{PA}$ and $\overline{PB}$ be two perpendiculars to ℓ as shown in Figure 1.29b. Then in $\triangle PAC$ the exterior angle PAC and its remote interior angle PBA are right angles. This contradicts the Exterior Angle Theorem. Consequently, more than one perpendicular to ℓ through P cannot exist. □

Now Solve This 1.5

In Figure 1.29, construct a circle with center at P and intersecting ℓ in two points. Next construct a point Q equidistant from these two points. Why is $\overleftrightarrow{PQ}$ perpendicular to ℓ?

The Exterior Angle Theorem or the uniqueness of the perpendicular to a line through a point can be used to prove the following theorem.

Theorem 1.11

Hypotenuse–Acute Angle Congruence Condition

If the hypotenuse and an acute angle of one right triangle are congruent to the hypotenuse and an acute angle of another right triangle, then the triangles are congruent.

Restatement

Given $\triangle ABC$ and $\triangle A_1B_1C_1$ in which $\overline{AB} \cong \overline{A_1B_1}$, $\angle A \cong \angle A_1$, and $\angle C$ and $\angle C_1$ are right angles.

Prove $\triangle ABC \cong \triangle A_1B_1C_1$.

Proof

In Figure 1.30, the marked parts of $\triangle ABC$ and $\triangle A_1B_1C_1$ are congruent. If $\overline{AC} \cong \overline{A_1C_1}$, then the triangles are congruent by SAS. If $AC \neq A_1C_1$, then $AC < A_1C_1$ or $AC > A_1C_1$. In either case, we can find a point C_2 on $\overrightarrow{AC}$ such that $\overline{AC_2} \cong \overline{A_1C_1}$. Then $\triangle ABC_2 \cong \triangle A_1B_1C_1$

by SAS and, consequently, $\angle AC_2B$ is a right angle. Then $\overline{BC_2}$ and $\overline{BC}$ are two different perpendiculars from B to $\overleftrightarrow{AC}$. This contradicts the uniqueness of a perpendicular from a point to a line. Thus $\overline{AC} \neq \overline{A_1C_1}$ must be rejected and, therefore, $\overline{AC} \cong \overline{A_1C_1}$ and $\triangle ABC \cong \triangle A_1B_1C_1$ by SAS. □

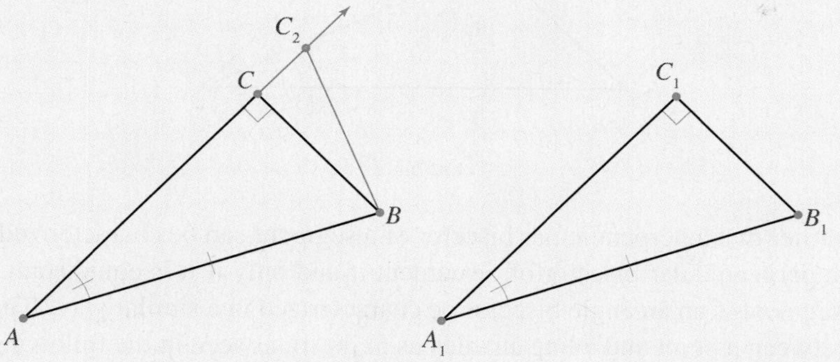

Figure 1.30

The existence of a unique line perpendicular to a given line through a given point enables us to define a concept important in finding the areas of triangles and other figures.

Definition of the Distance from a Point to a Line The distance from a point P to a line ℓ is the length of the segment connecting P with the foot of the perpendicular to ℓ through P. (The foot of the perpendicular is the point of intersection of the perpendicular and ℓ.)

In Figure 1.31, the length of $\overline{PQ}$ is the distance from P to ℓ.

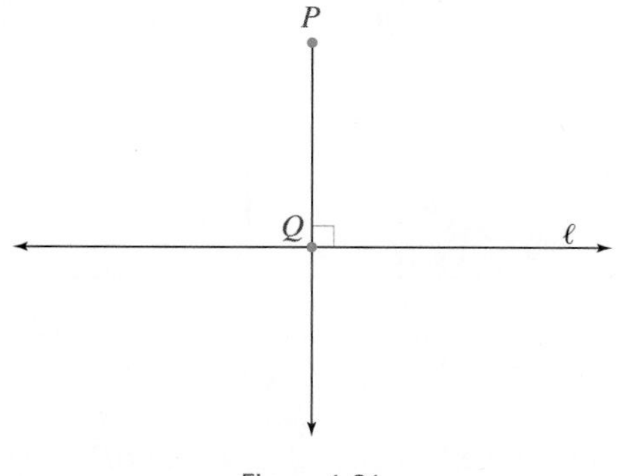

Figure 1.31

The distance from a vertex of a triangle to the line containing the opposite side is called a **height**. Thus a height is the length of an altitude. A triangle has three heights corresponding to its three sides. Figure 1.32 shows the three altitudes of an obtuse triangle and the corresponding heights h_1, h_2, and h_3. Notice that the lines containing the three altitudes intersect at a single point P in Figure 1.32. When three or more lines intersect in a single point, we say that they are **concurrent**. The proof that the lines containing the altitudes of a triangle are concurrent follows from the fact that the perpendicular bisectors of the sides of a triangle are concurrent. These and other concurrency theorems will be explored in the problem set at the end of this section.

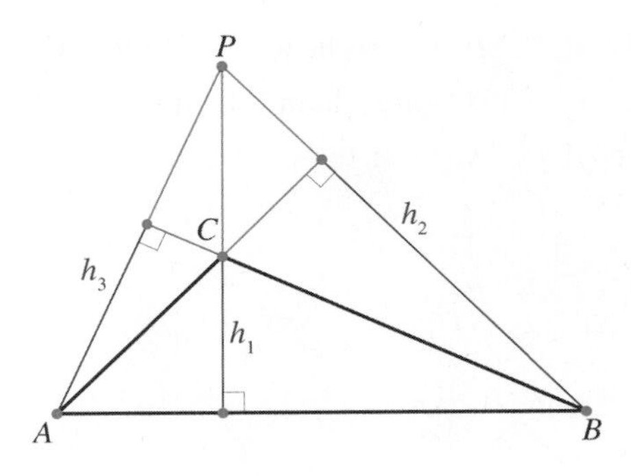

Figure 1.32

We saw earlier that a perpendicular bisector of a segment can be characterized as follows: A point is on the perpendicular bisector of a segment if and only if it is equidistant from the endpoints of the segment. Can an angle bisector be characterized in a similar way? The definition of the distance between a point and a line enables us to do so, as seen in the following theorem:

Theorem 1.12

A point is on the angle bisector of an angle if and only if it is equidistant from the sides of the angle.

Restatement

1. **Given** P on the angle bisector of $\angle A$ as in Figure 1.33.
   Prove $\overline{PB} \cong \overline{PC}$ ($\overline{PB}$ and $\overline{PC}$ are perpendiculars to the sides of the angle, respectively).
2. **Given** The distances from P to the sides of the angle are equal.
   Prove P is on the angle bisector of $\angle A$.

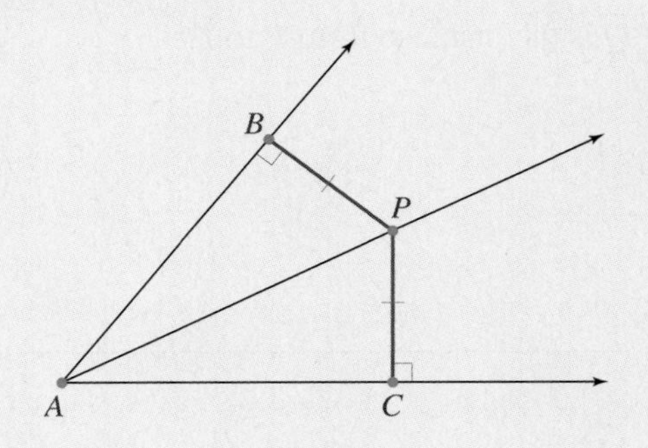

Figure 1.33

Proof 1 Because P is on the angle bisector of $\angle A$, $\triangle ABP \cong \triangle ACP$ by the Hypotenuse–Acute Angle Theorem. Consequently, $PB = PC$ (by CPCT). □

Proof 2 Since $\overline{PB} \cong \overline{PC}$ (given), $\triangle ABP \cong \triangle ACP$ by the Hypotenuse–Leg Condition. Consequently, $\angle PAB \cong \angle PAC$ (by CPCT). □

We can use the Exterior Angle Theorem to prove a theorem concerning the relative sizes of the angles opposite non-congruent sides of a triangle and, conversely, the relative sizes of the sides opposite non-congruent angles. The proofs of these theorems will be explored in the problem set at the end of this section.

Theorem 1.13

Given two non-congruent sides in a triangle, the angle opposite the longer side is greater than the angle opposite the shorter side.

Theorem 1.14

The Converse of Theorem 1.13

Given two non-congruent angles in a triangle, the side opposite the greater angle is longer than the side opposite the smaller angle.

Theorem 1.14 is instrumental in proving the intuitively well-known fact that the shortest path connecting any two points A and B is the segment $\overline{AB}$. This understanding is formalized in the following theorem, which is fundamental in most branches of mathematics:

Theorem 1.15

The Triangle Inequality

The sum of the lengths of any two sides of a triangle is greater than the length of the third side.

Restatement

Given $\triangle ABC$ with sides of length a, b, and c (shown in Figure 1.34a).

Prove $a + b > c$, $a + c > b$, and $b + c > a$.

Plan We prove the first inequality $a + b > c$. (The proofs of the other two inequalities are analogous.) Our plan is to construct a triangle with one side $a + b$ and another side c. Then we show that the angle opposite side $a + b$ is greater than the angle opposite side c and hence conclude from Theorem 1.14 that $a + b > c$.

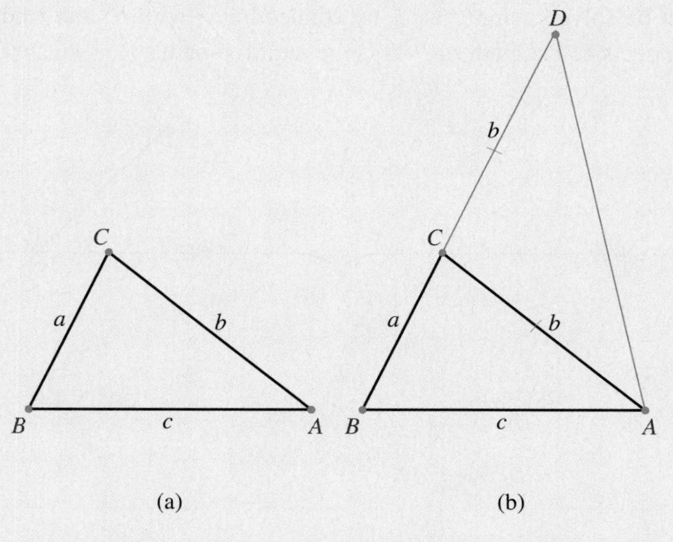

Figure 1.34

Proof

Extend side $\overline{BC}$ in Figure 1.34b so that $CD = b$. Next we show that $\angle BAD$ is greater than $\angle D$. We know that $\angle D \cong \angle CAD$ because $\triangle CDA$ is an isosceles triangle. Because C is in the interior of $\angle BAD$, we have $m(\angle BAD) > m(\angle CAD)$. Consequently, $m(\angle BAD) > m(\angle D)$. Applying Theorem 1.14 in $\triangle BAD$, we have $BD > BA$ or $a + b > c$. □

Now Solve This 1.6

1. Show that given three segments a, b, and c such that $a + b > c$ it is not always possible to construct a triangle whose sides are a, b, and c.
2. Construct three segments a, b, and c so that a triangle with these segments as sides will exist. Is the existence and therefore the construction of the triangle assured by Theorem 1.15? Justify your answer.

Example 1.4 **Hiker's Shortest Path.** Given line ℓ and points A and B not on the line, find the shortest path from A to a point on the line and then to B. (See also Chapter 0, "The Hiker's Path.")

Solution

Case 1. In Figure 1.35a, the points A and B are on opposite sides of the line. The segment $\overline{AB}$ intersects ℓ at P, and the "straight" path A–P–B is the shortest path. This is the case because if Q is any other point on ℓ, then by the triangle inequality $AQ + QB > AB$.

Case 2. Suppose that A and B are on the same side of ℓ, as in Figure 1.35b. We could reduce this case to the previous one by imagining that ℓ is a mirror and finding the shortest path connecting A with a point P on ℓ and the reflection B' of B in the mirror. The reflection of B in ℓ is obtained by dropping a perpendicular from B to ℓ and finding B' so that ℓ is the perpendicular bisector of $\overline{BB'}$. (A detailed study of reflections will be pursued in Chapter 5 on transformations.)

Notice that a property of a perpendicular bisector implies that if Q is any point on ℓ, then the length of the path A–Q–B is the same as the length of the path A–Q–B'. Consequently, this case reduces to finding the shortest path connecting A to some point on ℓ and B'. It can be solved using Case 1 by connecting A with B' and finding the point P where $\overline{AB'}$ intersects ℓ. The path A–P–B is the required path.

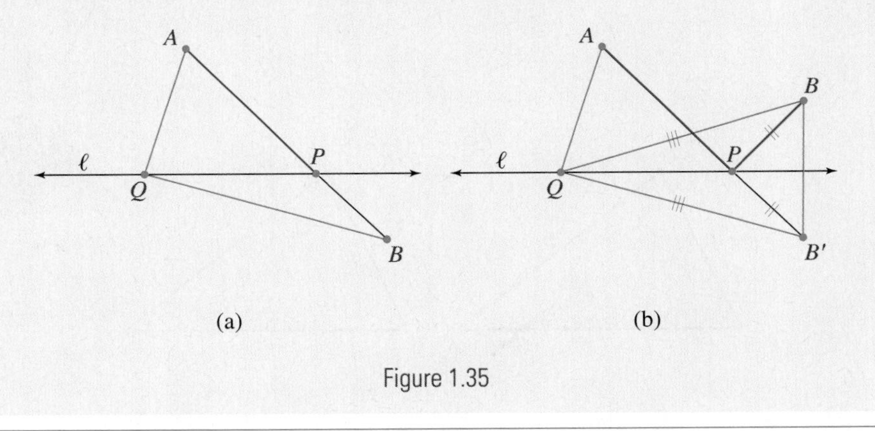

Figure 1.35

In the solution of Example 1.4, we used the concept of reflection in a line. For easy reference, here is the definition.

Reflection in a Line A **reflection in a line** ℓ assigns to each point P in the plane not on ℓ the point P', the *image of* P, in such a way that ℓ is the perpendicular bisector of $\overline{PP'}$. If P is on ℓ, then $P' = P$. (See Figure 1.36.)

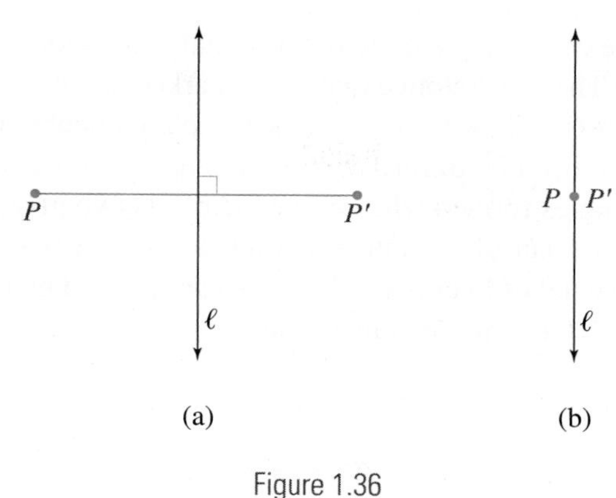

Figure 1.36

Remark

1. When P' is the image of P under reflection in a line ℓ, we also say that P' *is the reflection of* P *in* ℓ.
2. Reflection in a line is a function from the plane to the plane.

Properties of Parallel Lines

Through a point P not on a line ℓ, there exists a line parallel to ℓ. This fact follows from Theorem 1.16:

Theorem 1.16

If two lines in the same plane are each perpendicular to a third line in that plane, then they are parallel.

Restatement

Given k, ℓ, and m are three lines in the same plane as in Figure 1.37 such that $k \perp m$ and $\ell \perp m$.

Prove $k \parallel \ell$.

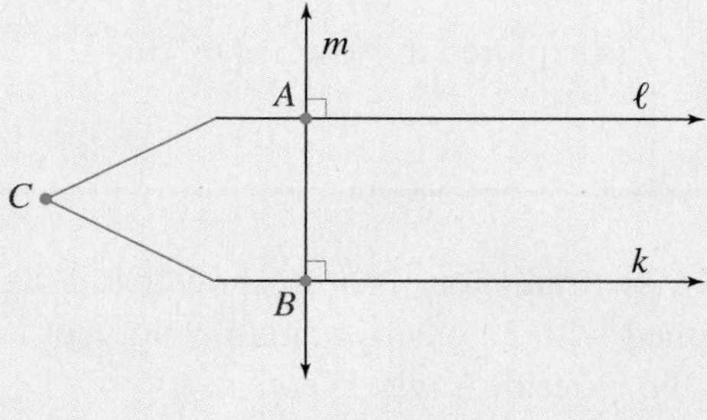

Figure 1.37

Proof

If k and ℓ are not parallel, then they must intersect at some point C. In that case, there will be two perpendiculars to m through C, which contradicts Corollary 1.5. Consequently, $k \parallel \ell$. □

We can also prove Theorem 1.16 by direct use of the Exterior Angle Theorem. The angles marked in Figure 1.37 are 90° each; hence $\angle CAB$ is also a right angle (property of vertical angles). Consequently, an exterior angle of $\triangle ABC$ is equal to a remote interior angle, which contradicts the Exterior Angle Theorem. Notice that if the marked angles in Figure 1.37 were only congruent but not necessarily 90° each, we could use the same argument to prove $k \parallel \ell$. This is stated in the next theorem (Theorem 1.17). Before we proceed, however, we need to define some common terminology for the angles formed when a line intersects two given lines.

Any line that intersects a pair of lines in two points is called a **transversal** of those lines. In Figure 1.38, line m is a transversal of lines k and ℓ. Angles are formed by these lines and are named according to their position relative to the transversal.

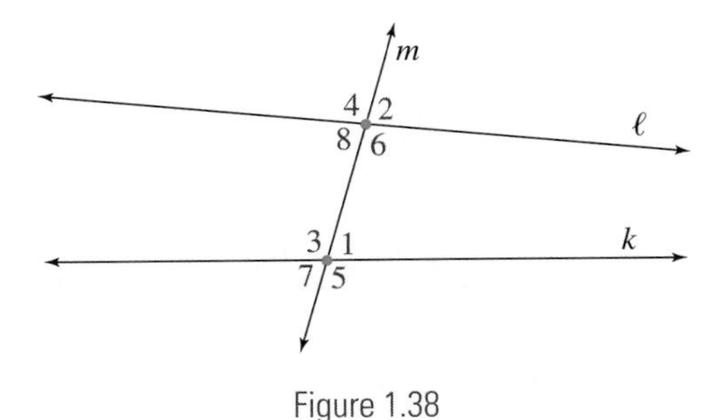

Figure 1.38

Interior angles: $\angle 1$, $\angle 3$, $\angle 6$, $\angle 8$

Corresponding angles: $\angle 1$ and $\angle 2$, $\angle 3$ and $\angle 4$, $\angle 5$ and $\angle 6$, $\angle 7$ and $\angle 8$

Alternate interior angles: $\angle 1$ and $\angle 8$, $\angle 3$ and $\angle 6$

Alternate exterior angles: $\angle 5$ and $\angle 4$, $\angle 7$ and $\angle 2$

Remark Two corresponding angles are congruent if and only if two alternate interior angles are congruent.

Theorem 1.17

If two lines are cut by a transversal and a pair of corresponding angles is congruent (or a pair of alternate interior angles is congruent), then the lines are parallel.

The proof of Theorem 1.17 is explored in Now Solve This 1.7.

Now Solve This 1.7

1. Prove Theorem 1.17 by contradiction. That is, show that if the lines were not parallel, a triangle would be formed with an exterior angle congruent to a remote interior angle, which will contradict the Exterior Angle Theorem.

2. Draw a line ℓ and a point P not on the line. Then construct a line m through the point P parallel to ℓ. Describe the construction and prove that $m \parallel \ell$.
3. Recall that we have defined a rhombus as a quadrilateral with all sides congruent. Prove that opposite sides of a rhombus are parallel.

4. Solve Problem 2 by constructing a rhombus.

You may wonder about the converse of Theorem 1.17: If two parallel lines are cut by a transversal, then a pair of corresponding angles is congruent. This statement is needed to prove that the sum of measures of the angles of any triangle is 180° and, therefore, many other theorems of

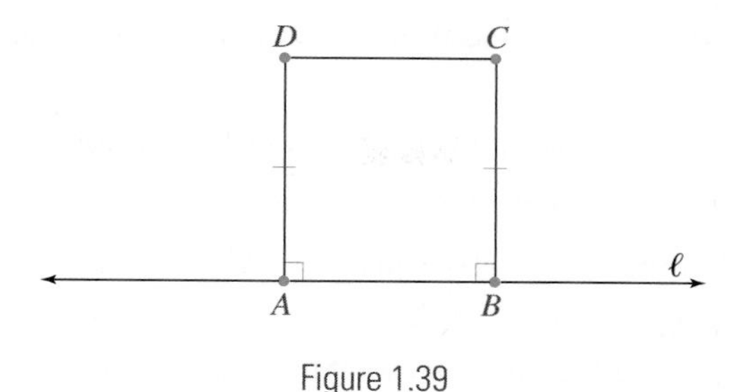

Figure 1.39

Euclidean geometry. In fact, the statement is equivalent to the famous Parallel Postulate discussed in Section 1.3 and hence distinguishes Euclidean geometry from non-Euclidean geometries. (See the Historical Note on Gerolamo Saccheri.) Even the existence of rectangles cannot be established using the axioms we have stated and the theorems we have proved so far. (A **rectangle** is a quadrilateral with four right angles.) If we try to construct a rectangle as shown in Figure 1.39 by constructing a line ℓ, marking two points A and B on the line, and then constructing two segments $\overline{AD}$ and $\overline{BC}$ each perpendicular to ℓ, we obtain what is called a **Saccheri quadrilateral**. The segment $\overline{AB}$ is called the *lower base* (at A and B, the angles are 90°) and the opposite side is called the *upper base*.

It is easy to prove that the diagonals of a Saccheri quadrilateral are congruent and that the upper base angles are congruent. Because we have not proved yet that the sum of the measures of the angles of a triangle is 180° (and hence that the sum of the measures in any quadrilateral is 360°), we cannot conclude that the Saccheri quadrilateral is a rectangle. This can be done with the Parallel Postulate (see the introduction to this chapter and Figure 1.1), as will be shown in the next section.

Historical Note: Gerolamo Saccheri

Gerolamo Saccheri (1667–1733), an Italian mathematician, was dissatisfied with the Parallel Postulate. Like most mathematicians at the time, he believed that it should be possible to prove the postulate as a theorem using previous axioms and theorems. He wrote a book entitled *Euclid Freed of All Blemish*, in which he tried to prove the Parallel Postulate. Although Saccheri's proof was wrong, in the attempt to find a proof he did prove many new theorems that are now considered a part of *absolute* or *neutral geometry*, a geometry independent of the Parallel Postulate. The theorems we have proved so far belong to the realm of absolute or neutral geometry.

Now Solve This 1.8

A **parallelogram** is defined as a quadrilateral in which each pair of opposite sides is parallel. Is it possible to prove that the diagonals of a parallelogram bisect each other using only the material covered so far? Explain.

Problem Set 1.2

Answer Problems 1 through 22 without using the Parallel Postulate.

1. The congruence $\triangle ABC \cong \triangle DEF$ can also be written $\triangle BAC \cong \triangle EDF$. How many such symbolic representations exist for two congruent triangles? Do different representations give different information about the triangles?

2. State and prove a congruence condition (other than the definition) for two quadrilaterals to be congruent.
3. State and prove a theorem analogous to Theorem 1.7 for non-convex kites.
4. Write a careful proof of Theorem 1.8 (SSS condition).
5. Write a proof of Theorem 1.9 (Hypotenuse–Leg Condition).
6. a. Draw an acute triangle (all angles less than 90°) and an obtuse triangle (one angle greater than 90°). In each case construct the three perpendicular bisectors of the sides of the triangle.

   b. In part (a) you must have noticed that the perpendicular bisectors of the sides of a triangle are concurrent—they intersect in a single point. Examine the following proof that this is the case and use a similar approach to prove that the angle bisectors of a triangle are concurrent.

   In $\triangle ABC$, let O be the point where the perpendicular bisectors of $\overline{AB}$ and $\overline{BC}$ intersect. We need to prove only that O is also on the third perpendicular. We have proved that a point is on the perpendicular bisector of a segment if and only if it is equidistant from the endpoints of the segment. Thus it suffices to prove that $AO = CO$. We have:

   $AO = BO$ (since O is on the perpendicular bisector of $\overline{AB}$)

   $BO = CO$ (since O is on the perpendicular bisector of $\overline{BC}$)

   Consequently, $AO = CO$ and, therefore, O must be on the perpendicular bisector of $\overline{AC}$.

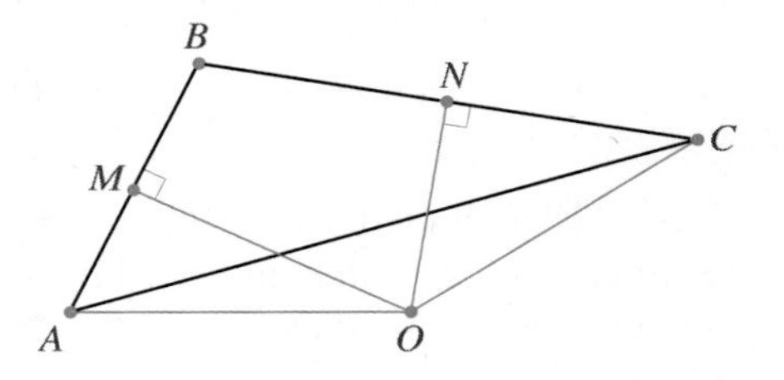

   c. The proof in part (b) is not completely rigorous; an assumption was made without proving it. What was the assumption?

7. Draw a line ℓ and a point P not on ℓ. Construct the perpendicular to ℓ through P. Describe the steps in the construction and prove that it is valid (i.e., that the line you constructed through P is perpendicular to ℓ).

8. A circle that passes through each vertex of a triangle is called a **circumscribed circle** or **circumcircle** as shown in figure (a). A circle that is tangent to each side of a triangle is called an **inscribed circle** or **incircle** as shown in figure (b).

   a. Assume that a tangent to a circle is perpendicular to the radius at the point of contact. Explain how to find the circumscribing circle and the inscribed circle for a given triangle. Justify your answers.

   b. Draw a triangle in which (i) all the angles are *acute* (less than 90°); (ii) one angle is *obtuse* (greater than 90°); and (iii) one angle is 90°. In each case construct the circumscribing and inscribed circles.

   c. What seems to be true about the centers of the circumscribed circles?

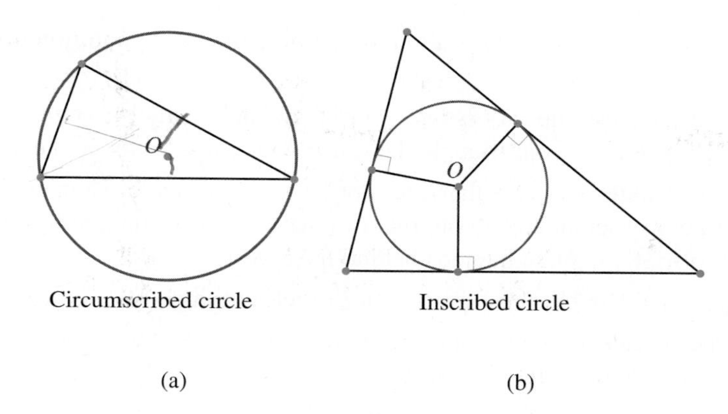

(a) (b)

9. **a.** Prove that the angle bisectors of a rhombus are concurrent (intersect in a single point). Then prove that the point where the angle bisectors intersect is equidistant from all sides of the rhombus.

   b. Construct a rhombus and the circle inscribed in the rhombus. Clearly identify the radius of the circle.

10. **a.** Prove that the angle bisectors of a convex kite are concurrent as follows. Consider the kite $ABCD$ in the figure below and its diagonal BD. Prove that $\overline{BD}$ is on the angle bisector of $\angle B$ as well as $\angle D$. Then construct the angle bisector of $\angle A$ and point O where that angle bisector intersects $\overline{BD}$. Next let d_1 be the distance from O to $\overline{AD}$, d_2 be the distance from O to $\overline{AB}$, d_3 be the distance from O to $\overline{BC}$, and d_4 be the distance from O to $\overline{DC}$. Argue that $d_1 = d_2$, $d_2 = d_3$, and $d_4 = d_1$. Conclude that $d_3 = d_4$ and hence that O is on the angle bisector of $\angle C$.

    b. Construct any kite and the circle inscribed in the kite. Clearly identify the radius of the circle. Describe the construction and prove that it is valid.

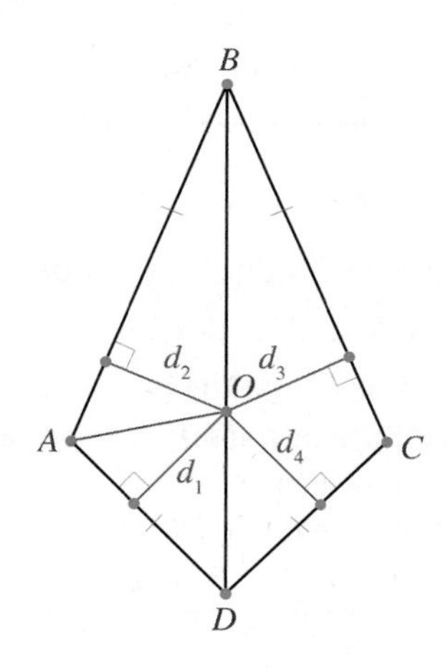

11. Use the Exterior Angle Theorem to prove that in any $\triangle ABC$, $m(\angle A) + m(\angle B) + m(\angle C) < 270°$.

12. Prove that in any quadrilateral, the sum of the lengths of any three sides is greater than the length of the fourth side.

13. Prove Theorem 1.14 (which is the converse of Theorem 1.13), which states that $\alpha > \beta \Rightarrow a > b$, as follows. Assume that $a > b$ is false (i.e., that $a \leq b$) and show that the

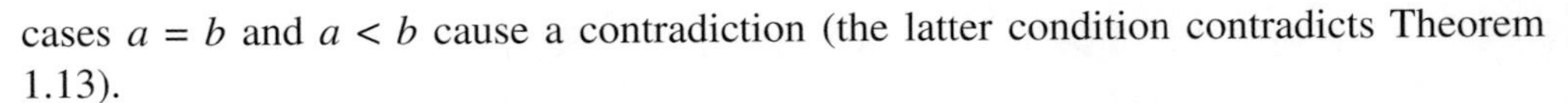

cases $a = b$ and $a < b$ cause a contradiction (the latter condition contradicts Theorem 1.13).

14. **a.** Construct a **scalene** (not isosceles) triangle and another triangle congruent to it using the three sides of your triangle. Describe your construction and explain why it is valid.

b. Draw an angle and a ray that does not intersect the angle. Then use your answer to part (a) to construct an angle congruent to the original angle that has the ray as one of its sides. Briefly explain the idea behind the construction.

c. Repeat part (b), but this time use an isosceles triangle to "duplicate" the angle.

15. Construct a scalene triangle and a triangle congruent to it using only two sides and the included angle of the original triangle.

•**16.** Construct a rhombus given its diagonals (that is, given two segments congruent to the diagonals). Describe and justify your construction.

17. Given three segments of length a, b, and c, what conditions must a, b, and c satisfy to form the sides of some triangle?

18. Prove that the sum of the distances from any point in the interior of a triangle to the three vertices is greater than half the **perimeter** (the sum of the length of the sides) of the triangle.

19. Prove that the perimeter of a quadrilateral is greater than the sum of the lengths of the diagonals.

20. In $\triangle ABC$, $\overline{OB}$ and $\overline{OC}$ bisect $\angle ABC$ and $\angle ACB$, respectively. Prove that if $AB > AC$, then $OB > OC$.

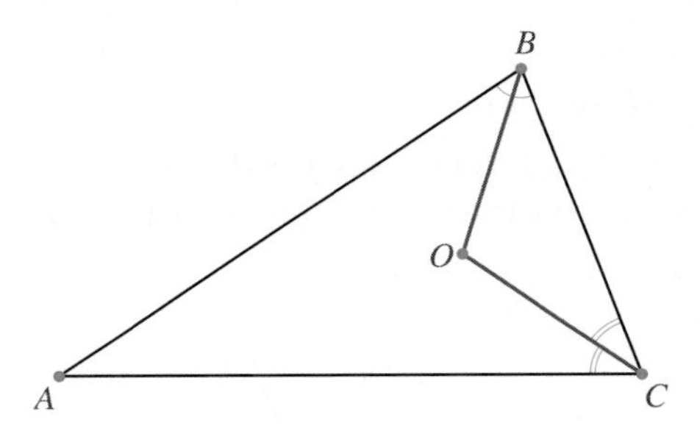

•**21.** Prove that if a, b, and c are the lengths of the sides of a triangle, then there exists a triangle with sides of lengths $\sqrt{a}$, $\sqrt{b}$, and $\sqrt{c}$.

22. $\angle AOB$ has vertex O, which is not on the paper. Construct the bisector of $\angle AOB$ (without extending the sides on additional paper). Describe your construction and prove that it produces the required angle bisector.

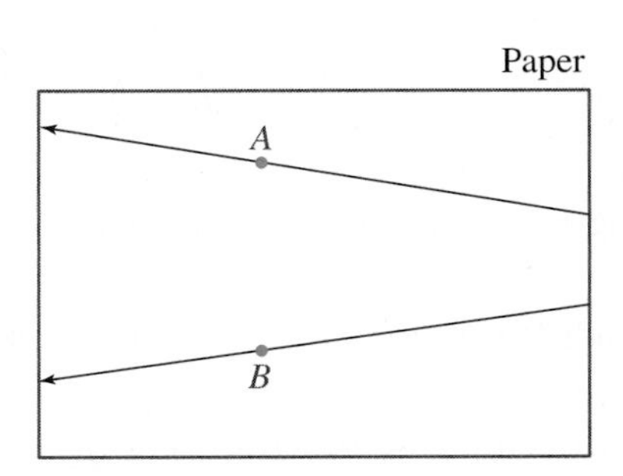

1.3 The Parallel Postulate and Its Consequences

Circa 300 B.C.E., Euclid arranged plane geometry based on five postulates. The fifth postulate, known as the Parallel Postulate, was stated by Euclid as follows:

> If a straight line falling on two straight lines make the interior angles on the same side less than two right angles, the two straight lines, if produced indefinitely, meet on that side on which the angles are less than the two right angles.

From ancient times this postulate was criticized for several reasons. Euclid proved 28 theorems before using the fifth postulate in a proof. The converse of the Parallel Postulate, equivalent to "The sum of two angles of a triangle is less than two right angles," was proved in Euclid's *Elements* as a theorem. In addition, mathematicians felt that because the postulate is so intuitively obvious, there must be a way to prove it. Indeed, for generations mathematicians believed that the Parallel Postulate could be proved based on Euclid's previous postulates. As a result of the many attempts to prove this postulate, several equivalent statements have been discovered. The most common substitute for Euclid's Parallel Postulate is known as **Playfair's axiom** after the Scottish mathematician John Playfair (1748–1819), although it appears earlier in the writing of Proclus (410–485 C.E.[1]) in his *Commentary on the First Book of Euclid's Elements.* Because of its simplicity Playfair's axiom is used today as *the* Parallel Postulate:

Axiom 1.2 The Parallel Postulate (Playfair's Axiom) *Given a line and a point not on the line, there exists a unique line through the point parallel to the given line.*

Restatement

Given ℓ_1 and P not on ℓ_1, there exists one and only one line ℓ_2 through P such that $\ell_1 \parallel \ell_2$ (see Figure 1.40).

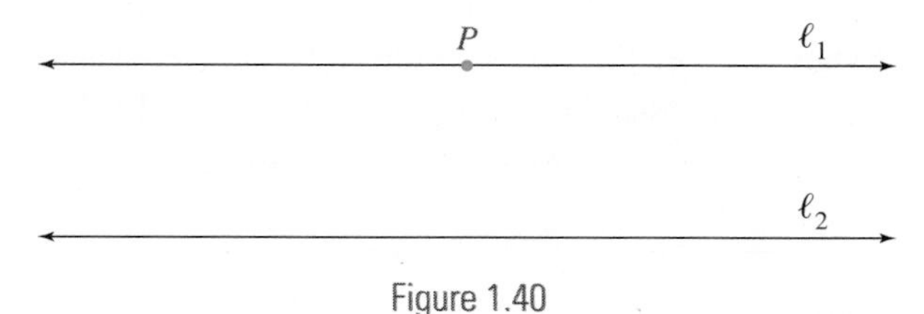

Figure 1.40

Historical Note: John Playfair

John Playfair (1748–1819), a Scottish mathematician and physicist, was educated at home until the age of 14, when he started his studies at the University of St. Andrews. He was ordained as a minister but continued his studies. In 1785, Playfair was awarded the chair of mathematics at the University of Edinburgh. In 1795, he wrote *Elements of Geometry*, which became a standard text in geometry and went through many editions. In that text, Playfair introduced his version of Parallel Postulate, which remains the standard today.

The Parallel Postulate is the turning point at which different geometries branch out. By accepting the Parallel Postulate, **Euclidean geometry** is developed. If we deny the postulate, there are two possibilities:

1. There is more than one parallel to a line through a given point not on the line.
2. There is no parallel to a line through a point not on the line.

---

1. C.E. stands for "Common Era."

In each case we obtain a **non-Euclidean geometry**. Statement 1 is the basis for **hyperbolic geometry**, while statement 2 with some modifications to previous axioms results in **elliptic geometry**.

Using the Parallel Postulate, the following converse of Theorem 1.17 can be proved as well as the subsequent theorems of Euclidean geometry:

Theorem 1.18

If two parallel lines are cut by a transversal, then a pair of corresponding angles is congruent.

Restatement

Given $\ell_1 \parallel \ell_2$ (as in Figure 1.41) and t (a transversal).

Prove $\angle 1 \cong \angle 2$

Plan We assume that the angles are not congruent and obtain a contradiction of the Parallel Postulate.

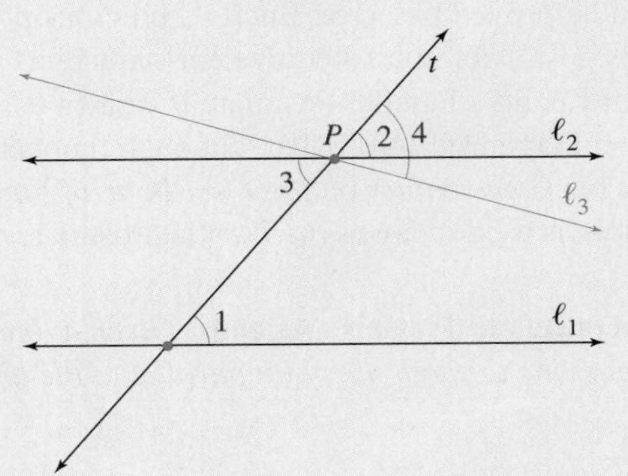

Figure 1.41

Proof

If $\angle 1$ and $\angle 2$ are not congruent, then there exists a line ℓ_3 through P such that $\ell_3 \neq \ell_1$ and so that $\angle 4$, which ℓ_3 makes with t, is congruent to $\angle 1$. Then by Theorem 1.17, $\ell_3 \parallel \ell_1$. Thus, through P, there are two lines ℓ_2 and ℓ_3 parallel to ℓ_1, which contradicts the Parallel Postulate. Consequently, $\angle 1 \cong \angle 2$. □

Theorem 1.18, its converse Theorem 1.17, and the fact that a pair of corresponding angles is congruent if and only if a pair of alternate interior angles is congruent imply Theorems 1.19 and 1.20:

Theorem 1.19

Two lines in a plane are parallel if and only if a pair of corresponding angles formed by a transversal is congruent.

Theorem 1.20

Two lines in a plane are parallel if and only if a pair of alternate interior angles formed by a transversal is congruent.

In Figure 1.42, $\ell_1 \parallel \ell_2$ and the angles are as marked. By Theorem 1.19, $\alpha = \gamma$. Because $\beta + \gamma = 180°$, we have $\alpha + \gamma = 180°$. Conversely, if $\alpha + \gamma = 180°$ (and it is not given that $\ell_1 \parallel \ell_2$), we can show that $\alpha = \gamma$ and, therefore, by Theorem 1.19 that $\ell_1 \parallel \ell_2$. These observations are stated in the following theorem:

Theorem 1.21

Two lines are parallel if and only if a pair of interior angles on the same side of a transversal is supplementary.

Restatement

In Figure 1.42, $\ell_1 \parallel \ell_2$ if and only if $\alpha + \beta = 180°$.

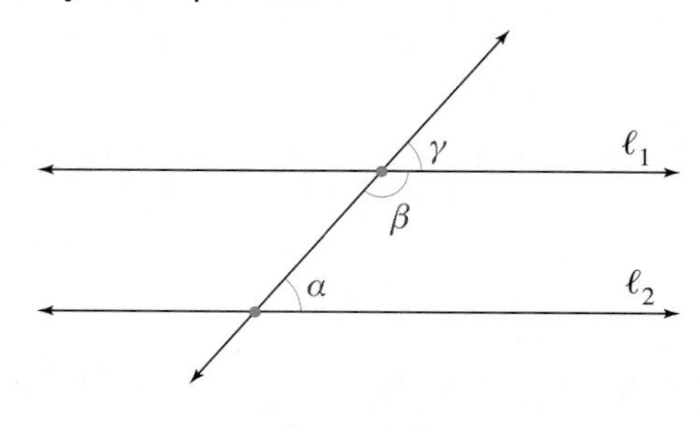

Figure 1.42

We can now prove a key theorem of Euclidean geometry:

Theorem 1.22

The sum of the measures of the interior angles of a triangle is 180°.

Restatement

Given $\triangle ABC$.

Prove $m(\angle A) + m(\angle B) + m(\angle C) = 180°$.

Proof

We could produce this result by showing that the sum of the measures of the angles of a triangle equals the measure of a straight angle. A straight line will create angles congruent to the angles of a triangle if it is parallel to one of the triangle's sides. Thus we construct in Figure 1.43 a line ℓ through C parallel to $\overline{AB}$. Because $\overleftrightarrow{AC}$ is a transversal for the parallel lines ℓ and $\overleftrightarrow{AB}$, $\alpha = \alpha_1$, as they are alternate interior angles. Similarly, because $\overleftrightarrow{CB}$ is a transversal for the same parallel lines, $\beta = \beta_1$. Consequently, $\alpha + \beta + \gamma = \alpha_1 + \beta_1 + \gamma = 180°$. □

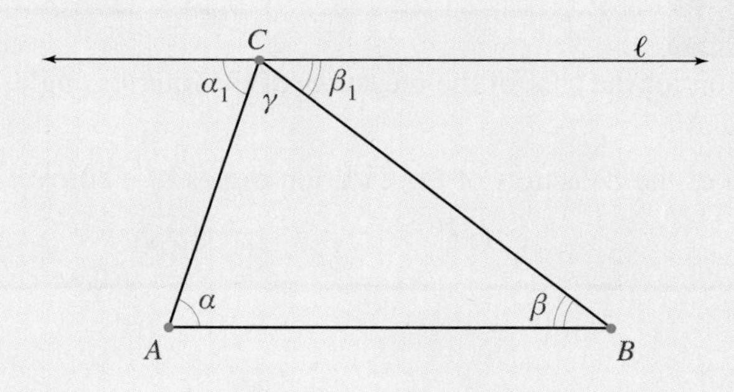

Figure 1.43

Example 1.5 In Figure 1.44, $\ell_1 \parallel \ell_2$ and k is a transversal. If $\overrightarrow{BC}$ and $\overrightarrow{AC}$ are angle bisectors of the interior angles as shown, find $m(\angle B)$.

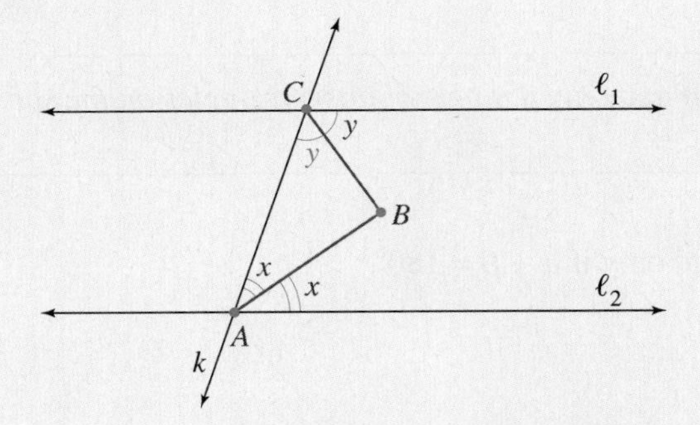

Figure 1.44

Solution

If we mark the measures of the angles as shown in Figure 1.44, we have

$$2x + 2y = 180^\circ \quad \text{(Theorem 1.21)}$$
$$x + y = 90^\circ$$

Now in $\triangle ABC$,

$$x + y + m(\angle B) = 180^\circ$$

Hence

$$m(\angle B) = 180^\circ - (x + y)$$
$$= 180^\circ - 90^\circ$$
$$= 90^\circ$$

Consequently, $\angle B$ is a right angle.

Now Solve This 1.9

1. Find an expression for the sum of the measures of the interior angles of a convex n-gon in terms of n.
2. What is the sum of the measures of the exterior angles of a convex n-gon? Prove your answer.

We are now able to prove the following:

Theorem 1.23

A Saccheri quadrilateral is a rectangle.

Recall that a rectangle is a quadrilateral with four right angles.

Restatement

Given In Figure 1.45, $\angle A$ and $\angle B$ are right angles, and $\overline{AD} \cong \overline{BC}$.

Prove $\angle D$ and $\angle C$ are right angles.

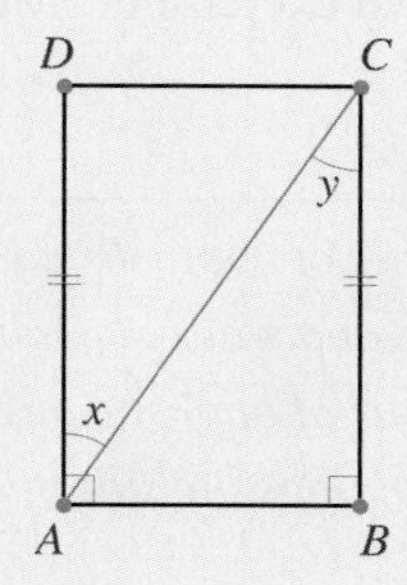

Figure 1.45

Proof

Construct the diagonal $\overline{AC}$. To prove that $\angle D$ is a right angle, we show that the diagonal divides $ABCD$ into congruent triangles. Notice that $x = y$ as alternate interior angles formed by the parallel lines AD and BC. Thus $\triangle ABC \cong \triangle CDA$ by SAS; consequently, $\angle D \cong \angle B$ and $\angle D$ is a right angle. Because the sum of the measures of the angles in any quadrilateral is 360° (see Now Solve This 1.9 or simply draw a diagonal and consider the sum of the angles in the two triangles), it follows that $m(\angle C) = 90°$. □

We define now some common terms of Euclidean geometry. We will investigate the properties of several quadrilaterals, such as the **trapezoid**, a quadrilateral with at least one pair of parallel sides; the **parallelogram**, a quadrilateral in which each pair of opposite sides is parallel; the **rhombus,** a parallelogram with two adjacent sides congruent (this definition is equivalent to the one given earlier, which stated that a rhombus is a quadrilateral with four congruent sides); and the **square**, a rhombus with a right angle. A **rectangle** can be defined as a parallelogram with a right angle. (It is easy to show that a quadrilateral in which all the angles are right angles is a rectangle, a property used in Theorem 1.23.)

The proofs of the following theorems are straightforward. You should prove them on your own.

Theorem 1.24

The measure of an exterior angle in a triangle is equal to the sum of the measures of its two remote angles.

Theorem 1.25

If a transversal is perpendicular to one of two parallel lines, it is also perpendicular to the other line.

Theorem 1.26

In a parallelogram:

1. *Each diagonal divides the parallelogram into two congruent triangles.*
2. *Each pair of opposite sides is congruent.*
3. *The diagonals bisect each other.*

It is useful to know which minimal properties characterize the various types of quadrilaterals. For example, which properties characterize a parallelogram? Before going any farther, list all conditions that are necessary and sufficient for a quadrilateral to be a parallelogram—without listing any more properties than needed. Then read on to compare your list of some of the useful conditions, and do the proofs as an exercise in Now Solve This 1.10.

Theorem 1.27

1. *A quadrilateral in which each pair of opposite sides is congruent is a parallelogram.*
2. *A quadrilateral in which the diagonals bisect each other is a parallelogram.*
3. *A quadrilateral in which each pair of opposite angles is congruent is a parallelogram.*
4. *A quadrilateral in which a pair of opposite sides is parallel and congruent is a parallelogram.*

The last condition in Theorem 1.27 may be the least well known but is often the most useful. Property 1 can be used for an efficient construction of parallel lines.

Now Solve This 1.10

1. Prove Theorems 1.24 through 1.27.
2. Part (1) of Theorem 1.27 can be written as follows: A necessary and sufficient condition for a quadrilateral to be a parallelogram is that each pair of opposite sides is congruent. Write the other parts of Theorem 1.27 using the phrase "necessary and sufficient."

Construction 1.5: Construction of a Line Through P (not on ℓ) Parallel to ℓ

In Figure 1.46a, a line ℓ and a point P are given. To construct through P a line parallel to ℓ, we make P a vertex of any parallelogram so that one of the sides of the parallelogram is on ℓ. By property 1 of Theorem 1.27, to achieve this goal we need simply make each pair of opposite sides congruent. We can do so by constructing all sides of the quadrilateral congruent as shown in Figure 1.46b. Notice that this construction yields a rhombus and, therefore, has opposite parallel sides.

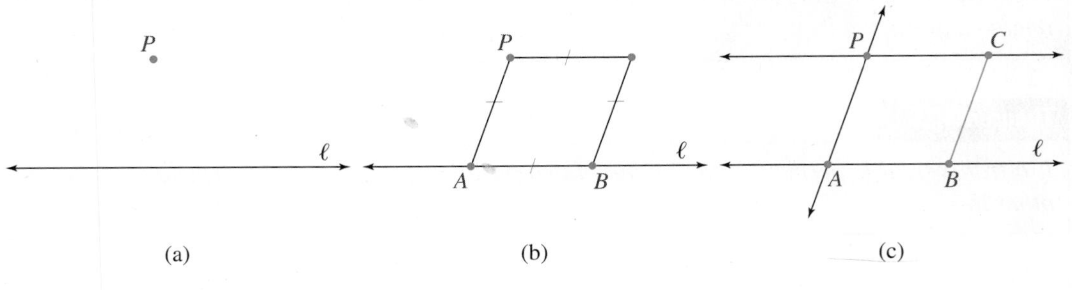

Figure 1.46

To accomplish the construction (see Figure 1.46c), draw any line through P that intersects ℓ and label the point of intersection A. (The labeling is, of course, not necessary for the actual construction.) Next make P and A be the first two vertices of a rhombus. Locate the next vertex B on ℓ so that $\overline{AB} \cong \overline{AP}$. Now find the fourth vertex C so that $\overline{PC} \cong \overline{AB} \cong \overline{AP}$. (For that purpose we use a compass to construct an arc with center B and radius AB and an arc with center P and radius AB. The intersection of the two arcs is the point C. Line PC is the required line.) □

Parallel Projections

If P is a point not on a line k, as in Figure 1.47a, then the point P', where the perpendicular through P to ℓ intersects ℓ, is called the **vertical projection** of P onto ℓ. We can actually project a point in any direction, not just the perpendicular one. In Figure 1.47b, k and ℓ are two lines and m is a transversal intersecting the lines at Q and Q', respectively. The line through P parallel to $\overline{QQ'}$ intersects ℓ at P'. The point P' is the **projection of P on ℓ parallel to $\overline{QQ'}$**. (The parallel projection of Q on ℓ is Q'.)

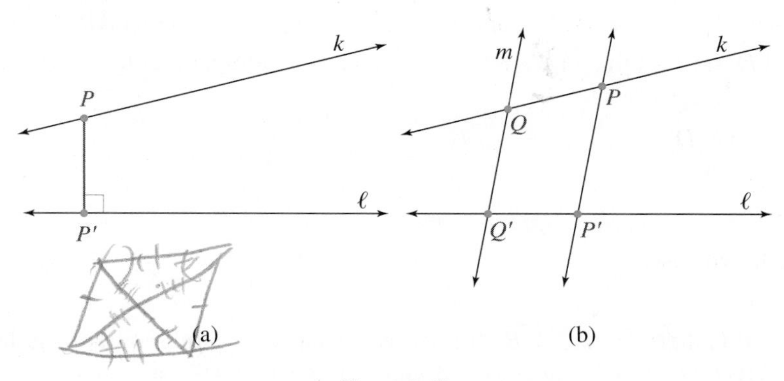

Figure 1.47

If we consider a line as a set of points, then a parallel projection is a function from k to ℓ that assigns to each point of k a corresponding point of ℓ. If we denote the function by f, then $f(P) = P'$ and $f(Q) = Q'$. Notice that the domain of this function is k and the range is ℓ, which we could denote as follows:

$$f: k \to \ell$$

This function is called the **projection of k on ℓ in the direction of m**. As shorthand, we refer to such functions as *parallel projections*. It is straightforward to show that a parallel projection function is one-to-one and onto and, therefore, a one-to-one correspondence. The next two theorems describe two useful properties of parallel projections. (See the Appendix for definition of *betweenness*.)

Theorem 1.28

A parallel projection preserves betweenness.

Restatement

Given $f: k \to \ell$, a parallel projection; A, B, C on k; $f(A) = A'$; $f(B) = B'$; $f(C) = C'$; and $A-B-C$ (i.e., B is between A and C).

Prove $A'-B'-C'$ (i.e., B' is between A' and C').

This theorem is illustrated in Figure 1.48. Because the result is so plausible, we will omit the proof here.

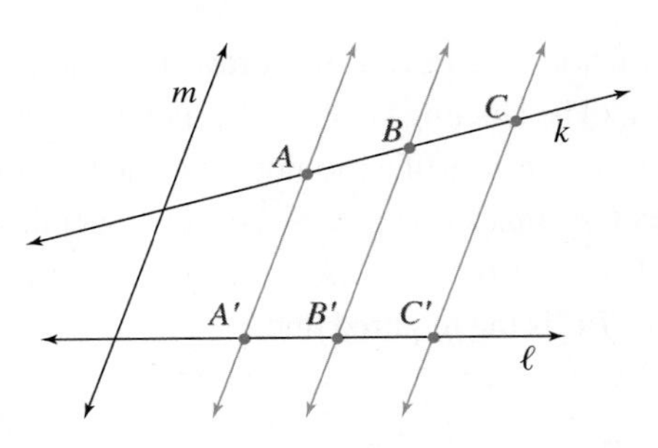

Figure 1.48

Theorem 1.29

A parallel projection preserves congruence of segments belonging to the same line.

Restatement

Given Figure 1.49 shows lines k, l, and m; A, B, C, and D on k so that $AB = CD$. Also A', B', C', and D' on ℓ so that $\overleftrightarrow{AA'}$, $\overleftrightarrow{BB'}$, $\overleftrightarrow{CC'}$, and $\overleftrightarrow{DD'}$ are parallel to m. (*Note:* It is not given that $AB = BC$.)

Prove $A'B' = C'D'$.

Proof

We distinguish two cases:

Case 1. If $k \parallel \ell$, then $\overline{AB}$ and $\overline{A'B'}$ are opposite sides of a parallelogram and hence $AB = A'B'$. Similarly, $CD = C'D'$. Since $AB = CD$, it follows that $A'B' = C'D'$.

Case 2. If k is not parallel to ℓ, to prove that $A'B' = C'D'$, we try to construct congruent triangles with corresponding sides AB and CD as well as sides congruent to $A'B'$ and $C'D'$. Keeping in mind the proved Case 1, this can be accomplished by constructing through A and C lines p and q parallel to ℓ, as shown in Figure 1.49. From Case 1, it follows that $A'B' = AE$ and $C'D' = CF$. Therefore it will suffice to prove that $AE = CF$. For that purpose we prove that $\triangle ABE \cong \triangle CDF$. Because each pair of similarly marked angles comprises corresponding angles between parallel lines, the similarly marked angles are congruent. It follows that $\angle ABE \cong \angle CDF$ and, by ASA, that $\triangle ABE \cong \triangle CDF$. Thus $AE = CF$ and, therefore, $A'B' = C'D'$. □

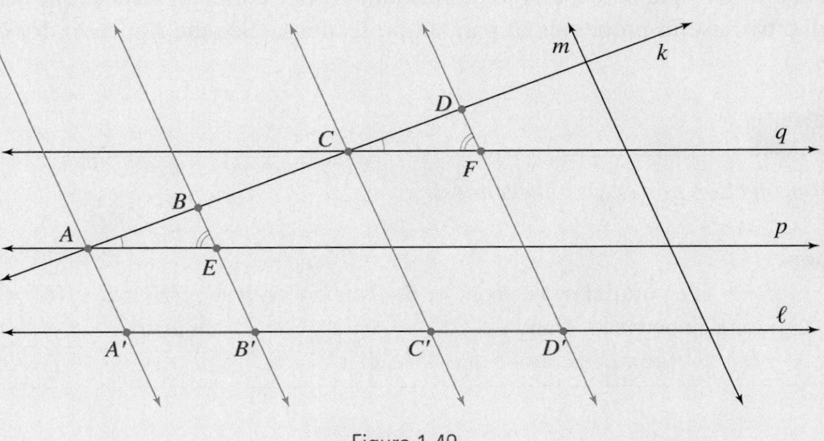

Figure 1.49

In Chapter 4, we will prove that parallel projections also preserve the ratio of the lengths of the segments. Functions preserving certain geometric properties will play a crucial role in later chapters.

An immediate and useful consequence of Theorem 1.29 arises when the congruent segments are adjacent—that is, when $B = C$ in Figure 1.49. We obtain the following corollary:

Corollary 1.6 *If three or more parallel lines intercept congruent segments on one transversal, then they intercept congruent segments on any other transversal.*

Restatement

If in Figure 1.50, $A_1A_2 = A_2A_3 = A_3A_4 = \cdots = A_{n-1}A_n$ and $\overline{A_1A'_1} \parallel \overline{A_2A'_2} \parallel \overline{A_3A'_3} \parallel \cdots \parallel \overline{A_nA'_n}$ then $A'_1, A'_2 = A'_2A'_3 = A'_3A'_4 = \cdots = A'_{n-1}A'_n$.

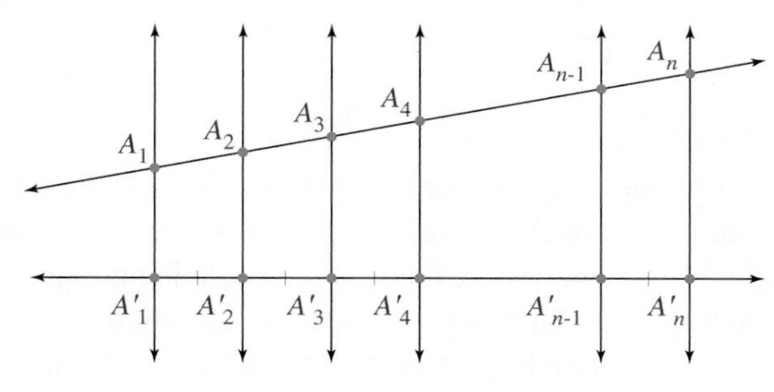

Figure 1.50

Corollary 1.6 can be used for the following construction.

Construction 1.6: Division of a Segment into Any Number n of Congruent Parts

Given $\overline{AB}$ as in Figure 1.51, we illustrate the construction for $n = 3$.

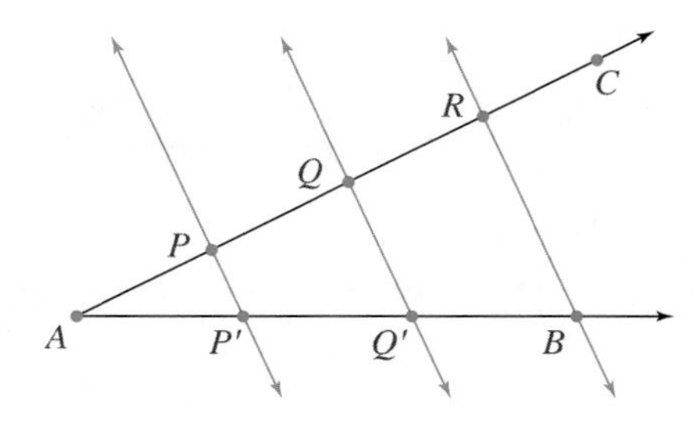

Figure 1.51

Through A, construct any ray $\overrightarrow{AC}$ and mark on $\overrightarrow{AC}$ any point P. Construct Q and R so that $\overline{AP} \cong \overline{PQ} \cong \overline{QR}$. Connect R with B. Through Q and P, draw lines parallel to $\overline{BR}$. The points of intersection P' and Q' accomplish the required construction since by Corollary 1.6, $\overline{AP'} \cong \overline{P'Q'} \cong \overline{Q'R}$. □

Figure 1.52 shows a special case of Corollary 1.6 for a triangle. In $\triangle ABC$, M is the midpoint of $\overline{AC}$, and a line is drawn through M parallel to $\overline{AB}$ that intersects $\overline{BC}$ at N. From Corollary 1.6, we can conclude that N is the midpoint of $\overline{BC}$. (In Figure 1.52, we use double arrows on $\overline{AB}$ and $\overline{MN}$ to indicate that the segments are parallel.) Thus we have the following theorem:

Theorem 1.30

A line through the midpoint of one side of a triangle and parallel to the second side bisects the third side.

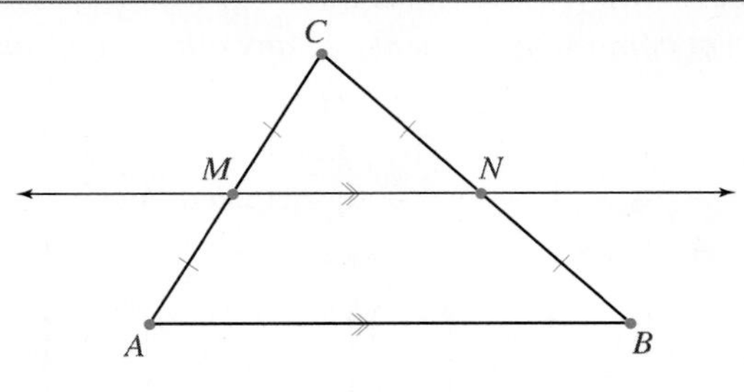

Figure 1.52

The segment connecting the midpoints of two sides of a triangle is called the **midsegment**. Each triangle has three midsegments. Because a segment has a unique midpoint, it follows from the last corollary that a midsegment of two sides of a triangle is parallel to the third side. How does the length of the midsegment in a triangle compare to the length of the parallel side? Given that any triangle is "half" of a parallelogram (a diagonal divides a parallelogram into two congruent triangles) we could turn $\triangle ABC$ in Figure 1.53a into a parallelogram by tracing $\triangle A_1B_1C_1$ over $\triangle ABC$ and then turning it upside down as shown in Figure 1.53b. (We will see later that this operation amounts to rotating $\triangle ABC$ about N by 180°.) We obtain a parallelogram if we make $B_1 = C$ and $C_1 = B$. It seems now that $\overline{MN} \cong \frac{1}{2}\overline{MM_1}$ and $\overline{MM_1} \cong \overline{AB}$ and consequently that $\overline{MN} \cong \frac{1}{2}\overline{AB}$. We state this result in the following theorem, along with a more rigorous argument following the idea presented above.

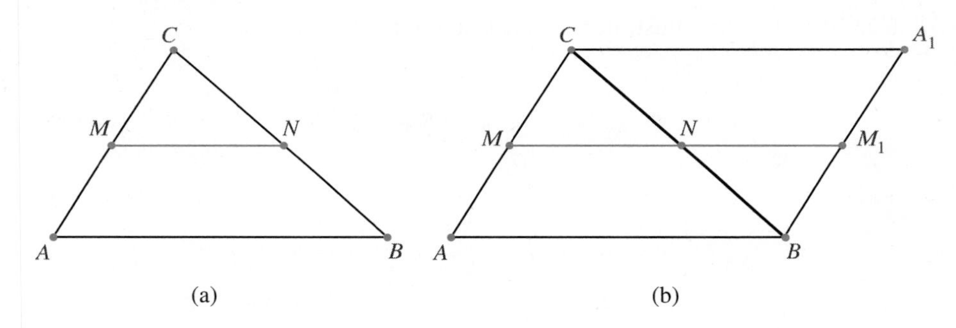

Figure 1.53

Theorem 1.31

The Midsegment Theorem

The segment connecting the midpoints of two sides of a triangle is parallel to the third side and half as long as that side.

Restatement

Given $\triangle ABC$ in Figure 1.53a; M and N are midpoints of $\overline{AC}$ and $\overline{BC}$.

Prove (1) $\overline{MN} \parallel \overline{AB}$; (2) $MN = \frac{1}{2}AB$.

Proof

Part (1) follows from Theorem 1.30. To prove part (2), we want to obtain a parallelogram and therefore draw through C a line parallel to $\overline{AB}$ and through B a line parallel to $\overline{AC}$ as shown in Figure 1.53b. The two lines intersect at A_1. From the definition, it follows that ABA_1C is a parallelogram. We extend $\overline{MN}$ until it intersects $\overline{BA_1}$ at M_1. At this point it seems obvious that $NM_1 = MN$ and that $MM_1 = AB$. Thus, if $MN = x$, then $2x = MM_1 = AB$ and hence $x = \frac{1}{2}AB$.

We can prove what we stated earlier as obvious in the following way. Theorem 1.29 implies that $A_1M_1 = BM_1$ and thus that $\overline{NM_1}$ is a midsegment in $\triangle A_1BC$. We can prove that $MN = M_1N$ by showing that $\triangle MCN \cong \triangle M_1BN$. We also know that $MM_1 = AB$ because AMM_1B is a parallelogram. □

Remark We could have postponed the Midsegment Theorem until Chapter 4 and deduced it from theorems concerning similar triangles. Instead, we chose to prove the theorem here because the methods used give us valuable practice and an opportunity to see different approaches.

Now Solve This 1.11

We have seen that the diagonals of a parallelogram bisect each other. Suppose they bisect each other at O.

1. If we draw a line through O and it intersects two parallel sides of the parallelogram at points P and Q, respectively, prove that $OP = OQ$.
2. The statement in Problem 1 is a generalization of a fact stated in the proof of Theorem 1.29. Which fact is it? Why is Problem 1 a generalization of that fact?

Medians of a Triangle

A **median** of a triangle is a segment connecting a vertex to the midpoint of the opposite side. The Midsegment Theorem can be used to prove a surprising property of the medians of a triangle. If you construct the three medians of an arbitrary triangle, you will notice that the medians are concurrent. In search of other properties of medians, we construct two medians $\overline{AM}$ and $\overline{CN}$ in $\triangle ABC$ as shown in Figure 1.54. As in Figure 1.53b, we trace $\triangle ACB$ to obtain $\triangle A_1C_1B_1$ and place it "upside down" so that $B_1 = C$ and $C_1 = B$. Rather than relying on this kind of informal tracing, we could also proceed formally as follows: We extend $\overline{AM}$ to A_1 so that $\overline{MA_1} \cong \overline{MA}$. We obtain a quadrilateral ABA_1C whose diagonals bisect each other. By Theorem 1.27, ABA_1C is a parallelogram.

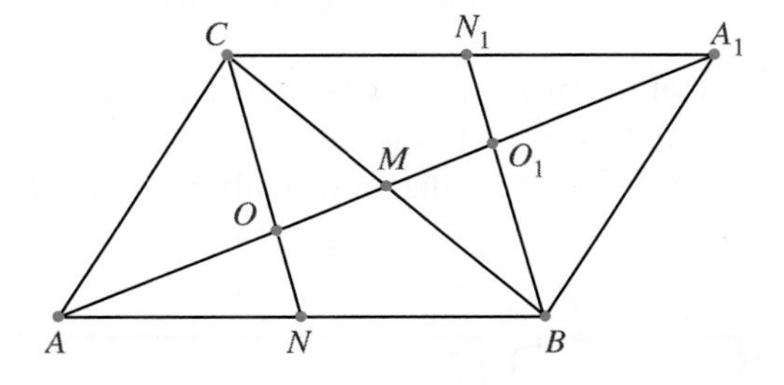

Figure 1.54

The corresponding medians in $\triangle A_1BC$ are $\overline{A_1M}$ and $\overline{BN_1}$. Because $\overline{CN_1} \cong \overline{N_1A_1}$ and $\overline{AN} \cong \overline{NB}$, the parallels $\overline{CN}$ and $\overline{N_1B}$ mark congruent segments on $\overline{OA_1}$ and on $\overline{AO_1}$. Thus

1. $\overline{OO_1} \cong \overline{O_1A_1}$
2. $\overline{AO} \cong \overline{OO_1}$
3. $\overline{AO} \cong \overline{OO_1} \cong \overline{O_1A_1}$

Because $AA_1 = 2\overline{AM}$, from statement (3) we have $AO = \frac{1}{3}AA_1 = \frac{1}{3} \cdot 2AM = \frac{2}{3}AM$. Thus $OM = \frac{1}{3}$AM. (Equivalently, from Now Solve This 1.11, we could argue that because $OM = MO_1$, $OM = \frac{1}{2}OO_1 = \frac{1}{2}AO$.)

The preceding investigation suggests the following theorem and its proof. (Two different proofs will be investigated in the problem set.)

Theorem 1.32

Property of Medians

The medians of a triangle are concurrent in a point whose distance to a vertex is two-thirds of the length of the median from that vertex.

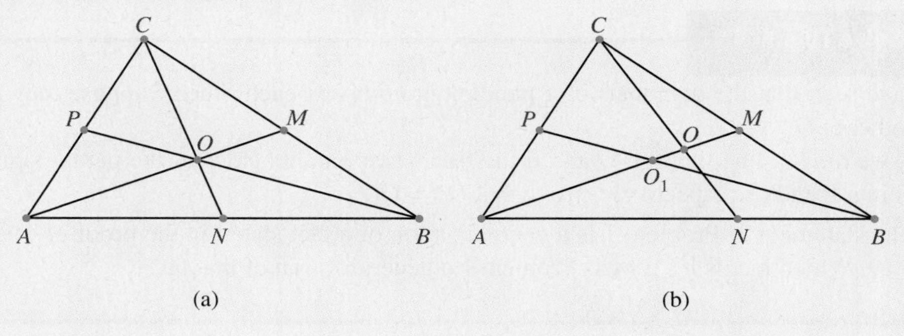

Figure 1.55

Restatement

If $\overline{AM}$, $\overline{CN}$, and $\overline{BP}$ are the three medians of $\triangle ABC$ as shown in Figure 1.55a, then they intersect in a single point O such that $AO = \frac{2}{3}AM$, CO $= \frac{2}{3}CN$, and $BO = \frac{2}{3}BP$. (Notice that each median is divided into segments whose lengths are in the ratio 2:1.)

Proof

Observe that because in the investigation preceding Theorem 1.32 we proved that any two medians intersect in a point whose distance to a vertex is two-thirds of the median from that vertex, the concurrency of the medians will follow immediately. Indeed, if the medians $\overline{AM}$ and $\overline{CN}$ intersect at O as shown in Figure 1.55b, then $\overline{AO} = \frac{2}{3}\overline{AM}$. If $\overline{AM}$ and $\overline{BP}$ intersect at O_1, then $AO_1 = \frac{2}{3}AM$. This implies that $AO = AO_1$ and hence that $O = O_1$. □

The point of concurrency of the medians of a triangle is called the **centroid**. The centroid is the center of gravity of the triangle when a triangle is considered as a thin plate of uniform density. (The center of gravity can be defined mathematically, and this definition can be found in most calculus texts.)

Now Solve This 1.12

The part of Theorem 1.32 saying that any two medians intersect at a point that divides each median into segments whose lengths are in the ratio 2:1 can be proved in a manner suggested in Figure 1.56. Write a proof based on this figure. (D is the midpoint of $\overline{CM}$, and E is the midpoint of $\overline{BM}$.)

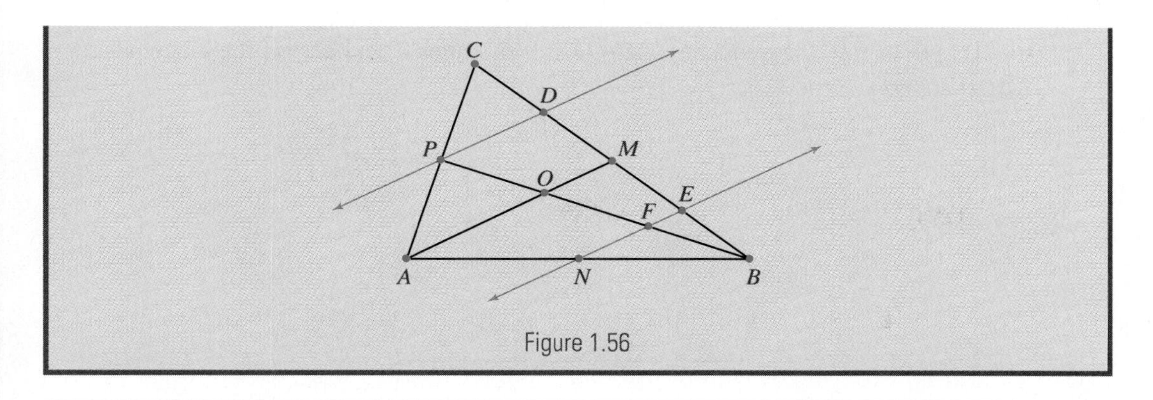

Figure 1.56

Example 1.6 What kind of quadrilateral is obtained when the midpoints of consecutive sides of the quadrilateral are connected?

Solution

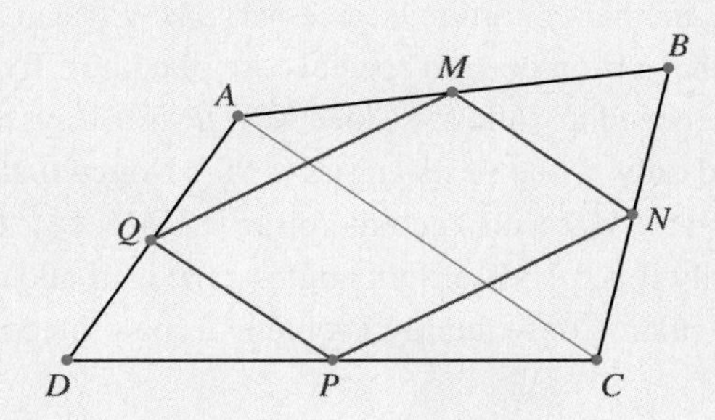

Figure 1.57

In Figure 1.57, the midpoints M, N, P, and Q of the sides of the quadrilateral $ABCD$ are connected to form the quadrilateral $MNPQ$. Starting with different quadrilaterals, accurate constructions reveal that in each case $MNPQ$ is a parallelogram. This fact can actually be discovered without any experimentation and proved for all quadrilaterals as follows.

The fact that M and N are midpoints suggests the use of the Midsegment Theorem. The side $\overline{MN}$ is a midsegment in $\triangle ABC$. Thus $\overline{MN} \parallel \overline{AC}$ and $MN = \frac{1}{2}AC$. Similarly, in $\triangle ADC$, $\overline{QP} \parallel \overline{AC}$ and $QP \cong \frac{1}{2}AC$. Both imply that $\overline{MN} \parallel \overline{PQ}$ and $\overline{MN} \cong \overline{PQ}$. This happens if and only if $MNPQ$ is a parallelogram.

Example 1.7 Complete each of the following statements assuming that the quadrilateral $MNPQ$ is the quadrilateral defined in Example 1.6. (Prove your answers.)

1. $MNPQ$ is a rhombus if and only if $ABCD$ is a quadrilateral such that _________.
2. $MNPQ$ is a rectangle if and only if $ABCD$ is a quadrilateral such that _________.

Solution

1. Using the result of Example 1.6, $MNPQ$ in Figure 1.58 is a rhombus if and only if $\overline{MN} \cong \overline{MQ}$. We need to write an equivalent condition involving characteristics of $ABCD$. We have $MN = \frac{1}{2}AC$ and $MQ = \frac{1}{2}BD$. Thus $MN = MQ$ if and only if $\frac{1}{2}AC = \frac{1}{2}BD$;

that is, $\overline{AC} \cong \overline{BD}$. Consequently, $MNQP$ is a rhombus if and only if the diagonals of $ABCD$ are congruent.

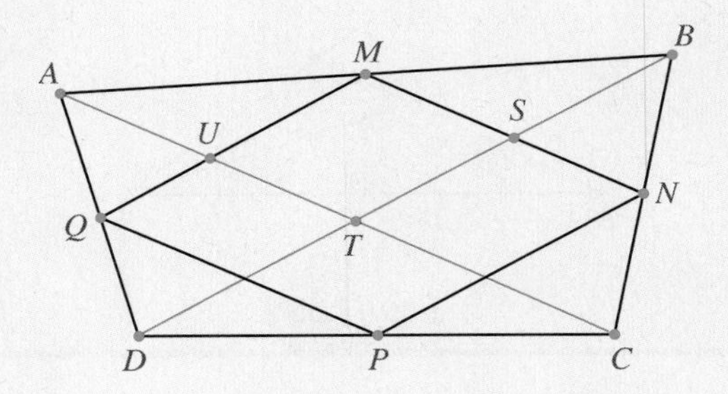

Figure 1.58

2. It is straightforward to show that if $ABCD$ is a rhombus or even a kite, then $MNPQ$ is a rectangle. However, neither condition is necessary: $MNPQ$ can be a rectangle if $ABCD$ is not a kite (recall that a rhombus is a special case of a kite). To find an "if and only if" condition, we can proceed as follows. Since $MNQP$ is always a parallelogram, it will be a rectangle if and only if one of its angles is 90°. Notice that $MSTU$ is also a parallelogram ($\overline{MS} \parallel \overline{UT}$ and $\overline{MU} \parallel \overline{ST}$) and therefore $\angle UMS \cong \angle UTS$. Thus $\angle UMS$ is a right angle if and only if $\angle UTS$ is a right angle—that is, if and only if the diagonals of $ABCD$ are perpendicular. Consequently Problem 2 could be completed "its diagonals are perpendicular."

Example 1.8

1. State and prove the Midsegment Theorem for trapezoids.
2. Prove that the segment connecting the two midpoints of the diagonals of a trapezoid is parallel to the bases of the trapezoid and find the length of this segment if the lengths of the bases are a and b.

Solution

1. Use the Midsegment Theorem for triangles to deduce that the midsegment of a trapezoid (the segment connecting the midpoints of the sides) is parallel to the bases and its length is one-half the sum of the lengths of the bases.

Given In Figure 1.59, $\overline{BC} \parallel \overline{AD}$, and M and N are the midpoints of $\overline{AB}$ and $\overline{CD}$, respectively.

Prove $\overline{MN} \parallel \overline{AD}$ and $MN = \frac{1}{2}(AD + BC)$.

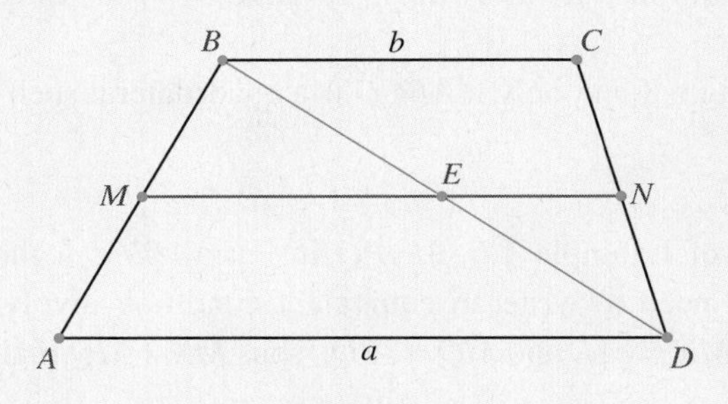

Figure 1.59

Proof

Through M, construct line ℓ (not shown in Figure 1.59) parallel to the bases. A property of parallel projections implies that ℓ intersects $\overline{CD}$ at its midpoint. Because the midpoint of a segment is unique, it follows that ℓ intersects $\overline{CD}$ at N. Consequently, $\overleftrightarrow{MN}$ is the line ℓ and, therefore, is parallel to the bases.

To find the length of $\overline{MN}$ given the lengths of the bases a and b, consider the two triangles created by the diagonal $\overline{BD}$ (the other diagonal could be used equally well). We have $ME = \frac{a}{2}$ and $EN = \frac{b}{2}$. Thus $MN = \frac{a+b}{2}$. □

2. **Given** In Figure 1.60, P and Q are the midpoints of the diagonals.

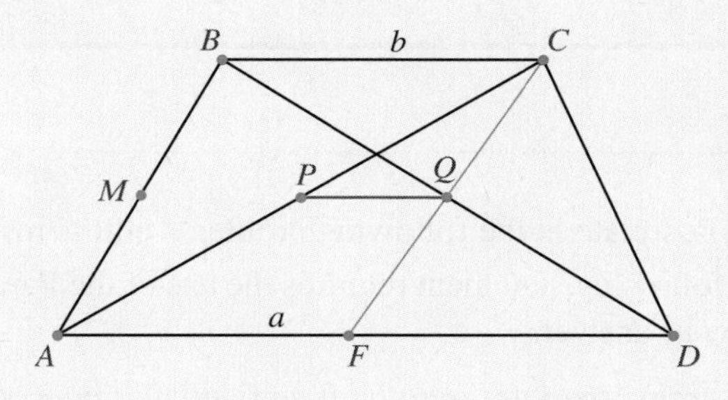

Figure 1.60

Prove $\overline{PQ}$ is parallel to the bases of the trapezoid. Also, express PQ in terms of a and b.

Proof

We try to make $\overline{PQ}$ a midsegment of a triangle. One such triangle can be obtained by connecting C with Q. The line CQ intersects $\overline{AD}$ at F. We show that $\overline{PQ}$ is a midsegment in $\triangle ACF$. Because P is the midpoint of $\overline{AC}$ (given), it remains to be shown that $\overline{CQ} \cong \overline{QF}$. For that purpose, notice that by ASA (why?) we have $\triangle BCQ \cong \triangle DFQ$. Thus $\overline{CQ} \cong \overline{QF}$ and, therefore, $\overline{PQ}$ is a midsegment in $\triangle ACF$. Consequently, we have $\overline{PQ} \parallel \overline{AF}$ and $PQ = \frac{1}{2}AF$. We now express $\overline{AF}$ in terms of a and b. $PQ = \frac{1}{2}AF = \frac{1}{2}(AD - FD)$. We have $AD = a$ and $FD = BC = b$. Therefore $PQ = \frac{1}{2}(a - b)$. □

Now Solve This 1.13

You can now prove your conjecture about the location of the treasure described in "The Treasure Island Problem" in Chapter 0 (see also Problem 1 in Problem Set 0). You should have conjectured that the treasure is independent of the location of the gallows Γ in Figure 1.61.

1. Show that if the conjecture is true, then by choosing the gallows Γ at T_2 (tree number 2), it follows that the treasure M is on the perpendicular bisector of the segment T_1T_2 connecting the trees T_1 and T_2, half the distance between the trees and above the lines connecting the trees.
2. Use the ideas in Figure 1.61 to prove the assertion in part (1). (We labeled $\Gamma C = h$, $S_1A = a$, and $S_2B = b$. You need to show that $T_1A = T_2B = h$, $CT_1 = a$, and $CT_2 = b$. Also show that $MN = \frac{1}{2}T_1T_2$ and $T_2N = T_1N$.)

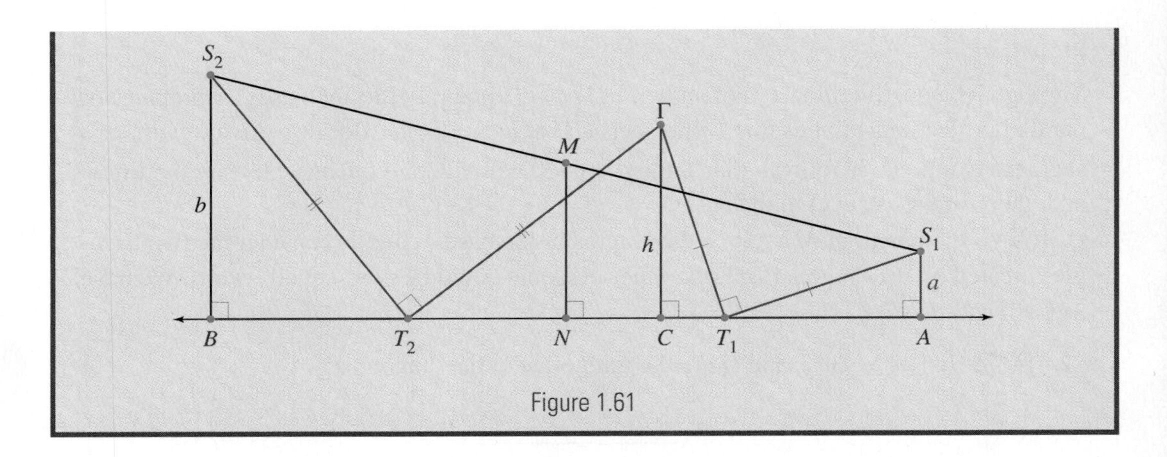

Figure 1.61

Problem Set 1.3

You may use the Parallel Postulate in the following problems unless instructed otherwise.

1. Which part of the following statement requires the use of the Parallel Postulate and which does not? Justify your answer.

   Two lines cut by a transversal are parallel if and only if a pair of corresponding angles is congruent.

2. **a.** In Now Solve This 1.4 you found the sum of the measures of the interior angles in a convex quadrilateral. Does your answer apply to concave quadrilaterals as well?

   b. Find the sum of the measures of the interior angles of any convex n-gon in terms of n by drawing all the diagonals from one vertex as shown below. (You may have used this approach in Now Solve This 1.4.)

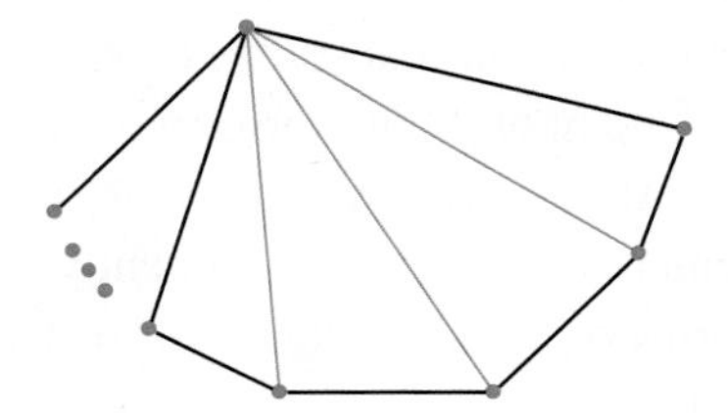

   ⋆**c.** Prove your answer to part (b) by mathematical induction.

3. Justify your conjecture in Now Solve This 1.4 about the sum of the measures of the angles of any convex n-gon by choosing any point P in the interior and connecting P with each of the vertices as shown.

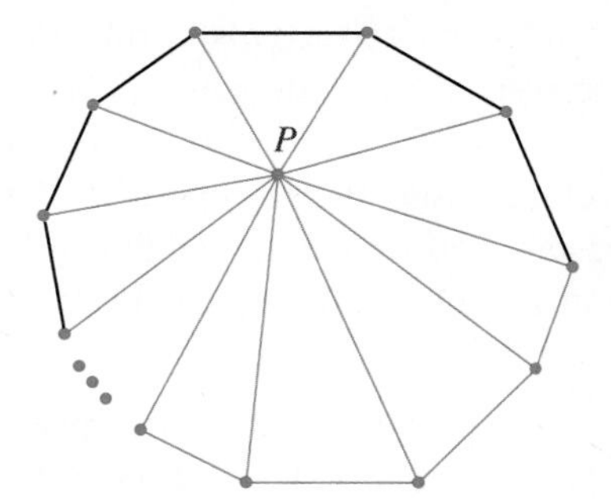

4. If $m(\angle BAC) = \alpha$ and $m(\angle B'A'C') = \beta$, $\overrightarrow{BA} \parallel \overrightarrow{B'A'}$, $\overrightarrow{AC} \parallel \overrightarrow{A'C'}$ as in the following figure, how are α and β related? Prove your answer.

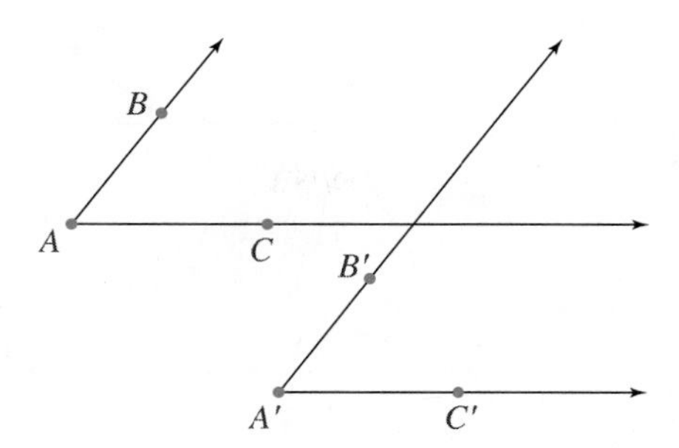

5. **a.** Let m, n, a, and b be lines such that $m \parallel n$, $m \perp a$, and $n \perp b$. Prove that $a \parallel b$.

b. Use your work from part (a) to prove that any two perpendicular bisectors of two sides in a triangle must intersect.

c. Did the proof of part (a) [and therefore of part (b)] depend on the Parallel Postulate? Justify your answer.

6. **a.** In the accompanying figure, the measures of the indicated angles are as shown. Complete the following and prove the completed statement:

$a \parallel b$ if and only if x, y, and z satisfy the condition _________.

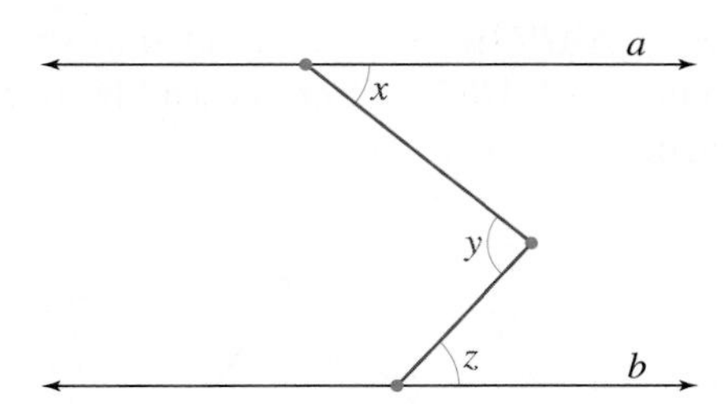

b. Write and prove a statement similar to the one in part (a) that corresponds to the figure below.

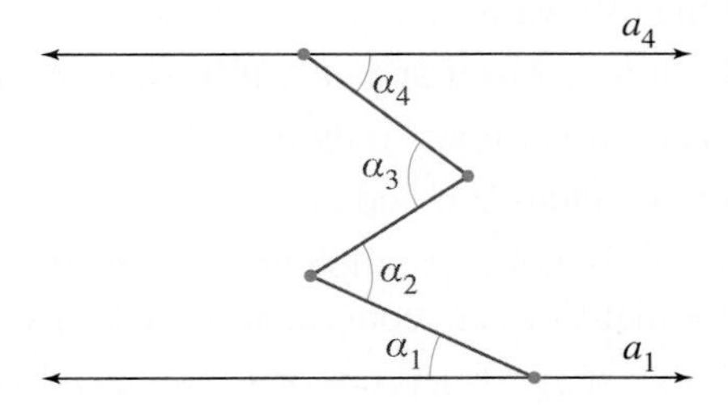

c. Write and prove a statement similar to the one in part (b) for

(i) Five angles

(ii) Six angles

d. Generalize the statement in part (c) for n angles $\alpha_1, \alpha_2, \ldots, \alpha_n$.

★**e.** Prove your generalization in part (d) by mathematical induction.

•**7.** In the accompanying figure, $a \parallel b$. The angles at P and Q have been trisected. The trisection rays form a quadrilateral $ABCD$. What is the most that can be said about the angles of $ABCD$? Prove your answer.

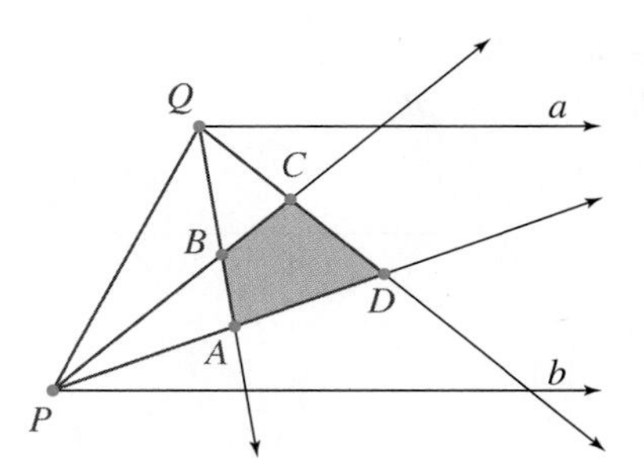

•**8.** In $\triangle ABC$, the angle bisectors at B and C form $\triangle BDC$. If $m(\angle A) = \alpha$, can $m(\angle D)$ be found in terms of α? If so, find it. If not, prove that it cannot be found.

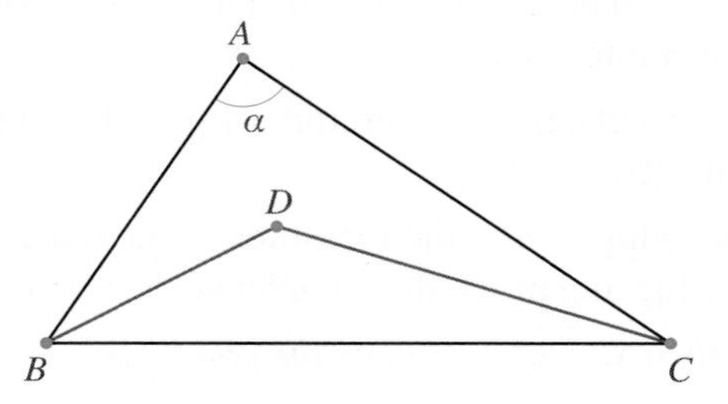

9. Two exterior angles in $\triangle ABC$ have been bisected as shown. Can $m(\angle D)$ be found if the measures of the angles of $\triangle ABC$ are not known? If so, find it. If not, show that $m(\angle D)$ cannot be determined.

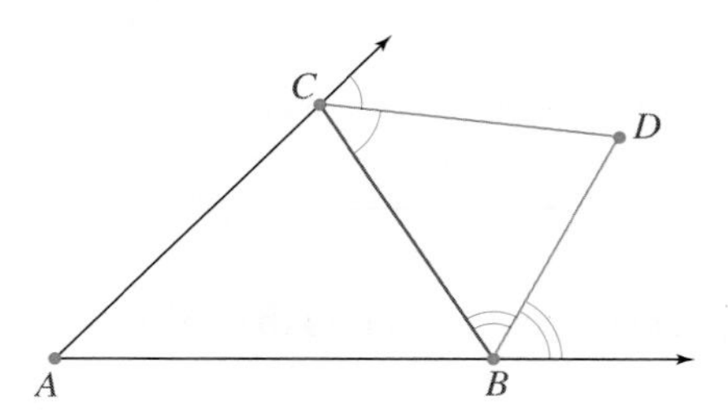

10. Complete each of the following statements and then prove the completed statements.

a. A quadrilateral is a *rhombus* if and only if its diagonals __________.

b. A quadrilateral is a *square* if and only if its diagonals __________.

•**c.** A quadrilateral is a *rectangle* if and only if its diagonals __________.

d. A quadrilateral that is not a parallelogram is an *isosceles trapezoid* (a trapezoid in which the nonparallel sides are congruent) if and only if its diagonals __________.

11. The following diagram suggests a proof of Theorem 1.13 that $a > b \Rightarrow \alpha > \beta$. Based on these diagrams, write a proof of Theorem 1.13.

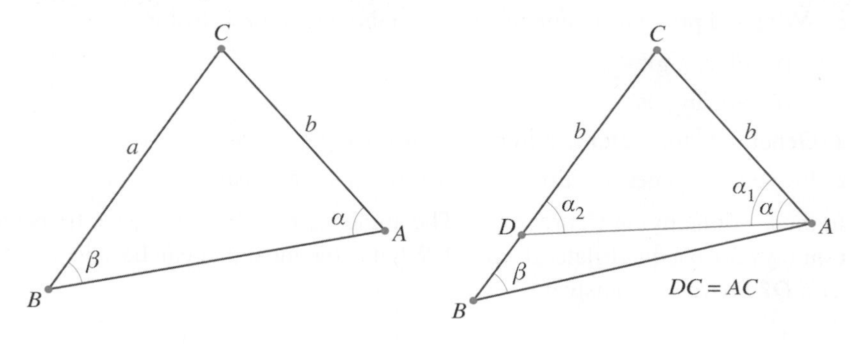

12. Prove that in a 30°–60°–90° triangle, the side opposite to the 30° angle is half as long as the hypotenuse. (*Hint:* Such a triangle is "half" an equilateral triangle.)

13. Devise as many different methods as possible for constructing a line through a given point parallel to a given line. Justify each method.

14. If k and ℓ are lines with transversal m and $f: k \to \ell$ is a parallel projection, explain why f is a one-to-one and onto function.

15. Divide a segment into five congruent parts. Justify your construction.

16. Give an alternative proof of the Midsegment Theorem by extending $\overline{MN}$, drawing through C line k parallel to $\overline{AB}$, labeling P as the intersection of $\overleftrightarrow{MN}$ with k, and proving that $MPCB$ is a parallelogram.

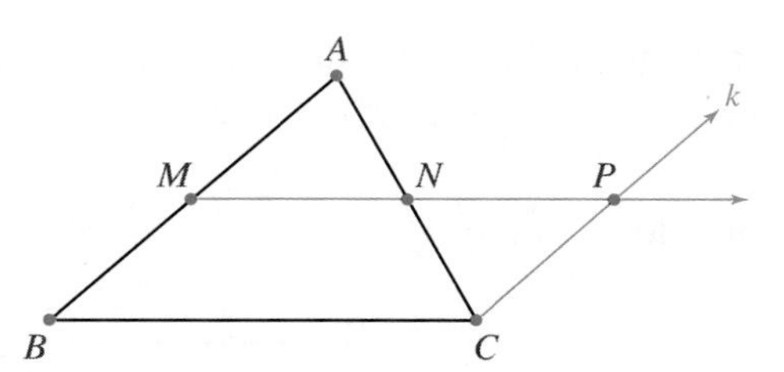

17. Give an alternative proof of the fact that any two medians divide each other in the ratio 2:1 by proving first that if P and Q are the midpoints of $\overline{BO}$ and $\overline{CO}$, respectively, then $MNQP$ is a parallelogram.

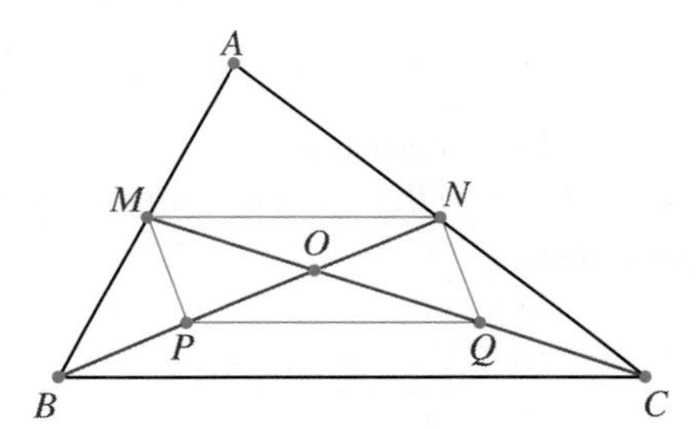

• **18.** Let $ABCD$ be a quadrilateral and $MNQP$ be the quadrilateral obtained by connecting the consecutive midpoints of the sides of $ABCD$. Complete the following statement and prove the completed statement:

MNQP is a square if and only if ABCD has the property _________.

19. Give an alternative proof of the Midsegment Theorem for trapezoids (in Example 1.8) by considering a proof suggested by tracing the trapezoid $ABCD$, turning the traced trapezoid "upside down," and attaching the two trapezoids so that their bases are collinear.

20. What, if anything, is missing (or wrong) in the following proof of Example 1.8, part 2? Through P, draw a line parallel to $\overline{BC}$ and intersecting $\overline{AB}$ at S. A property of parallel projection implies that S is the midpoint of $\overline{AB}$. Applying the Midsegment Theorem in $\triangle ABD$ and $\triangle ABC$, we have $PQ = QS - PS = \frac{1}{2}AD - \frac{1}{2}BC = \frac{1}{2}(a - b)$.

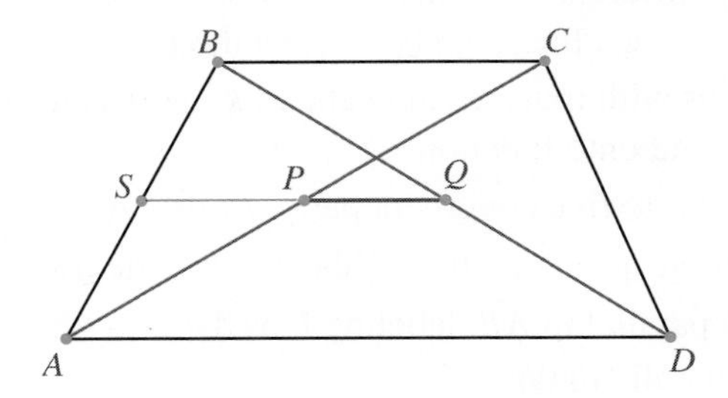

21. **a.** Let P and Q be the midpoints of the diagonals as in Problem 20. Draw $\overline{BP}$ and $\overline{CQ}$ as shown, and express UV in terms of a and b where $\overline{AD} = a$ and $\overline{BC} = b$. Justify your answer.

b. Under what condition will $U = V$?

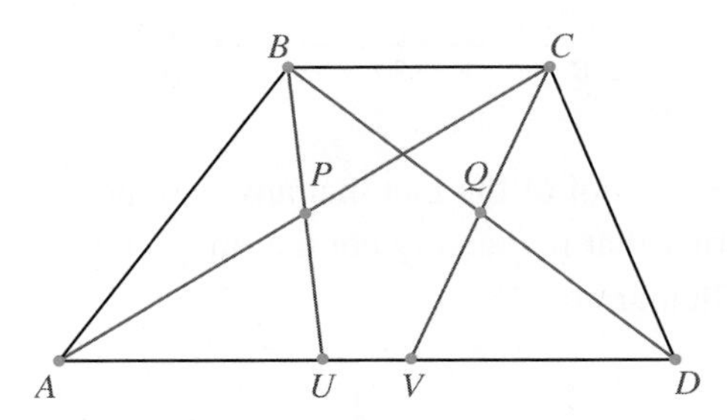

22. What kind of quadrilateral is obtained by connecting the intersection points of the angle bisectors of a rectangle? Provide the most information you can about the obtained quadrilateral and prove your answer.

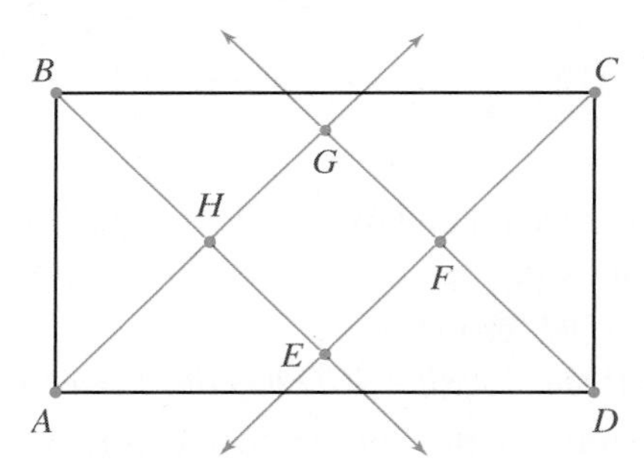

23. For the following figures, what kind of quadrilateral is always obtained by connecting the intersection points of the angle bisectors? Provide the most information you can about the quadrilaterals obtained and prove your answers.

a. A parallelogram

b. An isosceles trapezoid

24. • **a.** Prove that in a right triangle the median to the hypotenuse is half as long as the hypotenuse.

b. State and prove the converse of the statement in part (a).

25. In the figure below, $ABCD$ is a square and all marked angles are congruent. What is the most that can be said about quadrilateral $EFHG$? Prove your answer.

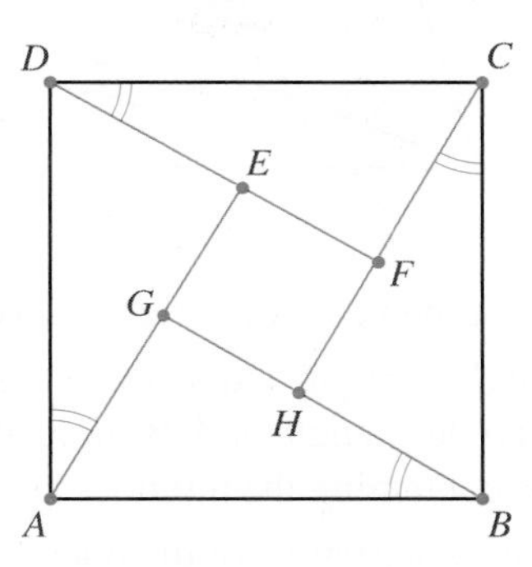

26. Prove the converse of the Midsegment Theorem: If a segment connects two points on two sides of a triangle in such a way that it is parallel to the third side and is half the length of that side, it is a midsegment.

27. $ABCD$ is a parallelogram whose diagonals intersect at E. The distances of the points A, B, C, D, and E from ℓ are a, b, c, d, and e, respectively.

a. Prove that $b + d = a + \mathrm{c}$.

b. Find e in terms of a, b, c, and d.

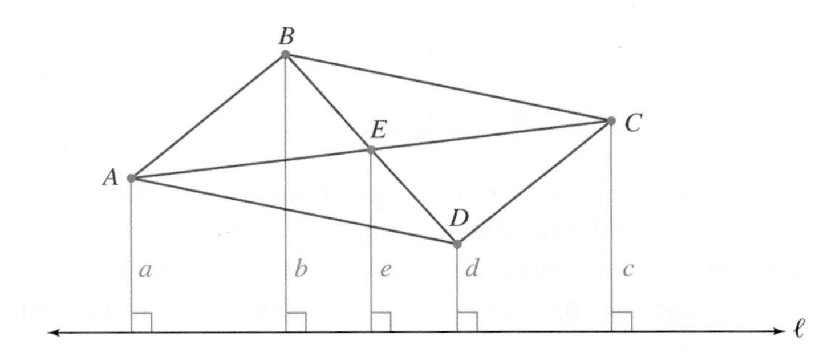

28. The segment $\overline{AB}$ (of fixed length) is moving so that the endpoints A and B are on the sides of a right angle. What is the locus (set of points) of all the midpoints of $\overline{AB}$? Prove your answer.

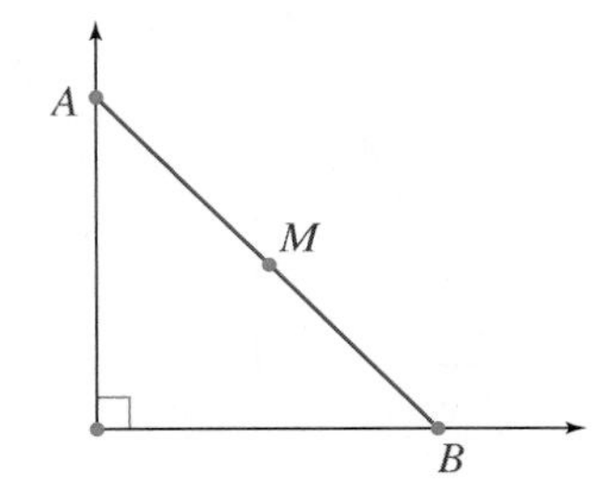

29. $ABCD$ is a parallelogram. Its diagonals divide the parallelogram into four triangles. Points F, G, H, and I are the incenters (the centers of the inscribed circles) of the corresponding triangles. What is the most that can be concluded about which type of quadrilateral $FGHI$ is? Prove your answer.

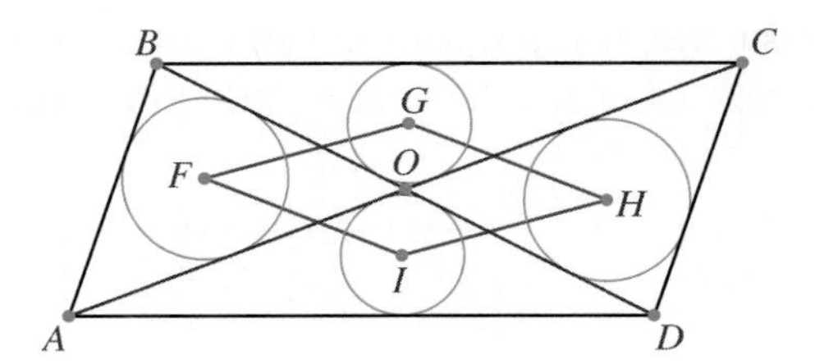

30. **a.** Prove that the three midsegments of a triangle divide it into four congruent triangles.

b. Construct an acute angle triangle and its three altitudes.

c. Construct an obtuse angle triangle and its three altitudes. Extend some of the altitudes to show that the lines containing the altitudes are concurrent.

d. Prove that the altitudes of a triangle or the lines containing the altitudes are concurrent. (*Hint:* Through each vertex of the triangle, construct a line parallel to the opposite side. The three lines intersect in three points, which determine a new triangle. Show that the perpendicular bisectors of the sides of the new triangle contain the altitudes of the original triangle.)

★• **31.** In the following figure, three adjacent squares form a rectangle. Prove (using only the materials covered) that $x + y + z = 90°$.

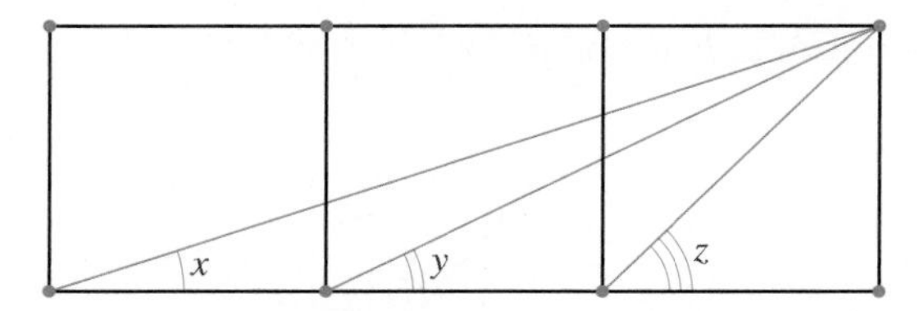

★• **32.** For which point P in the interior of an equilateral triangle $\triangle ABC$ will the sum of the distances from the three sides be at its minimum? (The sum of the distance is $x + y + z$.) *Hint:* Try various points and measure the corresponding values of $x + y + z$. Make a conjecture and prove the conjecture first for points on a side of a triangle. (Use only the concepts and theorems in this chapter.)

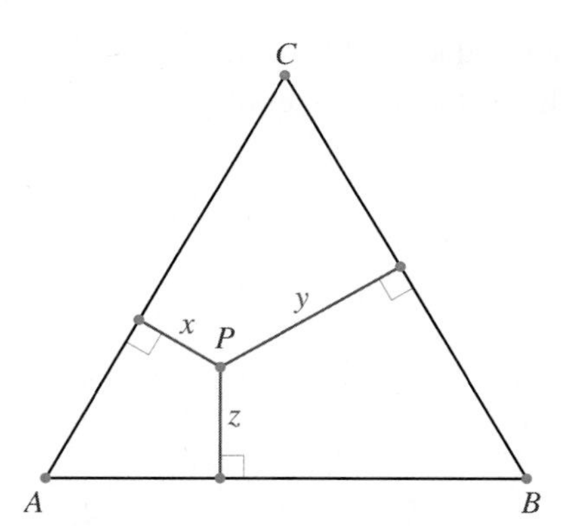

33. Describe how to perform each of the following constructions with "obstructions."

a. Construct the perpendicular bisector of a segment that is near the bottom of a page such that no drawing below the segment is allowed.

b. Construct the perpendicular to segment AB at A if an extension of the segment is not allowed.

34. Is the Parallel Postulate (Axiom 1.2) equivalent to the following? (Justify your answer.)

Two intersecting straight lines cannot both be parallel to the same straight line.

★ **35.** The following is a sketch of a proof showing that the Parallel Postulate implies Euclid's Fifth Postulate. Complete the details of the proof.

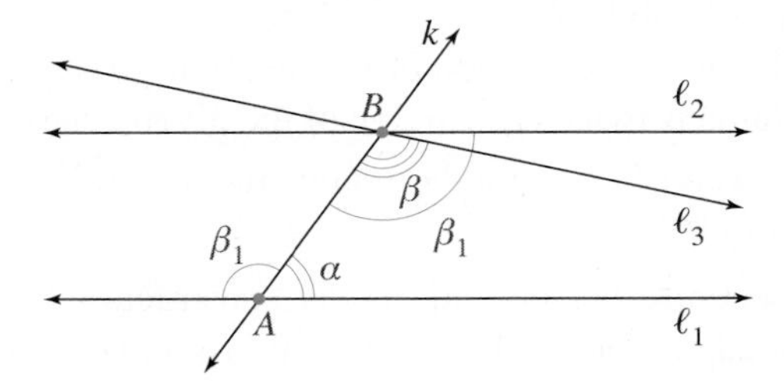

Lines ℓ_1 and ℓ_2 intersect a transversal k at points A and B, respectively, as shown in the figure. Assume that $\alpha + \beta < 180°$. If β_1 is supplementary to α, then we can conclude that $\beta_1 > \beta$. Thus there exists line ℓ_3 through B that forms with k an angle whose measure is β_1. Hence $\ell_3 \parallel \ell_1$. Because $\beta_1 > \beta$, $\ell_2 \neq \ell_3$. Consequently, ℓ_1 and ℓ_3 are not parallel and must meet. This will contradict the Exterior Angle Theorem.

★ **36.** Prove that if $ABCD$ is a quadrilateral in which $MN = \frac{1}{2}(a + b)$, where M is the midpoint of $\overline{AB}$, N is the midpoint of $\overline{CD}$, $AD = a$, and $BC = b$, then $ABCD$ is a trapezoid.

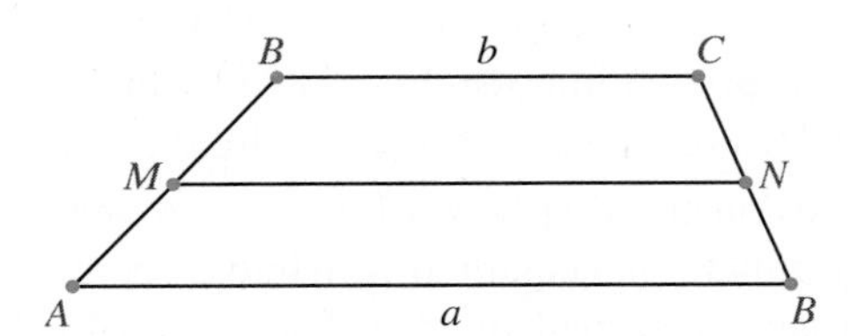

37. In the Treasure Island Problem, prove that the location of the treasure is independent of the location of the gallows by considering two positions for the gallows: one at one of the trees Γ^* and the other Γ in an arbitrary position. (*Hint:* Prove that the segments S_1S_2 and $S_1^*S_2^*$ bisect each other.)

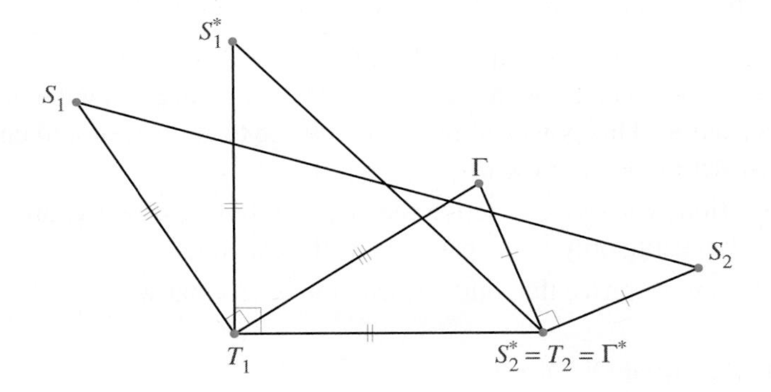

1.4 More on Construction

Three famous construction problems (using only a straight edge and a compass) have occupied mathematicians and amateurs since antiquity:

1. **Trisecting an Angle:** Is it possible to trisect any given angle—that is, to divide the angle into three congruent angles?
2. **Doubling the Cube:** Given a side of a cube, is it possible to construct a side of a new cube whose volume is twice the volume of the given cube?
3. **Squaring the Circle:** Is it possible to construct a square whose area is the same as the area of a given circle?

In the preceding sections, we saw that many constructions can be performed with only a compass and a straight edge. These three constructions, however, eluded the best mathematical minds for generations. In the nineteenth century, several mathematicians suggested that the famous three constructions couldn't be performed with straight edge and compass alone.

The first two were proved impossible by the French mathematician Pierre Wantzel (1814–1848) in 1837. The key to Wantzel's proof was the conversion of the geometric construction problems to their equivalent algebraic equations. Wantzel obtained criteria for solutions of polynomial equations with integer coefficients to be constructible with only straight edge and compass and showed that the first two problems involve equations whose solutions are not constructible. A more modern version of the proofs can be found in many abstract algebra textbooks (e.g., Beachy and Blair, pp. 283–288). For a more elementary approach, see the beautifully written *What Is Mathematics* (Courant and Robbins, pp. 134–138).

The third problem was proved impossible in 1882 by the German mathematician Ferdinand Lindmann (1852–1939). He noted that squaring the circle of radius 1 is equivalent to solving the equation $x^2 = \pi$ and hence constructing $\sqrt{\pi}$. Lindmann proved that π is transcendental—that is, it is not a solution of any polynomial equation with integer coefficients. His proof requires a considerable knowledge of advanced mathematics. The impossibility of squaring the circle or trisecting an angle has not stopped well-meaning amateurs from coming up with their own "constructions," which, of course, are wrong.

In contrast to these impossible constructions, we will focus in this section on some possible constructions using the theorems introduced in this chapter. More constructions will follow in forthcoming chapters after new theorems will be introduced. From time to time—and especially when a construction is not immediately apparent—we will discuss how a construction can be discovered and its correctness proved by using the following structure:

1. **Investigation** (sometimes also referred to as *analysis*), in which we imagine the given problem as solved and search for properties of the figure that will enable us to accomplish its construction. This powerful problem-solving technique should enable you to discover the construction on your own.
2. **Construction**, where we describe the steps in the construction and actually perform the construction using only a compass and a straight edge.
3. **Proof**, where we prove that our construction does what was asked for.

Construction 1.7: Equal Distances

Given two lines k and m and the point A on line k, construct points X and Y on k such that the distance from X to A equals the distance from X to m, and similarly such that the distance from Y to A equals the distance from Y to m.

Investigation To discover the location of the desired points X and Y, we first assume that the points have been found and then search for properties of Figure 1.62a that will help us construct these points. As $\overline{XX_1}$ and $\overline{YY_1}$ are perpendicular to m, it seems reasonable to construct the perpendicular from A to m. That perpendicular intersects line m in A_1, as shown in Figure 1.62b. (Notice that $\overline{AA_1}$ can be easily constructed using the basic construction of drawing the perpendicular from a point to a line.)

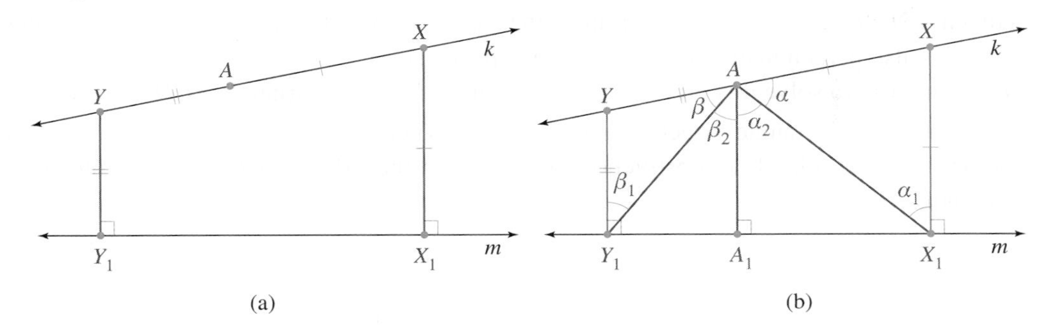

Figure 1.62

Next, since $AX = XX_1$ and $AY = YY_1$, we connect A with X_1 and with Y_1 to obtain the isosceles triangles $\triangle AXX_1$ and $\triangle AYY_1$. Because base angles in an isosceles triangle are congruent, we denote the pairs of congruent angles by α and α_1 and by β and β_1. Anticipating additional angles to be congruent, we name two more angles with vertices at A as α_2 and β_2, as shown in Figure 1.62b. Since lines YY_1, AA_1, and XX_1 are perpendicular to m, they are parallel to one another. Consequently, $\alpha_1 = \alpha_2$ (alternate interior angles created by the parallel lines AA_1 and XX_1 and the transversal AX_1) and $\beta_1 = \beta_2$. It follows that $\alpha = \alpha_1 = \alpha_2$ and $\beta = \beta_1 = \beta_2$; therefore $\overrightarrow{AX_1}$ and $\overrightarrow{AY_1}$ are angle bisectors of $\angle XAA_1$ and $\angle YAA_1$, respectively. Notice that these angle bisectors can be constructed because the angles formed by line k and $\overrightarrow{AA_1}$ are known. The intersections of the angle bisectors and line m are the points X_1 and Y_1. The intersections of the perpendiculars to m through X_1 and Y_1 with line k determine the required points X and Y.

Construction The steps of the construction were described toward the end of the investigation. For the sake of completeness, we repeat them here:

1. Construct the perpendicular from A to m. The foot of the perpendicular is A_1.
2. Construct the angle bisectors of the angles formed by lines k and $\overrightarrow{AA_1}$ intersecting line m at X_1 and Y_1, respectively.
3. At X_1 and Y_1, erect perpendiculars to m. These perpendiculars intersect line k at the required points X and Y.

Proof

We refer to Figure 1.62b. By Step 2, $\beta = \beta_2$ and $\alpha = \alpha_2$. Also, because $\overline{XX_1} \parallel \overline{AA_1} \parallel \overline{YY_1}$, $\beta_1 = \beta_2$ and $\alpha_1 = \alpha_2$. Thus $\alpha = \alpha_1$ and $\beta = \beta_1$. Consequently, $\triangle AXX_1$ and $\triangle AYY_1$ are isosceles triangles and, therefore, $AX = XX_1$ and $AY = YY_1$ as required. □

Construction 1.8: Extending a Line Beyond an Obstruction

Extend line AB beyond the small obstruction shown in Figure 1.63, without the straight edge touching the obstruction.

at its length. Next, through E we draw the line parallel to k and through C we draw the line parallel to ℓ. These lines intersect ℓ and k at X and Y, respectively, and intersect each other at Q. The line PQ is the required line.

Construction

1. Through P, construct lines parallel to the sides of the angle and intersecting these sides at B and D.
2. Construct points C and E in the interior of the angle by extending $\overline{BP}$ at its length and $\overline{DP}$ at its length.
3. Through C, construct a line parallel to ℓ and through E a line parallel to k. The point where the constructed lines intersect is Q.
4. The line PQ is the required line.

Proof

From the preceding construction, it follows that both $ABPD$ and $AXQY$ are parallelograms (X and Y, respectively, are the intersection points of the lines through E and C parallel to ℓ and k). Because $PE = PD$ (construction) and $\overline{DE} \parallel \overline{AX}$, $\overline{BC}$ is a midsegment in the parallelogram $AXQY$. Therefore the midpoint P of $\overline{BC}$ is also the point of intersection of the diagonals of $AXPY$. □

Now Solve This 1.15

In Problem 22 of Problem Set 1.2, we were asked to construct the angle bisector of an angle whose vertex is inaccessible (actually part of the angle bisector). Here we ask you to explore three approaches that might differ from yours.

1. Construct an isosceles triangle ABC and then construct the perpendicular bisector of $\overline{BC}$. You will need to construct the angle first (α in Figure 1.67). The construction of the angle is suggested in Figure 1.67.
2. Bisect the angle α whose vertex is B and construct the perpendicular to that angle bisector.
3. Use the fact that the three angle bisectors of a triangle are concurrent. Then construct any triangle with one (obstructed) vertex A and arbitrary points B and C on the sides of the angle (like in Figure 1.67 except that $\triangle ABC$ does not need to be isosceles). Construct the point where the angle bisectors intersect. Repeat this process for a different triangle.

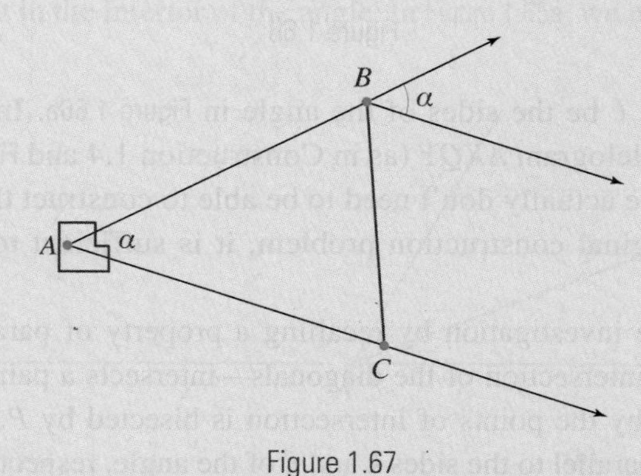

Figure 1.67

Construction 1.11: SSA Construct a Triangle Given Two Sides and an Angle Opposite One of the Sides

Investigation In Figure 1.68a, we are given angle B and segments b and c congruent to sides AC and AB of $\triangle ABC$, respectively. We need to construct $\triangle ABC$.

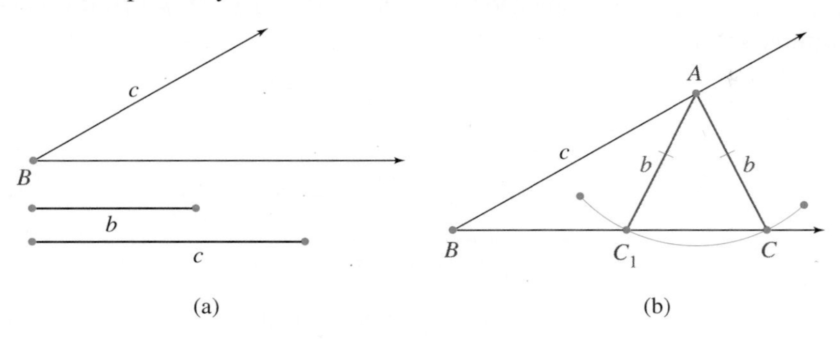

Figure 1.68

In Figure 1.68b, we first construct $\angle B$ and then place point A on one of the sides such that $BA = c$. To find the third vertex C, notice that C is on a ray, on the other side of $\angle B$, and at the distance b from A (i.e., on the circle with center A and radius b). That circle intersects the ray in two points, C and C_1, and hence the problem has two solutions: $\triangle ABC$ and $\triangle ABC_1$.

Because the construction and its proof are embedded in the investigation, they are not provided here separately. However, you are invited to further explore the construction by finding when the construction has two solutions, one solution, or no solution. □

Now Solve This 1.16

Referring to Construction 1.10 and Figure 1.68a, construct figures and explore a number of solutions for each of the following problems:

1. When the side b opposite the given angle is shorter than the other side c (Figure 1.68b)
2. When $b > c$
3. When $b = c$
4. When b is equal to the distance from A to the opposite side
5. When b is less than the distance from A to the opposite side

Problem Set 1.4

1. The figure below suggests how to construct a trapezoid given four of its sides. Solve this construction problem by completing the three stages: investigation, construction, and proof.

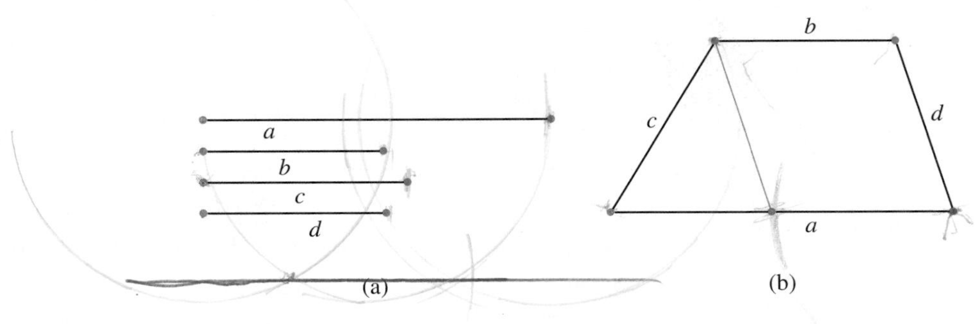

2. Construct a right triangle given its hypotenuse c and a leg a.

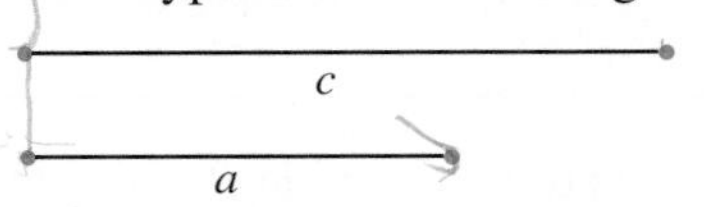

3. Construct an isosceles triangle ABC with $AB = AC$, given $\angle A$ and its perimeter (the sum of its three sides).

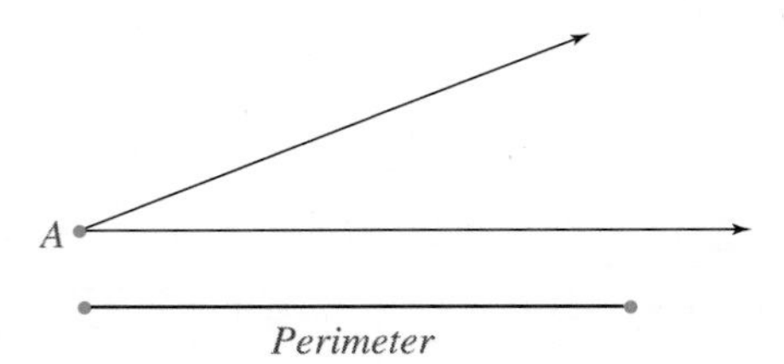

4. For each of the following, construct all the points that are equidistant from three given lines. In each case draw the lines.

a. The lines do not intersect in a single point (that is, the lines are not concurrent) and no two lines are parallel.

b. Exactly two lines are parallel.

c. The three lines are concurrent.

5. Construct a triangle, given one side a, the altitude h to the side, and the median m to the side.

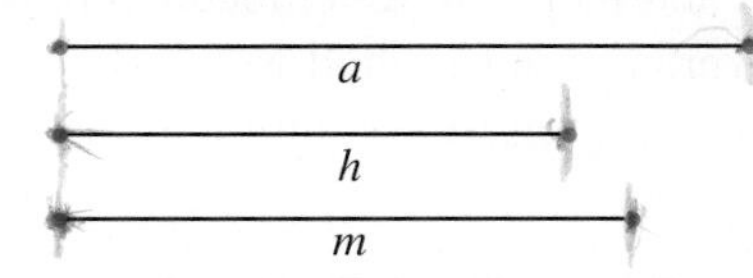

•**6.** Construct your own arbitrary $\triangle ABC$ and inscribe in it a rhombus such that one of the angles of the rhombus is an angle of the triangle.

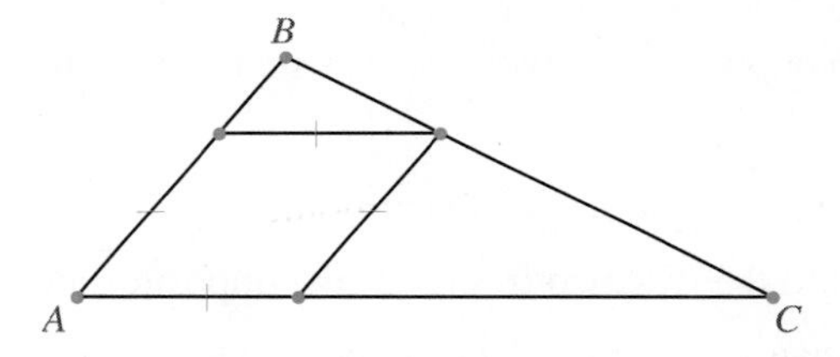

7. Construct an equilateral triangle given its height h.

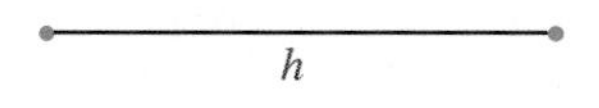

8. P is on one side of a given $\angle A$. Find point X on the other side of the angle such that $AX + PX = s$, the length of a given segment.

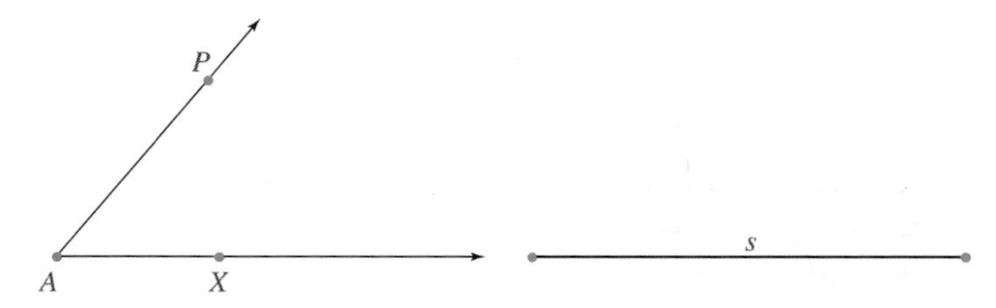

•**9.** Divide a given segment into three congruent parts using a theorem about medians of a triangle.

In some of the following construction problems the given segments or angles are not drawn. When this is the case (like in the next problem) construct your own figure and extract (copy) the given data from the figure and then use only the extracted data to construct the figure.

10. Construct a quadrilateral given three of its sides and two diagonals.

11. Construct a triangle given two of its sides and a median to the third side.

• **12.** Construct a triangle given three of its medians.

13. **a.** If $\overline{AB}$ is a diameter of a semicircle and C is any point on the semicircle, prove that $\angle ACB$ is 90°.

b. Use your result from part (a) to construct a right triangle given its hypotenuse c and one side (labeled a in the figure below).

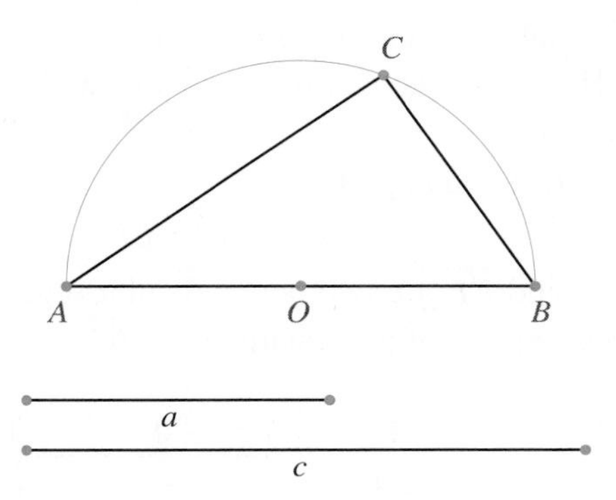

c. Solve the problem in part (b) without using the theorem in part (a).

14. Construct a right triangle given the hypotenuse c and the sum of the remaining sides a and b.

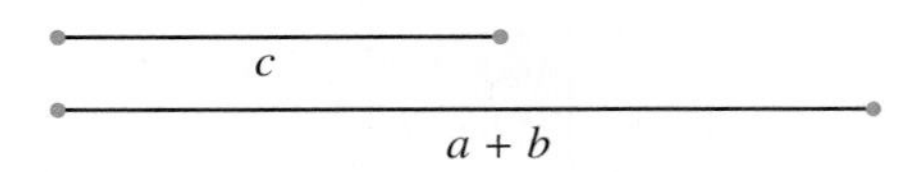

15. Construct a triangle given a side a, angle α opposite a, and the altitude h to one of the other sides.

• **16.** Construct a square inscribed in a given rhombus.

• **17.** In $\triangle ABC$, construct points P and Q on sides $\overline{AB}$ and $\overline{BC}$, respectively, such that $\overline{PQ} \parallel \overline{AC}$ and $PQ = AP + QC$.

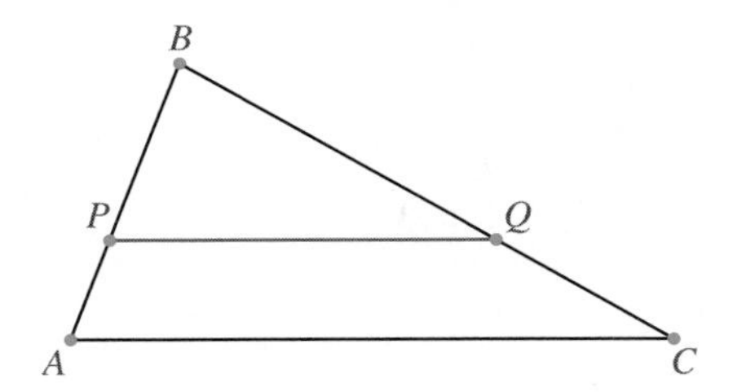

18. The point of intersection of an altitude with a side of a triangle is called the **foot of the altitude**. Given the three feet of the altitudes of a triangle, construct the triangle.

19. **a.** Construct a hexagon $ABCDEF$ (see the figure below), which is not regular but for which each of the interior angles is 120°.

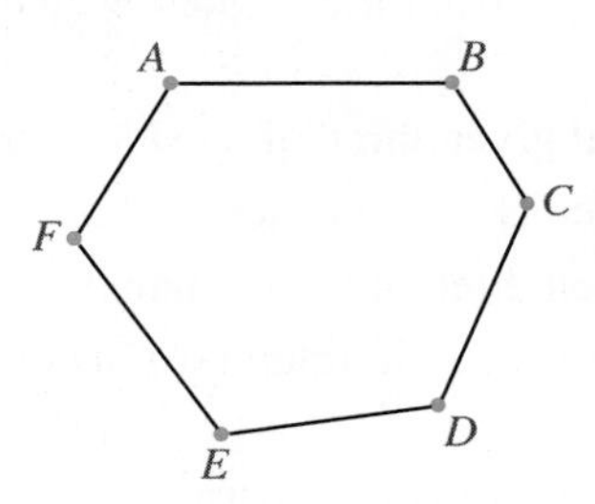

b. If two polygons have corresponding angles that are congruent and corresponding sides that are proportional (the ratio of corresponding sides is a constant), the polygons are called *similar*. How many nonsimilar hexagons with all interior angles 120° are there? Justify your answer.

c. Prove that in the hexagon in part (a),

$$AF + AB = ED + DC$$

20. • **a.** Prove that if $\overline{HG}$ and $\overline{EF}$ are two perpendicular segments connecting arbitrary points on pairs of opposite sides of the square, then $HG = EF$.

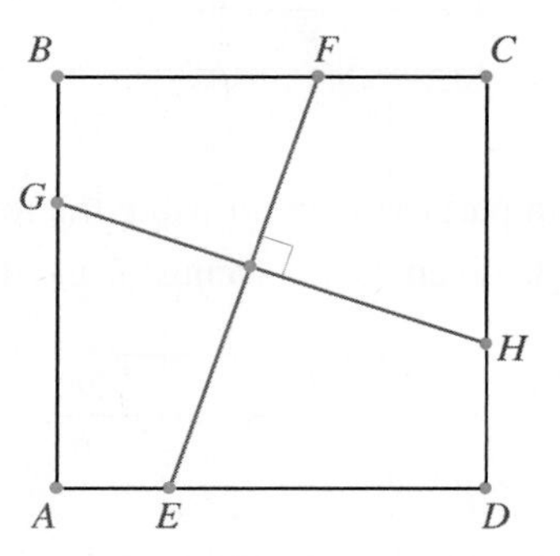

b. Suppose X, Y and Z, W are pairs of points on opposite sides of a square that remain after the sides of the squares have been erased. Reconstruct the square. Is the solution unique?

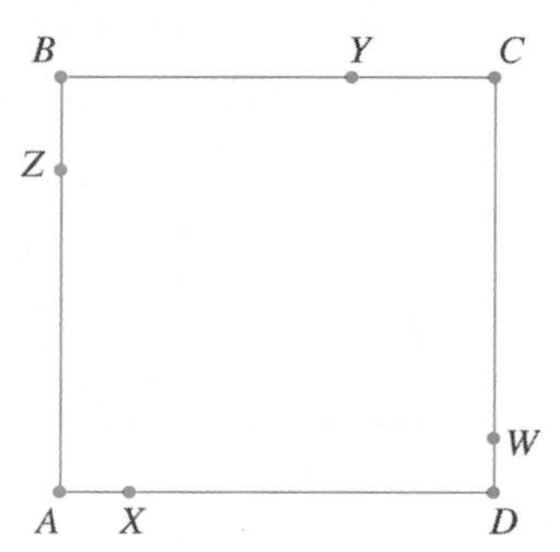

★ **21.** Given points A and B on the same side of line ℓ, find (construct) point X on ℓ such that $\angle AXC$ is twice the size of $\angle BXD$. (Points C and D are arbitrary points on ℓ such that X is between C and D. These points appear only for the purpose of naming the angles.)

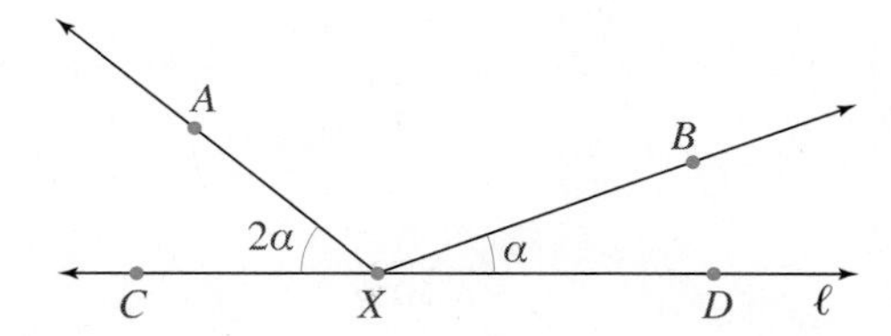

Circles

2

It is impossible to be a mathematician without being a poet in soul.
—Sonia Kovalevskaya (1850–1891), a famous Russian mathematician

Introduction

A **circle** is the set of all points in the plane, at a given distance (the **radius**) from a fixed point in the plane (the **center**). Two circles are **congruent** if and only if their radii are equal. The circle is one of the most commonly encountered figures. Wheels, balls, cross sections of trees, the shape of the full moon, and the equator—all suggest circles. Some of the most important concepts in mathematics, such as π and the circular or trigonometric functions, are related to the circle. Circles have many fascinating properties in their own right. We will explore some of these properties in this chapter.

2.1 Basic Properties of Arcs and Central and Inscribed Angles

A line that intersects a circle in two points is called a **secant**; the segment connecting those two points is called a **chord**. In Figure 2.1a, $\overleftrightarrow{PQ}$ is a secant and $\overline{PQ}$ is a chord. A chord containing the center of the circle is a **diameter**, which is twice the radius.

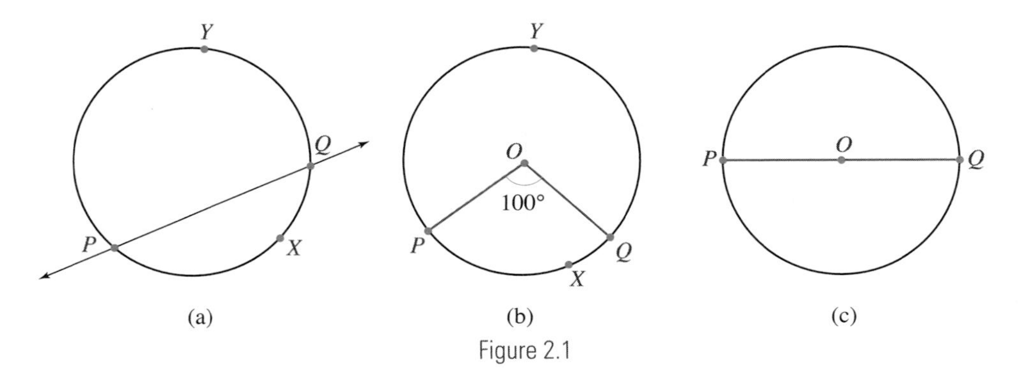

Figure 2.1

Problem 2.1 Find a relationship between the measure of an inscribed angle and the measure of its intercepted arc.

Definition of a Tangent to a Circle A **tangent to a circle** O at point P on the circle is the line through P, perpendicular to the radius OP.

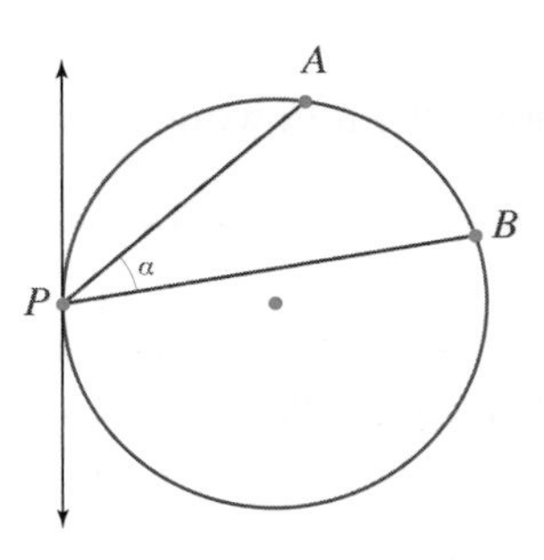

Figure 2.8

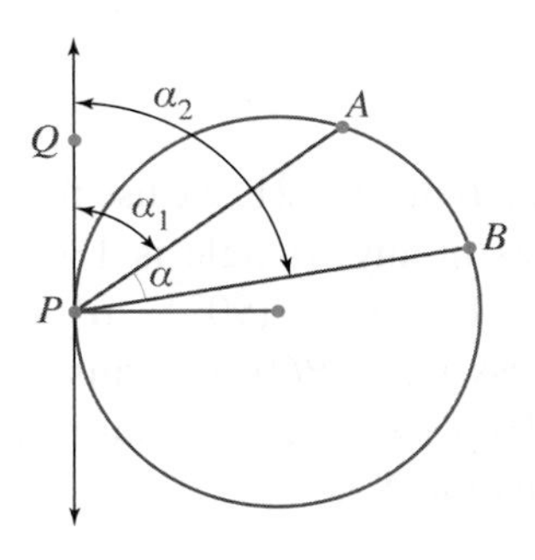

Figure 2.9

Solution

In Figure 2.9, we need to find a relationship between α and $m(\overset{\frown}{AB})$. To use Theorem 2.1, we draw the tangent $\overleftrightarrow{PQ}$. Each of the angles that the tangent makes with $\overline{PA}$ and $\overline{PB}$ can be expressed now in terms of the corresponding measures of the intercepted arcs. In addition, α is the difference of these angles. Referring to Figure 2.9, we have

$$\alpha_2 = \frac{1}{2}m(\overset{\frown}{PAB})$$

$$\alpha_1 = \frac{1}{2}m(\overset{\frown}{PA})$$

$$\alpha = \alpha_2 - \alpha_1 = \frac{1}{2}m(\overset{\frown}{PAB}) - \frac{1}{2}m(\overset{\frown}{PA})$$

$$\alpha = \frac{1}{2}[m(\overset{\frown}{PAB}) - m(\overset{\frown}{PA})]$$

$$\alpha = \frac{1}{2}m(\overset{\frown}{AB})$$

□

In Figure 2.9, $\overset{\frown}{PAB}$ is a minor arc, so it can also be written as $\overset{\frown}{PB}$. Notice that the solution to Problem 2.1 does not depend on the size of α and hence is completely general. We summarize the result of Problem 2.1 in the following useful theorem:

Theorem 2.2

The Inscribed Angle Theorem

The measure of an inscribed angle equals half the measure of the intercepted arc.

The proof of Theorem 2.2 is given in the solution to Problem 2.1. Figure 2.10 illustrates the theorem in the case where the center of the circle lies in the interior of the inscribed angle. A different proof of Theorem 2.2 (commonly found in textbooks) that does not use Theorem 2.1 will be explored in the problem set at the end of this section.

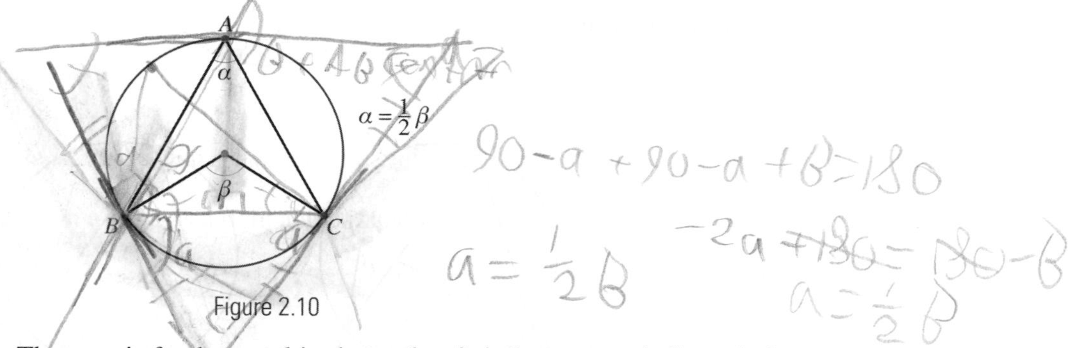

Figure 2.10

The Inscribed Angle Theorem is fundamental in the study of circles, so we shall use it frequently. Its full significance is perhaps more evident in the following corollary:

Corollary 2.1 *In any circle, all the inscribed angles intercepting the same arc are congruent.*

Proof

By Theorem 2.2, the measure of each of the inscribed angles in Figure 2.11 is half the measure of $\overset{\frown}{BC}$. □

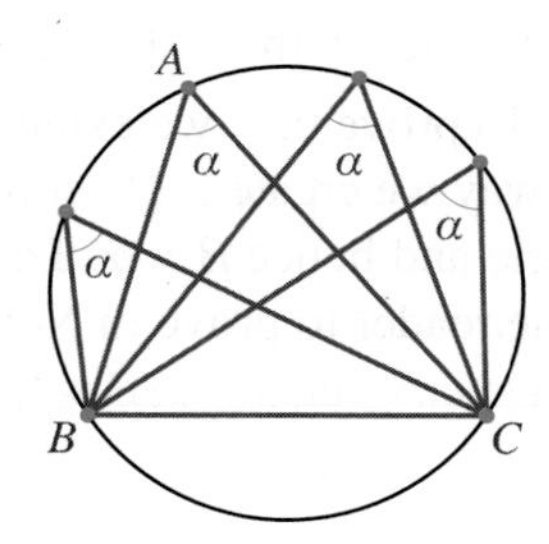

Figure 2.11

Theorem 2.2, along with Corollary 2.1, is usually referred to as the Inscribed Angle Theorem. We will frequently describe the chord of the intercepted arc as being "seen" by the inscribed angle. In Figure 2.11, this would mean that $\overline{BC}$ is seen from A by $\angle BAC$ or by α. Another way to say this is "α is the angle by which the chord BC is seen from A." Using this terminology,

> The Inscribed Angle Theorem says that from all points on $\overset{\frown}{BAC}$, the chord $\overline{BC}$ is seen by the same angle, whose measure is half the measure of the intercepted arc.

In Figure 2.11, the measure of $\angle BAC$ is designated by α. We have shown that from every point on $\overset{\frown}{BAC}$, the chord $\overline{BC}$ is seen by angle α. Nevertheless, we have not proved that the points on that $\overline{DE}$ is the perpendicular bisector of $\overline{OC}$. This information implies that D is equidistant from O and C—that is, $OD = DC$. Because OD and OC are radii, $\triangle ODC$ is equilateral and

Proof

Earlier, we showed that the condition is necessary; we need to show that the condition is sufficient. Suppose that $ABCD$ in Figure 2.18 is a quadrilateral in which

$$m(\angle A) + m(\angle C) = 180° \tag{2.3}$$

We assume that $ABCD$ is not cyclic and obtain a contradiction. We construct a circle with center O that contains any three of the vertices. In Figure 2.18, O is the circle through B, C, and D.

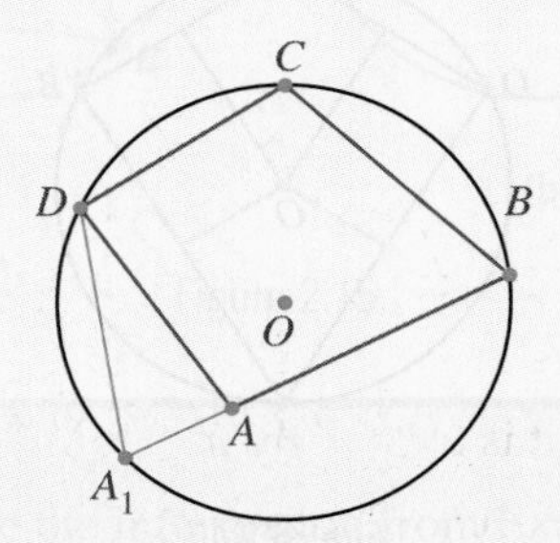

Figure 2.18

Because we assumed that $ABCD$ is not cyclic, A is not on the circle. We distinguish two cases and for each obtain a contradiction. If A is in the interior of the circle, we extend $\overline{AB}$ so that it intersects the circle at A_1.[3] In this way, we get quadrilateral A_1BCD inscribed in a circle. By the already proved first part of the theorem, we have

$$m(\angle A_1) + m(\angle C) = 180° \tag{2.4}$$

Equations 2.3 and 2.4 imply that $m(\angle A) = m(\angle A_1)$, which is impossible because $\angle A$ is an exterior angle in $\triangle DAA_1$ and, therefore, greater than any non-adjacent interior angle of the triangle. A similar contradiction is obtained for the case when A is in the exterior of the circle and is left as an exercise. □

Now Solve This 2.5

If available, use a computer and software such as the Geometer's Sketchpad to perform the following constructions. (A compass and straightedge will also suffice.)

1. Construct a circle and a quadrilateral $ABCD$ with perpendicular diagonals inscribed in the circle as shown in Figure 2.19a. Through P (the point of intersection of the diagonals), draw a line perpendicular to one of the sides. The line divides the opposite side into two segments of length a and b. Measure the segments, and find the ratio between their lengths. Repeat the experiment for a few other such quadrilaterals. Make a conjecture about your results and prove it.
2. Draw any quadrilateral $ABCD$ and its angle bisectors as shown in Figure 2.19b. The points of intersection of the four angle bisectors form a new quadrilateral $WXYZ$. Construct a circle through any three of the vertices of the new quadrilateral. What do you observe about the circle? Repeat the experiment for two other quadrilaterals. Make a conjecture about your results and prove it.

3. Notice the strategy of relating the given problem to a familiar theorem (or a problem).

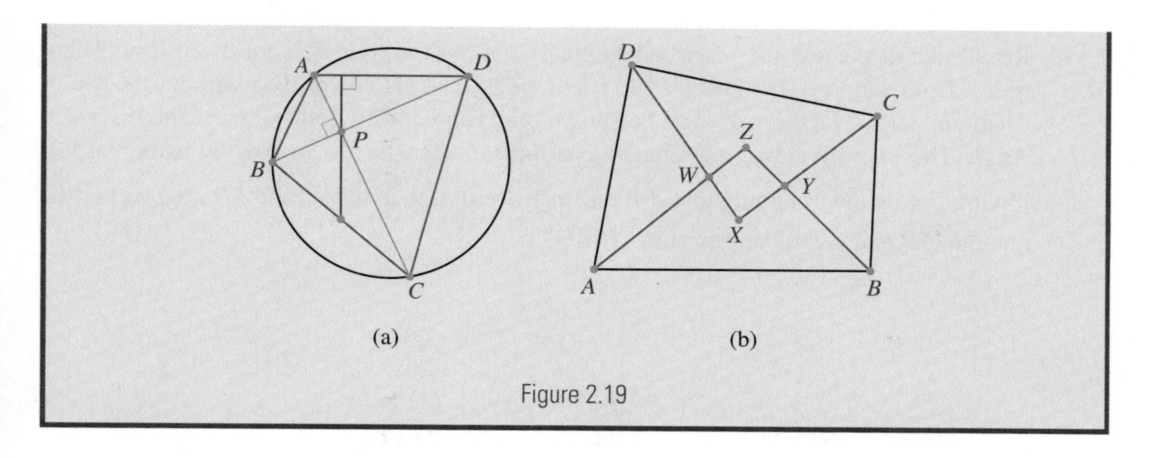

Figure 2.19

Problem Set 2.1

In the following problems, if a point is labeled by the letter O, it is the center of the circle.

1. **a.** State and prove an "if and only if" condition relating two congruent chords in a circle to their distances from the center of the circle.

   b. Given two chords in a circle, prove that the longer chord is closer to the center of the circle.

   c. State and prove the converse of the statement in part (b).

2. State and prove the converse of Proposition 2.3.

3. **a.** Prove that a tangent to a circle intersects the circle in only one point.

   b. Prove that a line that intersects the circle in exactly one point is a tangent to the circle.

4. Give an alternative proof of the Inscribed Angle Theorem, by first proving a special case stated in part (a) below and then using your result from part (a) to prove the cases in parts (b) and (c). In each case show that $\alpha = \frac{1}{2}\beta$.

   a. One side of the inscribed angle contains the diameter of the circle.

   b. The center of the circle is in the interior of the angle. [*Hint:* Construct the diameter through the vertex of the angle and apply your result from part (a).]

   c. The center of the circle is in the exterior of the angle.

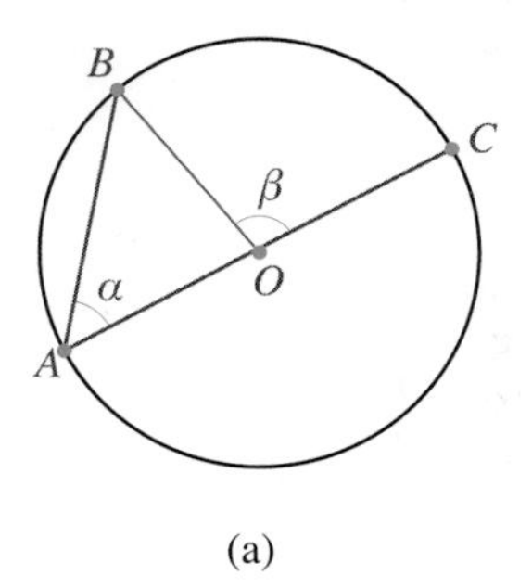

(a)

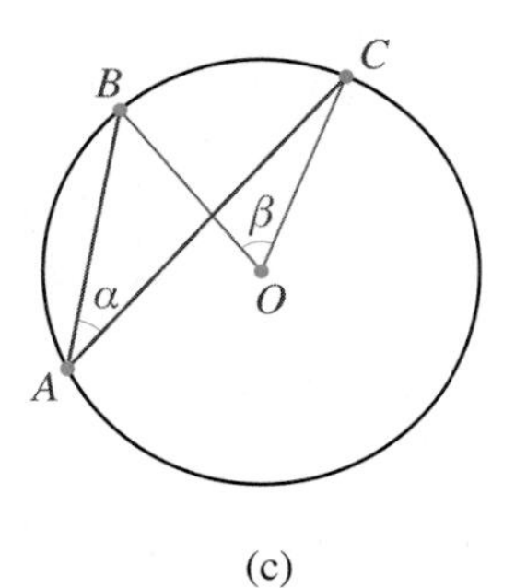

(b)

(c)

5. Which proof of the Inscribed Angle Theorem do you like better—the one in the text or the one outlined in Problem 4? Explain why.

6. Recall that in the text the Inscribed Angle Theorem (Theorem 2.2) followed from Theorem 2.1. As suggested in Problem 4, Theorem 2.2 can also be proved without the use of Theorem 2.1. In fact, it can even be shown that Theorem 2.1 follows from the Inscribed Angle Theorem by viewing a tangent as a limit of secants. Do so now by using the following figure and assuming that if P moves toward A, then the secant $\overleftrightarrow{AP}$ approaches the tangent $\overleftrightarrow{AB}$ and $\angle PAQ$ approaches $\angle BAQ$.

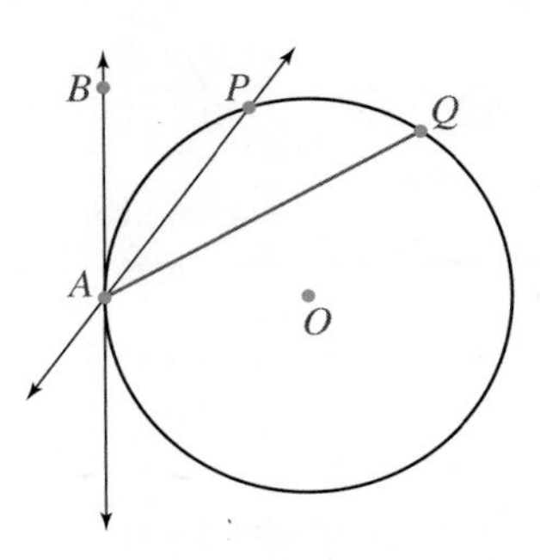

•**7.** In each of the following cases, if possible find unique values for x or for x, y, and z. If not possible prove that there are infinitely many solutions. (O is the center of each circle.)

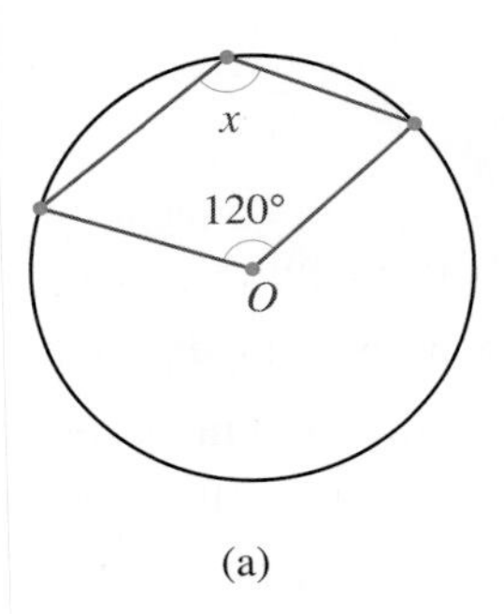

(a)

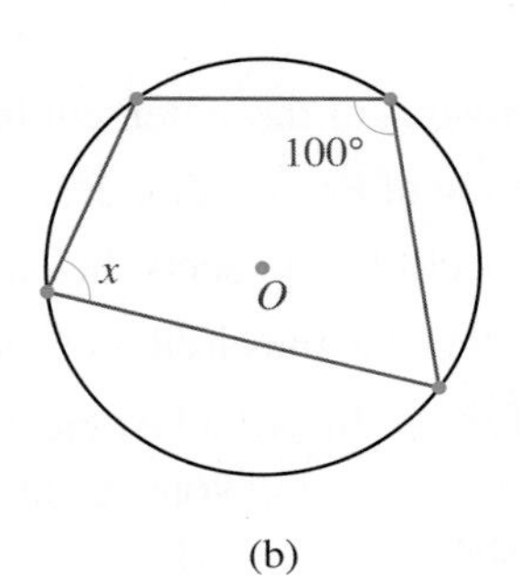

(b)

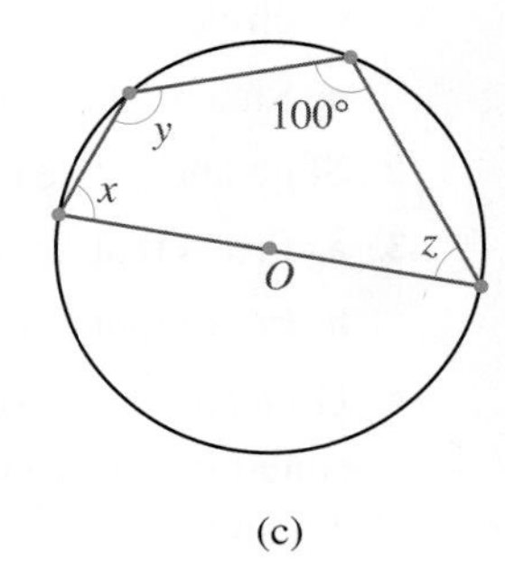

(c)

•**8.** $\triangle ACB$ is inscribed in a circle. H is the orthocenter (the point where the altitudes intersect) of $\triangle ACB$. Point K is the intersection of $\overrightarrow{AD}$ and the circle. Prove that $\overline{HD} \cong \overline{DK}$.

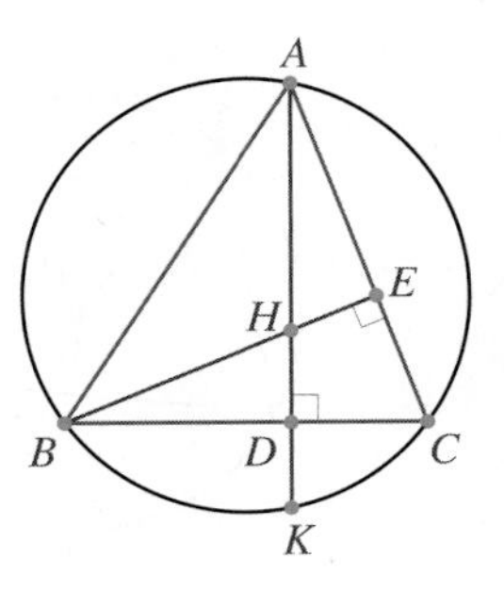

9. a. In the accompanying figures, two secants intersect in the interior and the exterior of the circle, respectively. In each case, find the angle α formed by the secants in terms of the measures of the intercepted arcs. State and prove the corresponding results. (*Hint:* Connect some of the points in the figures to create inscribed angles.)

b. Explain why the Inscribed Angle Theorem is a special case of each of the results in part (a).

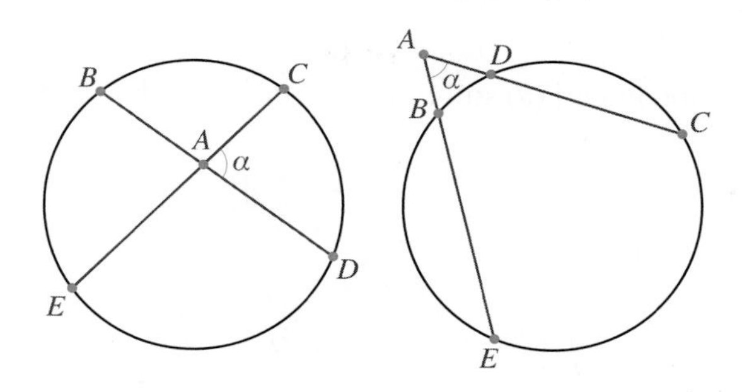

10. Prove that two parallel chords in a circle create congruent arcs. (If $\overline{BC} \parallel \overline{AD}$, show first that $\overline{AB} \cong \overline{CD}$.)

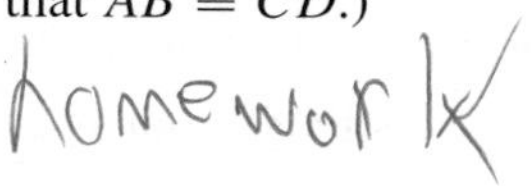

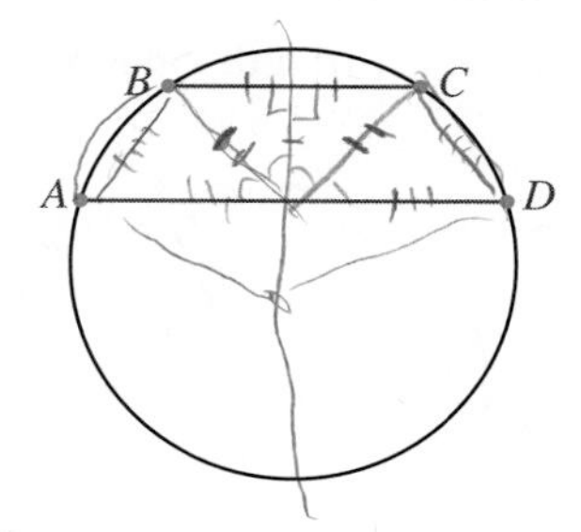

11. Problem 9(a) can now be answered in a different way using the result of Problem 10. Use the accompanying figure to answer the first part in Problem 9(a) and sketch an analogous figure to answer the second part of Problem 9(a).

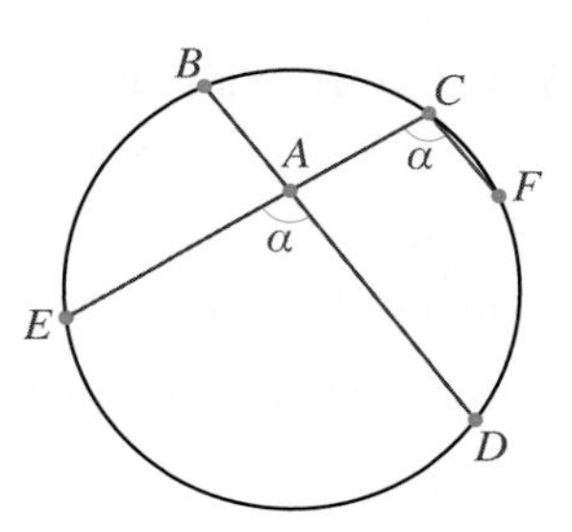

12. Answer each of the following questions and for each write a related "if and only if" statement. Justify your answers.

a. What kind of parallelograms are cyclic?

b. What kind of trapezoids are cyclic?

• **13.** $\triangle ABC$ is inscribed in a circle. Line m is tangent to the circle at C, $n \parallel m$. Line n intersects $\triangle ABC$ at D and E. Is $ABED$ cyclic? Either prove or disprove it.

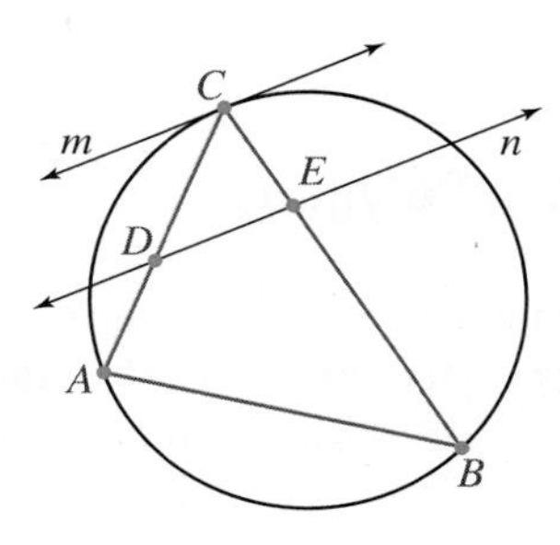

14. $\triangle ABC$ is inscribed in circle O as shown. The altitude $\overline{BH}$ intersects the diameters through A and C at D and E, respectively.

a. Prove that $\triangle DOE$ is isosceles.

b. Is the assertion in part (a) still true if $\triangle ABC$ is an obtuse triangle? Why or why not?

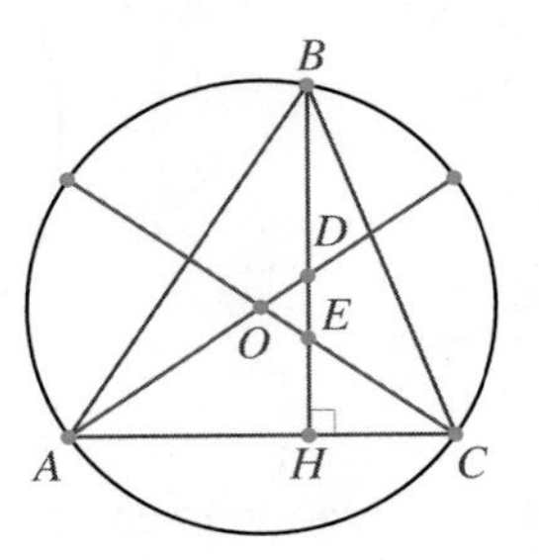

15. $\overline{AB}$ is a diameter of a semicircle with center O. Let S be on $\overrightarrow{AB}$, and let $\overleftrightarrow{SC}$ be a tangent to the circle. If $\overrightarrow{ST}$ bisects $\angle ASC$, find the measure of $\angle STC$. Justify your answer.

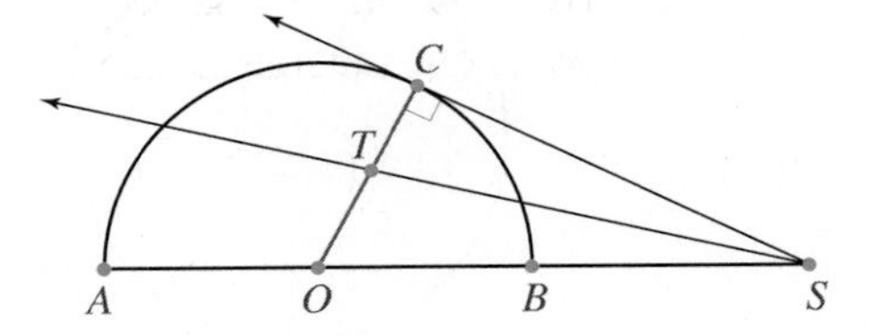

16. A is a point in the exterior of a circle. (Choose A close to the circle.) Let B and C be the points of tangency of the tangents drawn from A to the circle, and let $\overline{CP}$ be a diameter. The line $\overleftrightarrow{PB}$ intersects $\overleftrightarrow{CA}$ at point E. Prove that $m(\angle BAC) = 2m(\angle BEA)$. (*Note:* You need to draw your own figure.)

17. $ABCD$ is inscribed in circle O. $\overline{AB}$ is a diameter, and E is the intersection of $\overrightarrow{AD}$ and $\overrightarrow{BC}$.

a. Prove that $\angle ODC \cong \angle AEB$.

b. Prove that $\overline{OD}$ is the tangent at D to the circle that circumscribes $\triangle DEC$.

c. Which of the preceding statements, if any, are true if $\overline{AB}$ is not a diameter? Justify your answer.

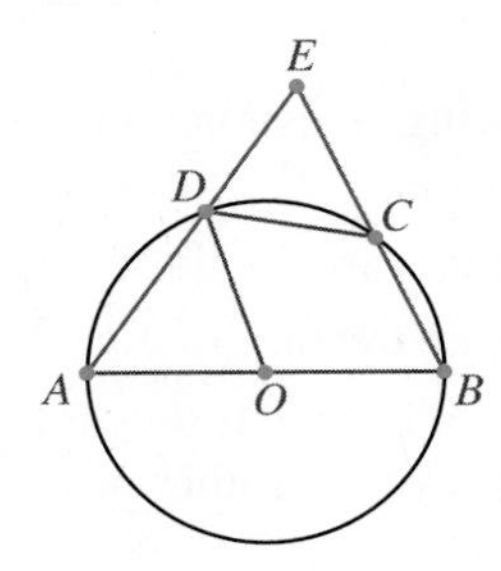

2.2 Circles Inscribed in Polygons

When each side of a polygon is tangent to a circle, the circle is said to be **inscribed in the polygon**; we also say that the **polygon circumscribes the circle**. Such a polygon is called **circumscribable**. Figure 2.20a and Figure 2.20b show a circumscribable triangle and a circumscribable quadrilateral, respectively.

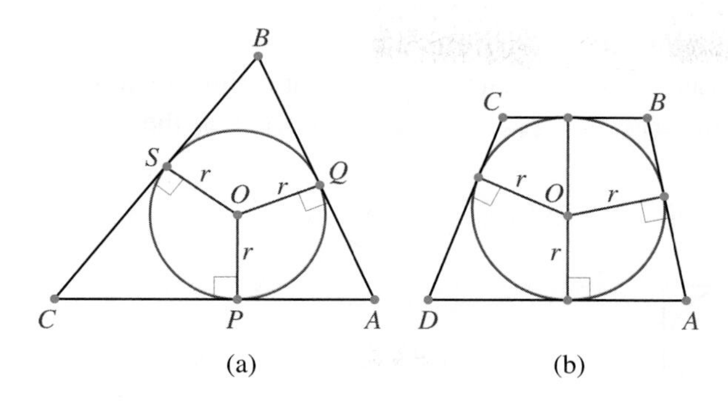

Figure 2.20

In Chapter 1, we learned that it is possible to inscribe a circle in every triangle. Let's reexamine why by looking at the procedure for constructing such a circle, if it exists. Notice that each of the distances OP, OQ, and OS from the center O to the corresponding sides must be equal to r, the radius of the circle. In particular, O is equidistant from the sides of $\angle A$ and hence must be on the angle bisector of $\angle A$. Similarly, O is on the angle bisector of $\angle B$ and of $\angle C$. This suggests the following theorem, which was both stated and proved in Chapter 1 (we restate it here for convenience):

Theorem 2.5

The angle bisectors of any triangle intersect at a single point O, which is the center of the inscribed circle. The distance from O to any of the sides of the triangle is the radius of the circle.

Notice in Figure 2.20a that P, Q, and S are the points of tangency. (Why?) It seems that $AP = AQ$. This follows from the fact that $\triangle AQO \cong \triangle APO$. Similarly, $BQ = BS$ and $CS = CP$. This suggests the following useful theorem:

Theorem 2.6

From a point in the exterior of a circle, the two tangent segments are congruent.

Proof

Figure 2.21 shows the two tangent segments $\overline{AP}$ and $\overline{AQ}$ from A. We need to show that $\overline{AP} \cong \overline{AQ}$. Notice that $\triangle AQO \cong \triangle APO$ by the hypotenuse–leg condition because the angles at P and Q are right angles (definition of a tangent), and the triangles share the same hypotenuse $\overline{AO}$ and $\overline{PO} \cong \overline{QO}$. ($O$ is on the angle bisector of $\angle A$ and hence equidistant from the sides of the angle.) Hence $\overline{AQ} \cong \overline{AP}$. □

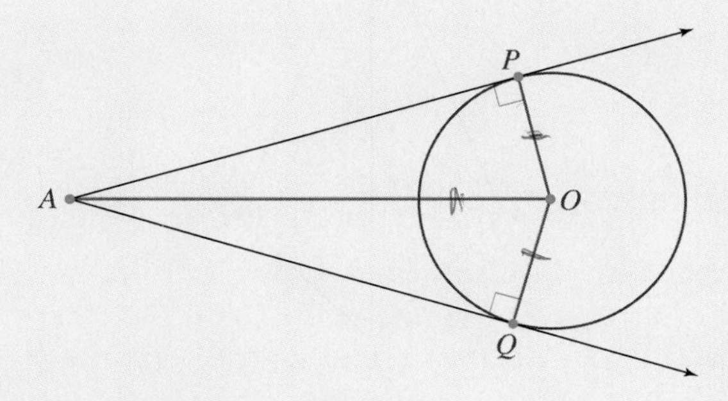

Figure 2.21

Circles Inscribed in Quadrilaterals

Figure 2.22a and Figure 2.22b show that a circle cannot be inscribed in every quadrilateral. What makes the quadrilateral in Figure 2.22c circumscribable whereas the others are not?

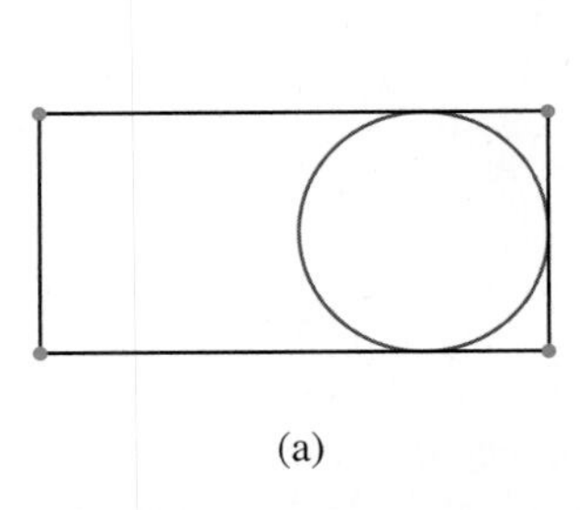

(a)

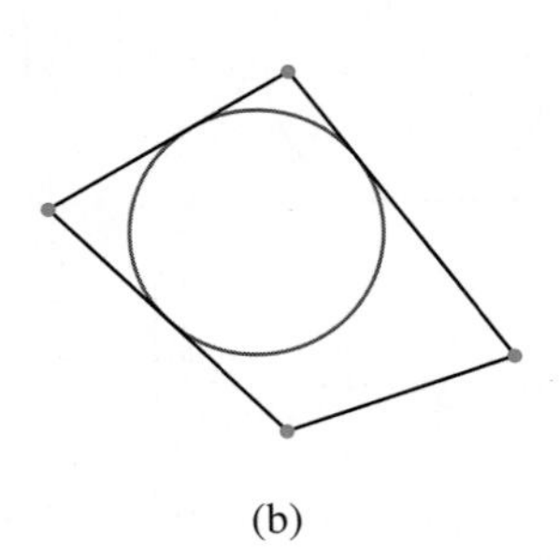

(b)

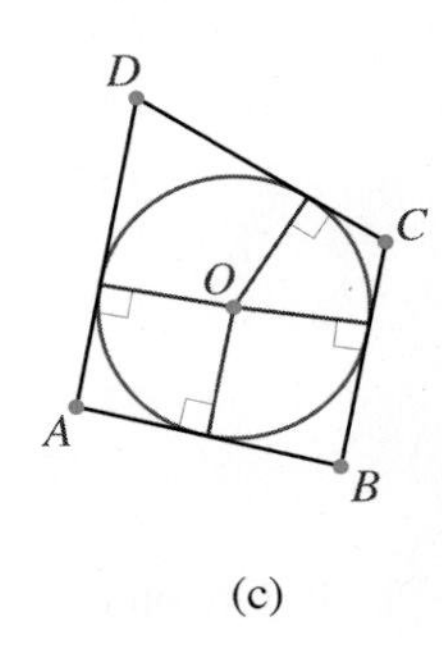

(c)

Figure 2.22

Notice that in Figure 2.22c, the distance from the center O to each of the sides of the quadrilateral is the radius of the circle. Thus O is equidistant from each side of any of the angles. Consequently, O must be on each of the angle bisectors—that is, the angle bisectors intersect at O. Conversely, suppose that the angle bisectors of the four angles of a quadrilateral are concurrent and intersect at point O. Can we conclude, then, that the quadrilateral is circumscribable? Because a point is on the angle bisector of an angle if and only if it is equidistant from the sides of the angle, it follows that O is equidistant from all sides of the quadrilateral. Thus a circle with center O and radius equal to the distance from O to any of the sides of the quadrilateral can be inscribed in the quadrilateral. We have proved the following theorem:

Theorem 2.7

A circle can be inscribed in a quadrilateral if and only if the angle bisectors of the four angles of the quadrilateral are concurrent.

Notice that a circle can be inscribed in every square because the diagonals of a square are also its angle bisectors. Can you think about other familiar quadrilaterals in which a circle can be inscribed? (Try to find one before reading on.) Such a quadrilateral is given in the following example.

Example 2.2 Prove that a circle can be inscribed in every kite.

Proof

By Theorem 2.7, it is sufficient to prove that the angle bisectors of a kite are concurrent. Consider the kite $ABCD$ in Figure 2.23, in which $AB = BC$ and $AD = DC$.

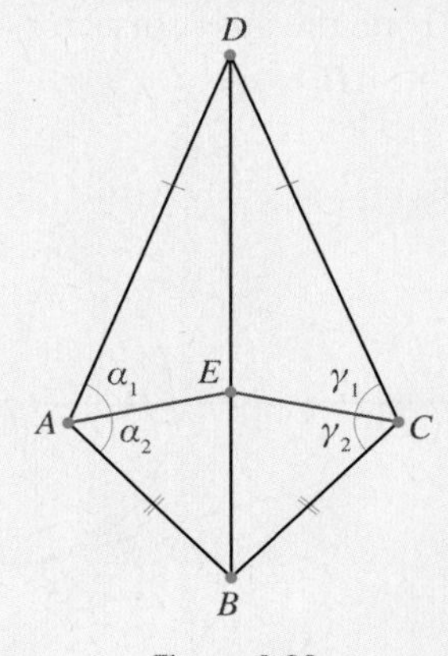

Figure 2.23

From the properties of a kite, we know that the diagonal $\overline{BD}$ bisects $\angle B$ and $\angle D$. We next construct the angle bisector of $\angle A$, which will divide $\angle A$ into two congruent angles α_1 and α_2, and will intersect $\overline{BD}$ in some point E. We will show that the angle bisector of $\angle C$ also intersects $\overline{BD}$ at E. One way to achieve this goal is to connect E with C and to prove that $\overline{CE}$ (which makes angles γ_1 and γ_2 with the sides $\overline{DC}$ and $\overline{BC}$, respectively) is on the angle bisector of $\angle C$. That is, we show that $\gamma_1 = \gamma_2$. To prove that $\gamma_1 = \gamma_2$, we show first that $\alpha_1 = \gamma_1$ and $\alpha_2 = \gamma_2$. We will reach these pairs of equal angles by observing that α_1, γ_1 and α_2, γ_2 are corresponding parts in congruent triangles. For that purpose, we show that each pair of these angles is congruent by noting that the angles in each pair are corresponding parts in congruent triangles. Notice that $\triangle ABE \cong \triangle CBE$ by SAS because $AB = BC$ (definition of a kite), $BE = BE$, and $\angle ABE \cong \angle CBE$ (property of a kite). Similarly, $\triangle ADE \cong \triangle CDE$. Consequently, $\angle BCE \cong \angle BAE$ and $\angle DCE \cong \angle DAE$. Thus $\alpha_1 = \gamma_1$ and $\alpha_2 = \gamma_2$. As $\alpha_1 = \alpha_2$, we conclude that $\gamma_1 = \gamma_2$. Consequently, $\overrightarrow{CE}$ is the angle bisector of $\angle C$ and, therefore, all the angle bisectors of a kite are concurrent. □

A Relationship Among the Sides of a Circumscribable Quadrilateral

In Example 2.2, we saw that a circle can be inscribed in any kite. We also saw, as in Figure 2.22b, that some quadrilaterals are not circumscribable. We can, however, always find a unique circle that is tangent to any three sides of a quadrilateral (Why?), though it may not be tangent to the fourth side as well. If a circle can be inscribed in a quadrilateral, as in Figure 2.24, then all the sides are tangent to the circle. Therefore, by Theorem 2.6, all tangent segments from the same vertex are congruent. This fact is indicated in Figure 2.24 by designating the lengths of the congruent tangent segments by the same letters.

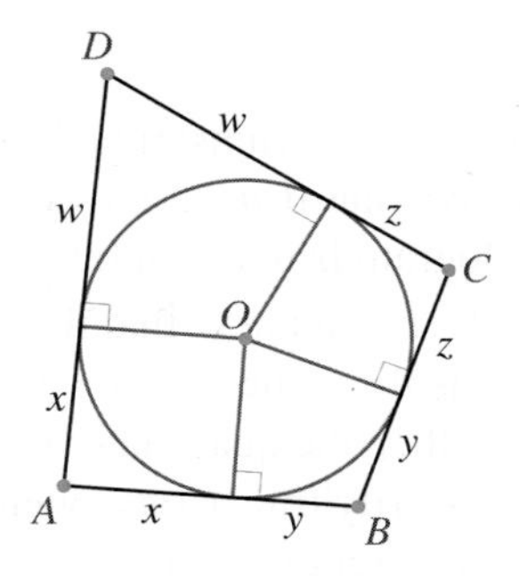

Figure 2.24

Consequently, we have

$$AB = x + y$$
$$BC = z + y$$
$$DC = w + z$$
$$DA = w + x$$

These equations suggest a relationship among the lengths of the four sides of the quadrilateral. Next notice that

$$AB + DC = (x + y) + (w + z) = x + y + z + w$$
$$BC + DA = (z + y) + (w + x) = x + y + z + w$$

sides of $\triangle DEF$, which we know are concurrent. Let O be the intersection of these three perpendicular bisectors, which are also the angle bisectors of $\angle A$, $\angle B$, and $\angle C$. From O, we construct $\overline{OP}$, $\overline{OQ}$, $\overline{OR}$, and $\overline{OS}$ (the perpendiculars to the four sides of $ABCD$). It remains to be shown that O is on the angle bisector of $\angle D$. This will be the case if and only if $OS = OR$.

Because O is on the angle bisector of $\angle A$, it is equidistant from the sides AB and AD; hence $OS = OP$. Similarly, because O is on the angle bisector of $\angle B$, $OP = OQ$. Because O is on the angle bisector of $\angle C$, $OQ = OR$. These three equations imply that $OS = OR$ and hence $OS = OP = OQ = OR$. Because these are the lengths of segments from O perpendicular to the sides of $ABCD$, O is the center of the circle inscribed in $ABCD$. □

Now Solve This 2.6

Why in the proof of Theorem 2.9 was it important to prove first the special case when the quadrilateral is a kite?

Notice that the last two theorems combined can be stated in one theorem:

Theorem 2.10

A quadrilateral is circumscribable if and only if the sum of the lengths of two of its opposite sides equals the sum of the lengths of the other two opposite sides.

Example 2.4 In Figure 2.27, $\overline{AB}$ is the diameter of circle O, and $\overleftrightarrow{BN}$ and $\overleftrightarrow{EF}$ are tangents (E is an arbitrary point on the circle). The tangent $\overleftrightarrow{EF}$ intersects $\overleftrightarrow{BN}$ at C, and $\overleftrightarrow{AE}$ intersects $\overleftrightarrow{BN}$ at D. Prove that $BC = CD$.

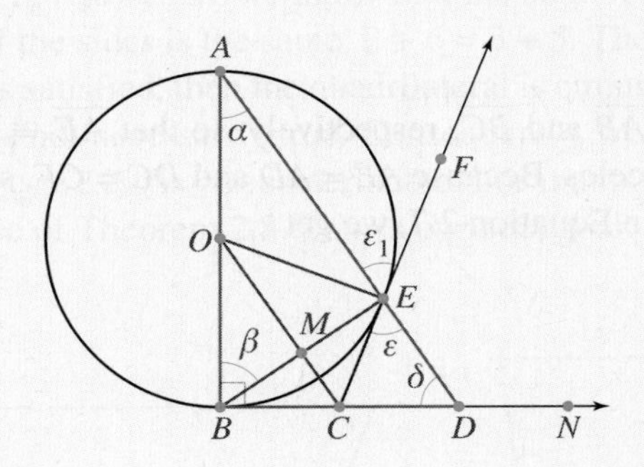

Figure 2.27

Investigation and a Proof What can we conclude about C?[5] Because $\overline{CB}$ and $\overline{CE}$ are tangent segments from point C, by Theorem 2.6 we have

$$BC = CE \tag{2.10}$$

If $BC = CD$ was true, then with Equation 2.10 it would imply that $CE = CD$. Conversely, if we could show that $CE = CD$, then with Equation 2.10 we would have $CD = BC$. It seems easier to prove that $CE = CD$, as they are sides in $\triangle ECD$ and we are more familiar with investigating triangles. In fact, it would suffice to show that the angles opposite these sides

5. Notice that we use the strategy, "What can you deduce from the 'given'?"

are congruent—in other words, referring to Figure 2.27, that $\varepsilon = \delta$. This is equivalent to what we need to prove and seems to be easier because we have substantial information about angles related to the circle. Thus our new subgoal is to show that $\varepsilon = \delta$.

Notice that $\varepsilon_1 = \varepsilon$ (vertical angles). Thus it suffices to show that $\varepsilon_1 = \delta$. As δ is neither an angle between a chord and a tangent nor an inscribed angle, we try to express δ in terms of such angles. In $\triangle ABD$, $\angle B$ is a right angle. (Why?) Therefore

$$\delta = 90° - \alpha \tag{2.11}$$

We will show that $\varepsilon_1 = 90° - \alpha$. For that purpose, we connect B with E to create a right angle BEA (which intercepts the diameter $\overline{AB}$). Now we have $\varepsilon_1 = \beta$ (both angles intercept the same arc AE). From the right triangle ABE, $\beta = 90° - \alpha$. Thus $\varepsilon_1 = 90° - \alpha$. This result, along with Equation 2.11, implies that $\varepsilon_1 = \delta$ and hence that $\varepsilon = \delta$. Consequently, the subgoal and the required results are proved.

We next give a proof based on the above investigation but omit the motivation for each step. The new proof is therefore shorter.

Proof 1

Because $\overline{BN}$ in Figure 2.27 is tangent to the circle at B, $\angle B$ is a right angle. From $\triangle ABD$, $\delta = 90° - \alpha$. Also $\varepsilon_1 = \varepsilon$ (vertical angles). Connect B with E. We have $\varepsilon_1 = \beta$ since both angles intercept the same arc AE (Corollary 2.2). Because $\angle AEB$ intercepts the diameter $\overline{AB}$, it is a right angle. Now from $\triangle ABE$, $\beta = 90° - \alpha$. Thus

$$\varepsilon_1 = 90° - \alpha$$

Consequently, $\varepsilon_1 = \delta$ and $\varepsilon = \delta$. This result implies $CE = CD$. Because $CE = BC$ (tangent segments from C), we have $CD = BC$. □

In what follows, we give a somewhat different proof of the result in Example 2.4. (The plan and the proof are combined.)

Proof 2

We need to show that C is the midpoint of $\overline{BD}$. This reminds us of the Midsegment Theorem. To apply that theorem, we need to create a triangle and a midsegment. For that purpose, we connect O with E and with C. Because $\overline{BC} \cong \overline{CE}$ (tangent segments) and $\overline{OB} \cong \overline{OE}$ (radii), it follows that $OECB$ is a kite. We know that the diagonals of a kite are perpendicular to each other, that $\overline{OC}$ bisects the angles at O and C, and that M is the midpoint of $\overline{BE}$. We focus now on $\triangle ABD$ and try to show that $\overline{OC} \parallel \overline{AD}$. This last relation will follow if a pair of corresponding angles created by $\overline{OC}$, $\overline{AD}$, and a transversal are congruent. Because $\angle BMO$ is a right angle, it suffices to show that $m(\angle BEA) = 90°$. Indeed, $\angle BEA$ is a right angle: It subtends the diameter $\overline{AB}$. Consequently, $\overline{OC} \parallel \overline{AD}$. Because O is the midpoint of $\overline{AB}$ and $\overline{OC} \parallel \overline{AD}$, it follows that C is the midpoint of $\overline{BD}$ (property of parallel projections). □

Now Solve This 2.7

1. The following is yet another idea for a proof of Example 2.4—that is, the statement that in Figure 2.27, C is the midpoint of $\overline{BD}$. Complete the proof:

   We prove that $\overline{OC} \parallel \overline{AD}$ by showing that $\alpha = m(\angle BOC)$.

2. Prove that $ABCE$ in Figure 2.27 is not circumscribable.

Example 2.5 If $\triangle ABC$ is an equilateral triangle inscribed in a circle and P is any point on $\overset{\frown}{AC}$, prove that $AP + PC = PB$. (See Figure 2.28.)

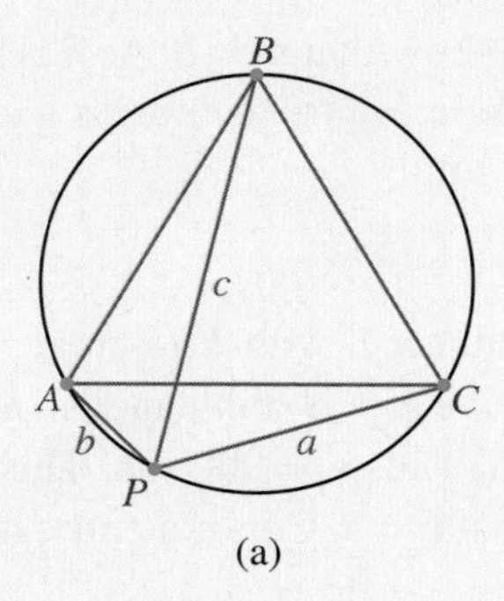

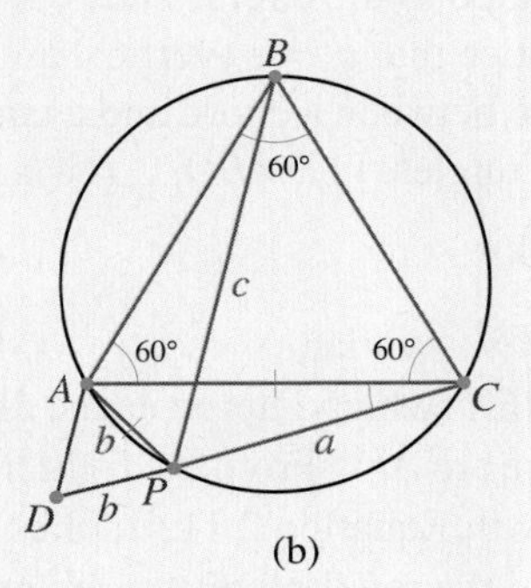

Figure 2.28

Plan Referring to Figure 2.28a, we need to show that $a + b = c$. Notice that c is a side in $\triangle ABP$. Because of our experience in working with triangles, we try to make $a + b$ be a side in a triangle. One way to achieve this goal is to extend $\overline{PC}$ so that $DP = b$, as shown in Figure 2.28b. Then $a + b$ is a side in $\triangle ACD$. We want to prove that $\overline{DC} \cong \overline{BP}$, so it makes sense to try to show that $\triangle ABP$ and $\triangle ACD$ are congruent. This will be investigated next in Now Solve This 2.8.

Now Solve This 2.8

Complete the proof that $AP + PC = PB$ (in Figure 2.28b) by showing that

1. $\angle APD \cong \angle ABP$
2. $m(\angle ACD) = 60°$ and hence that $m(\angle DAP) = 60°$
3. $\angle BAP \cong \angle DAC$

Remark Figure 2.28 is a vivid example of the fact that congruence of two corresponding sides and an angle does not assure congruence of triangles. In $\triangle ABP$ and $\triangle ACP$, $\overline{AP}$ is a common side, $\overline{AC} = \overline{AB}$, and $\angle ABP \cong \angle ACP$, yet the triangles are not congruent because $m(\angle APB) = 60°$ and none of the angles in $\triangle ACP$ is 60°. (Why?) Because this condition is true for any P on $\overset{\frown}{AC}$, we have exhibited infinitely many such pairs of triangles.

Problem Set 2.2

• **1.** $\triangle ABC$ is inscribed in a circle. Point D is the center of the inscribed circle. Prove that $\triangle DAE \cong \angle ADE$.

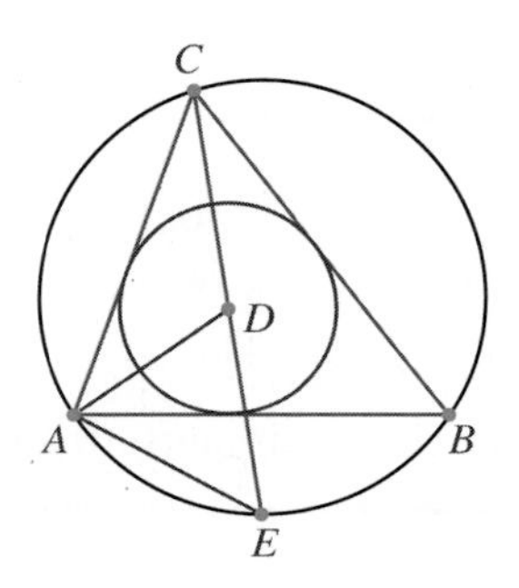

2. If a circle with center O is inscribed in a trapezoid whose bases are tangent to the circle at points P and Q (see figure), prove each of the following statements:

a. O, P, and Q are collinear.

b. The diameter of the circle equals the height of the trapezoid.

c. $m(\angle COB) = 90°$.

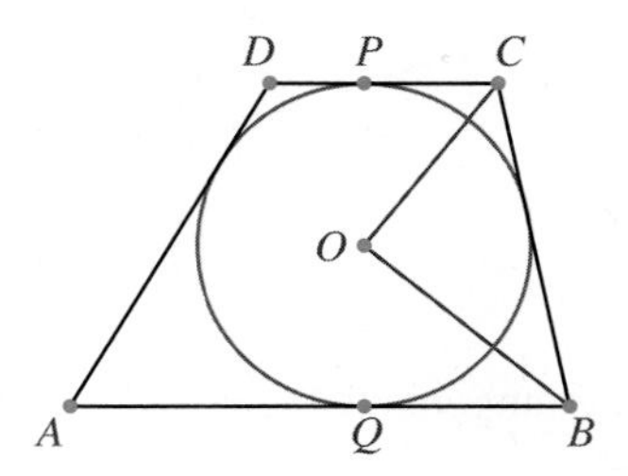

3. Two congruent circles intersect at P and Q as shown. A line is drawn through Q that intersects the circles at A and B. Another line is drawn through P that intersects the circles at A and C.

a. Prove that $PB = PA$.

b. If $\overline{AQ}$ is a diameter, prove that $CP = AP$ and that $m(\angle CBA) = 90°$.

•**4.** The two circles shown below share a common center (they are *concentric circles*). If $\overline{AB}$ and $\overline{CD}$ are two chords in the larger circle tangent to the smaller circle, prove that $AB = CD$.

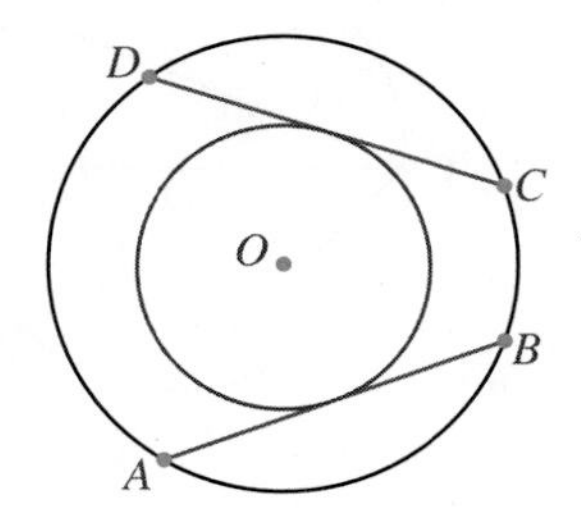

5. In the accompanying figure, $\overline{PA}$ and $\overline{PB}$ are tangents to circle O. If Q is on $\overset{\frown}{AB}$ and $\overline{CD}$ is tangent at Q, prove that

a. $m(\angle COD)$ is the same regardless of the position of Q on $\overset{\frown}{AB}$.

b. The perimeter of $\triangle PCD$ is the same regardless of the position of Q on $\overset{\frown}{AB}$.

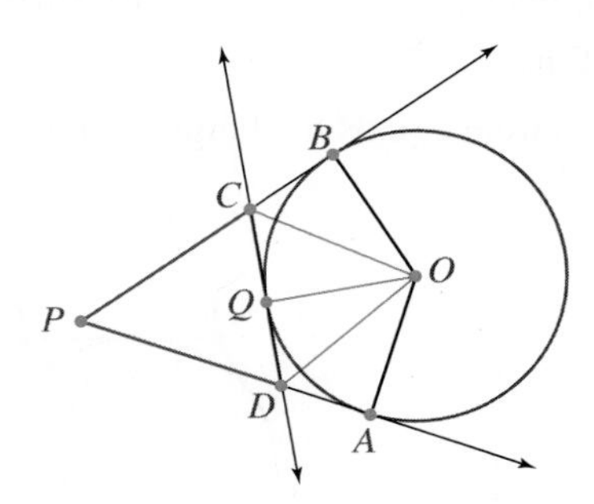

•**6.** Suppose $\overleftrightarrow{PA}$ and $\overleftrightarrow{PB}$ are tangents to a circle at A and B, respectively, and $m(\angle BPA) = \alpha$. Answer the following:

a. Express the marked angles γ and δ at C and D, respectively, in terms of α.

b. Prove that $\alpha = \gamma - \delta$.

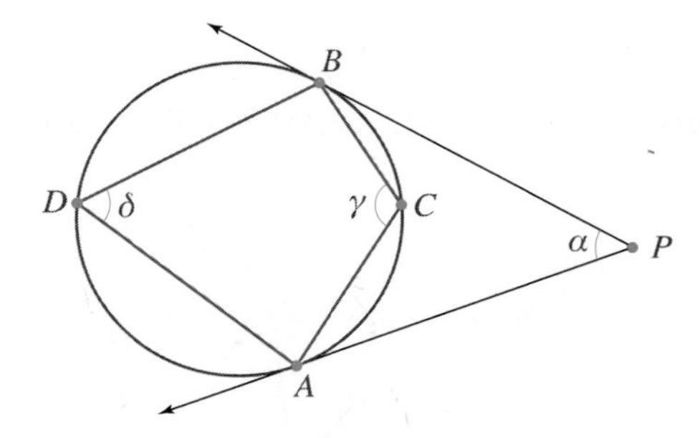

7. $ABCD$ is a quadrilateral with right angles at A and at D. Circle O is tangent to the sides $\overline{AB}$ and $\overline{AD}$ at B and E, respectively. The diagonal $\overline{AC}$ contains O. The side $\overline{CD}$ intersects the circle at P, and $\overline{PB}$ intersects $\overline{AC}$ at Q.

a. Find the angles of $\triangle CQP$.

b. Prove that $AQ = r$, where r is the radius of the circle. That is, prove that $\overline{AQ} \cong \overline{CO}$.

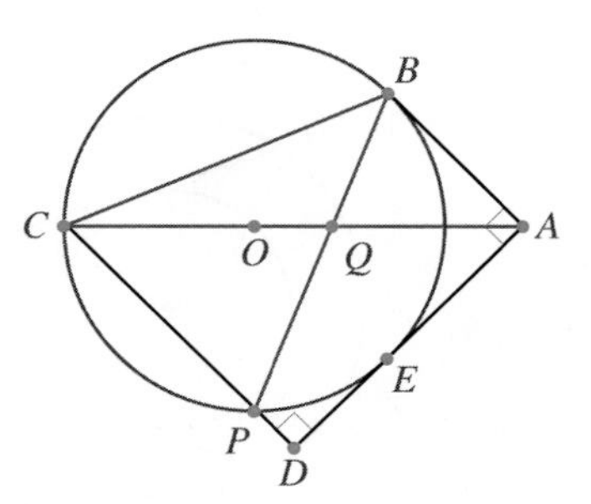

•**8.** A circle is tangent at P to the side $\overline{BC}$ of the square $ABCD$. The vertices A and D are on the circle as shown. The side $\overline{DC}$ intersects the circle at Q. Prove that

a. $\angle QPA$ is a right angle.

b. $\overline{AP}$ bisects $\angle QAB$.

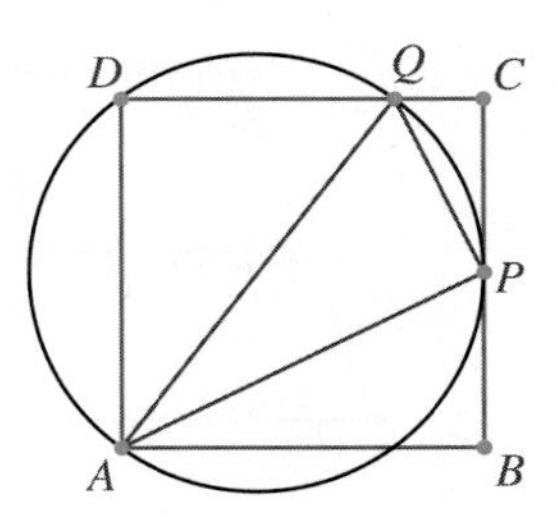

9. Determine whether it is possible to find a circumscribable, isosceles "true" trapezoid (in which the parallel sides are not congruent) that has each of the following properties. Justify your answers.

a. A diagonal bisects an angle of the trapezoid.

b. The trapezoid is cyclic.

c. The diagonals are perpendicular to each other.

10. a. Circle O is inscribed in a rhombus. $ABCD$ is a quadrilateral whose vertices are the four points of tangency. What kind of quadrilateral does $ABCD$ seem to be? Prove your answer.

b. State and prove the converse of the theorem suggested in part (a).

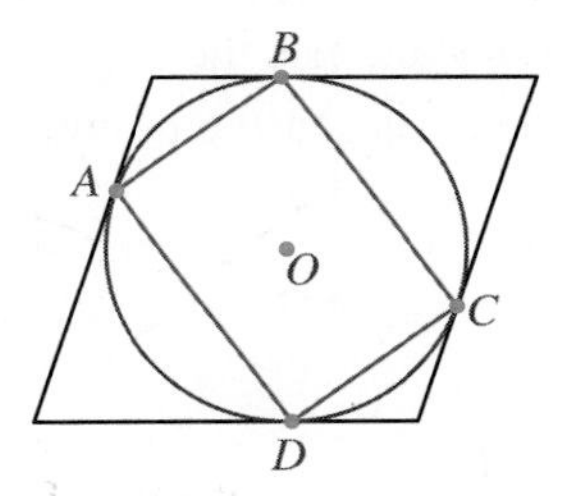

•**11.** Circle O is inscribed in trapezoid $ABCD$, and $\overline{MN}$ is the midsegment of the trapezoid. Prove the following:

•**a.** $\overline{MN}$ contains O.

•**b.** $ON = NB$ and $OM = MA$.

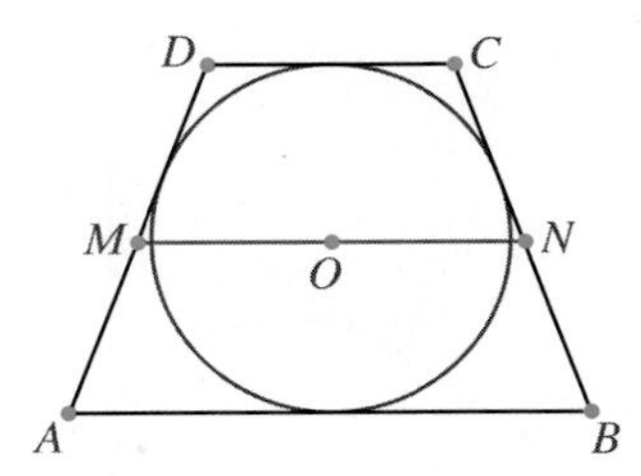

12. In a trapezoid $ABCD$, $\overline{MN}$ is the midsegment and P is a point on $\overline{MN}$ such that $PN = NB$ and $PM = MA$.

a. It is possible to inscribe a circle in $ABCD$? Justify your answer.

b. Prove that P is the center of the inscribed circle.

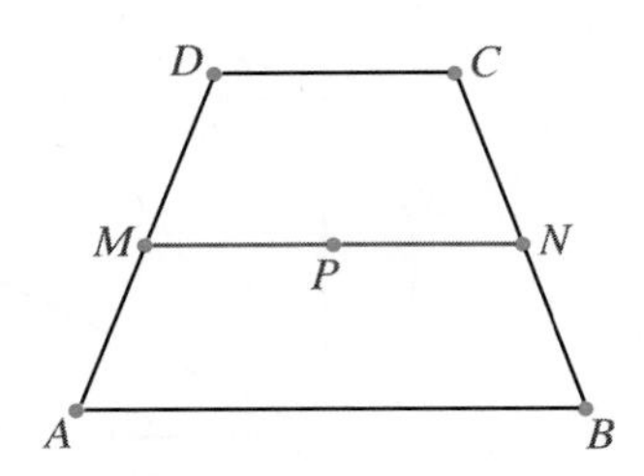

13. Prove that the centers O_1 and O_2 of two **disjoint circles** (circles that have no common points) and the point P, where the two tangents shown, intersect are collinear.

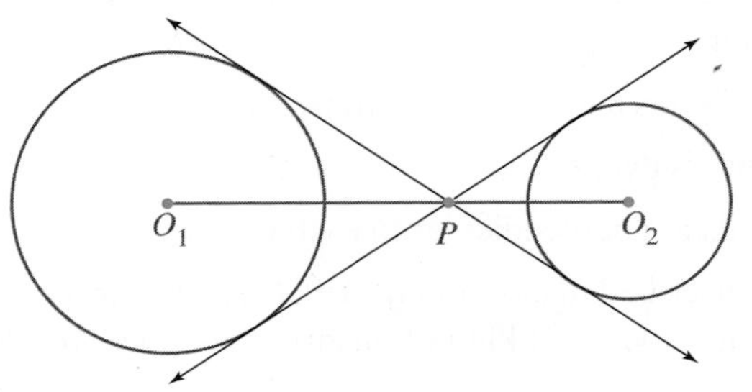

14. Two circles that have only one point in common are called **tangent circles.** Prove that

a. The common point of two tangent circles and the centers of the circles are collinear. (*Hint:* Assume the contrary and use the triangle inequality.)

b. The tangent to one of the circles at the point of contact is also tangent to the other circle.

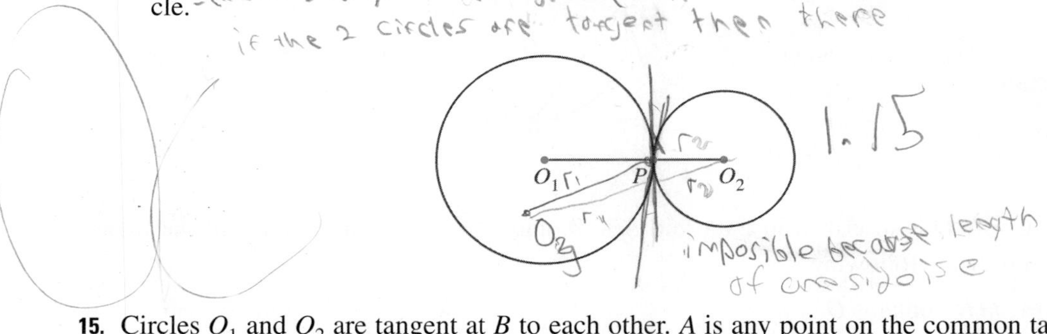

15. Circles O_1 and O_2 are tangent at B to each other. A is any point on the common tangent through B, and $\overrightarrow{AP}$ and $\overrightarrow{AQ}$ are tangents to the circles O_1 and O_2, respectively. Prove that $AP = AQ$.

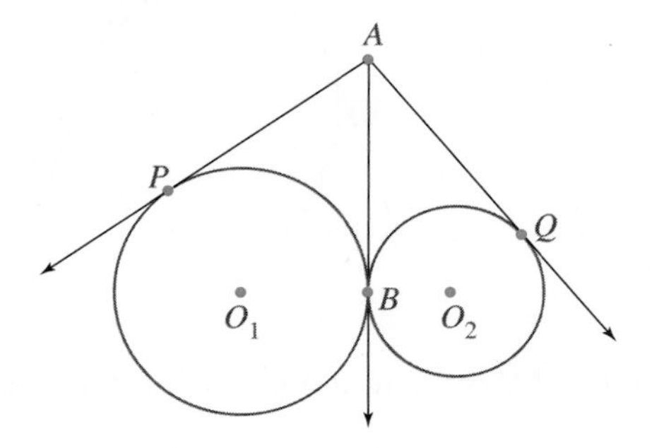

16. The circles O_1 and O_2 are tangent at P. The line AB is tangent to the circles at A and B, respectively. If C is on $\overline{AB}$ and $\overrightarrow{CP}$ is the tangent through P, prove that

a. $AC = CB$

b. $m(\angle APB) = 90°$

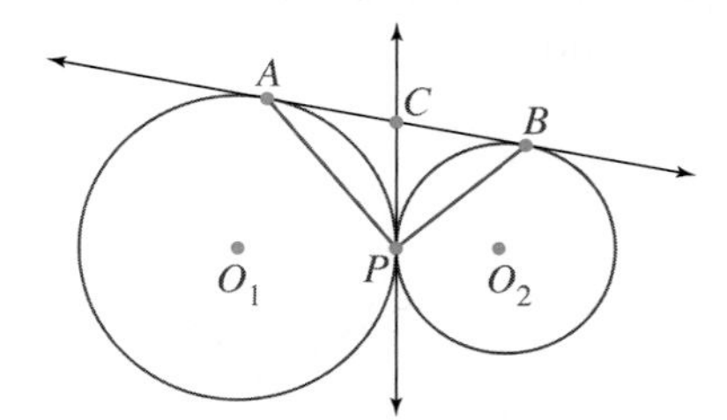

17. The circles O_1 and O_2 are tangent at P. A line through P intersects the first circle at A and the second circle at B. Prove that the tangent at A to the first circle is parallel to the tangent at B to the second circle.

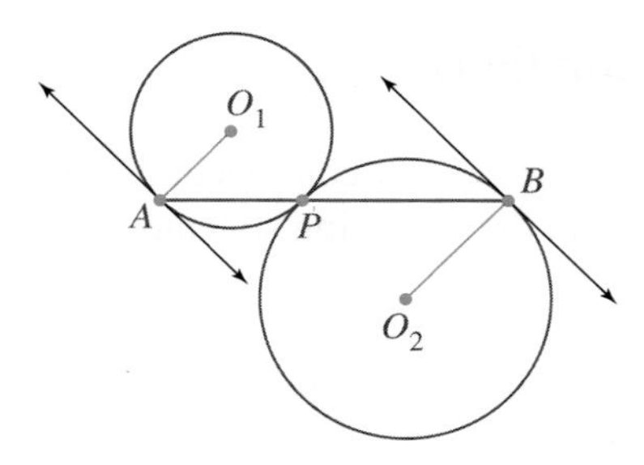

★ 18. Two congruent circles O_1 and O_2 intersect at A and B. A third circle with center at B intersects O_1 in C and O_2 in D. (These points are in the same half plane determined by $\overline{AB}$.) Prove that A, C, and D are collinear.

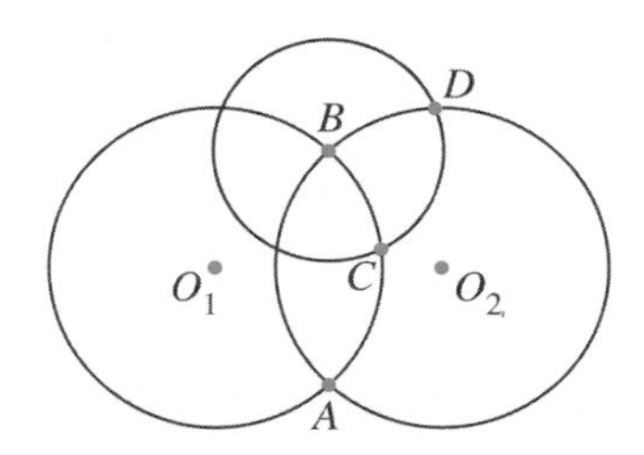

19. The vertices of $\triangle EFD$ are on the sides of $\triangle ABC$ as shown. Construct the three circles that circumscribe $\triangle ADE$, $\triangle DFC$, and $\triangle EBF$. Repeat the construction for differently positioned triangles EFD.

a. Based on what you observed, make a conjecture concerning the three circles.

b. Prove your conjecture in part (a).

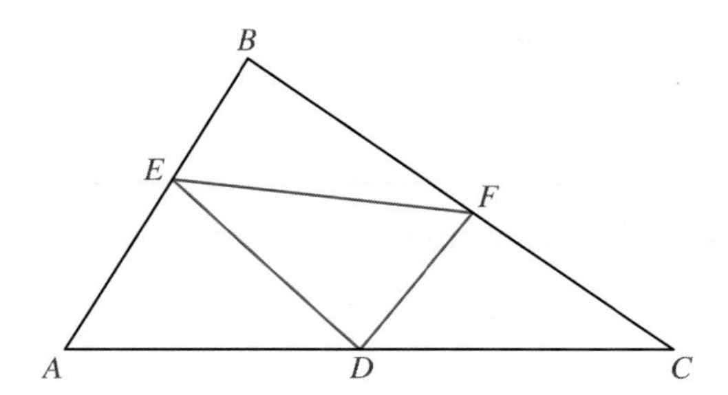

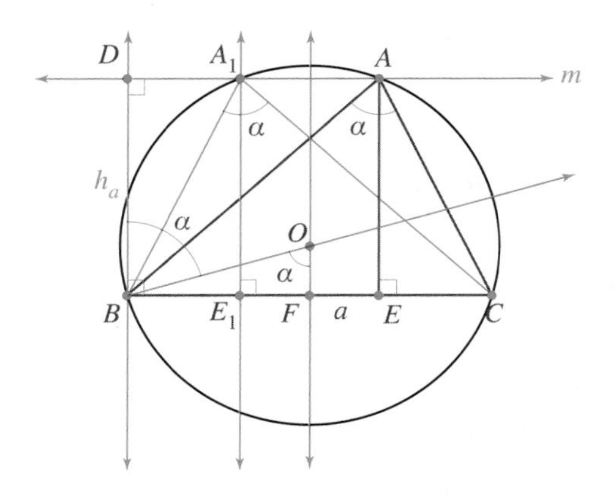

Figure 2.35

3. Construct the locus of all points (in the half plane of Step 2) from which $\overline{BC}$ is seen by α.
4. Determine the points of intersection (if any) of the loci of Steps 2 and 3. These points are A and A_1 in Figure 2.35.
5. $\triangle ABC$ and $\triangle A_1BC$ are the required triangles.

Proof

$\triangle ABC$ and $\triangle A_1BC$ each have side $\overline{BC}$ congruent to the given side a. The angles at A and A_1 are α by virtue of our construction (Construction 2.1). AE and A_1E_1 are the heights of the triangles; each equals h_a because the distance from every point on line m to $\overleftrightarrow{BC}$ is h_a. □

Remark The location of the vertex A in Construction 2.2 was found as the intersection of two loci: a line and an arc. A similar approach is useful in most construction problems. The number of points of intersection determines the number of solutions.

Now Solve This 2.10

Prove that in Figure 2.35, $\triangle BCA \cong \triangle CBA_1$.

Construction 2.3

Construct a triangle from a, α, m_b as shown in Figure 2.36.

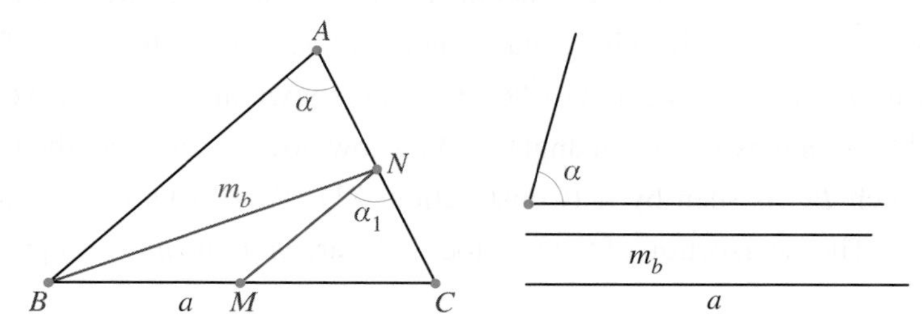

Figure 2.36

Investigation In $\triangle ABC$ in Figure 2.36, note that from A, the side $\overline{BC}$ is seen by α. We need another locus for A, but we do not see an obvious way to find it. We note that point N is on a circle with center B and radius m_b, but this information does not seem to help us in finding A or N. To find the location of N, we need another locus to which N belongs. For that purpose, we construct through N a line parallel to $\overleftrightarrow{AB}$ intersecting $\overline{BC}$ at M. It follows from the Midsegment Theorem that M is the midpoint of $\overline{BC}$. As $\overleftrightarrow{NM} \parallel \overleftrightarrow{AB}$ in Figure 2.36, $\alpha_1 = \alpha$. Now we know how to construct $\triangle BNC$ as N is on two loci: the circle with center B and radius m_b, and the locus of all points from which $\overline{MC}$ is seen by α_1 that equals α. We can now extend $\overline{NC}$ at its length to obtain A.

Construction

1. Construct $\overline{B_1C_1}$ congruent to a (Figure 2.37).
2. Construct the midpoint M_1 of $\overline{B_1C_1}$.
3. Construct the locus of all points from which $\overline{M_1C_1}$ is seen by α (the major arc M_1C_1).
4. Construct the locus of all points at distance m_b from B_1 (the circle with center at B_1 and radius m_b).
5. Find N_1, the point of intersection of the loci in Steps 3 and 4.
6. Extend $\overline{C_1N_1}$ at its length to locate A_1.
7. Connect A_1 with B_1. Now $\triangle A_1B_1C_1$ is the required triangle.

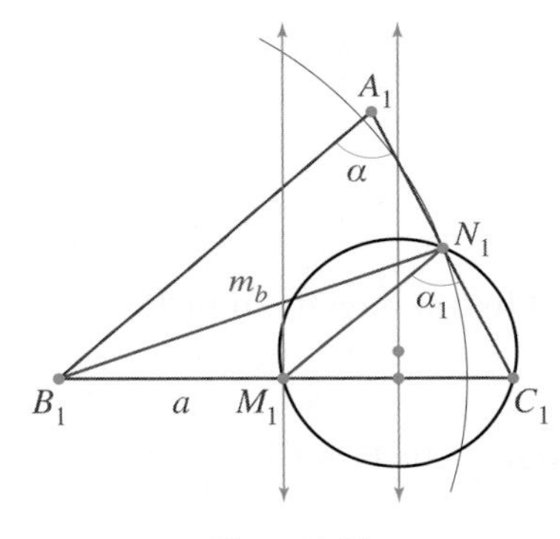

Figure 2.37

Proof

$\overline{B_1C_1}$ is congruent to a by construction. Because M_1 and N_1 were constructed to be midpoints of $\overline{B_1C_1}$ and $\overline{A_1C_1}$, respectively, $\overline{M_1N_1}$ is a midsegment in $\triangle A_1B_1C_1$ and hence $\overline{M_1N_1} \parallel \overline{A_1B_1}$. Thus $\angle B_1A_1C_1 \cong \angle M_1N_1C_1$. Notice that $\angle M_1N_1C_1$ was constructed to be congruent to α and hence $m(\angle B_1A_1C_1) = \alpha$. Also, $\overline{B_1N_1}$ was constructed to be congruent to m_b. By construction, N_1 is the midpoint of $\overline{A_1C_1}$, so B_1N_1 is the required median. □

Now Solve This 2.11

Figure 2.38 suggests a somewhat different approach to Construction 2.3. Extend $\overline{BN}$ by its length to create point D and hence the parallelogram $ABCD$. Complete the investigation, describe the construction, and prove that it satisfies the given requirements.

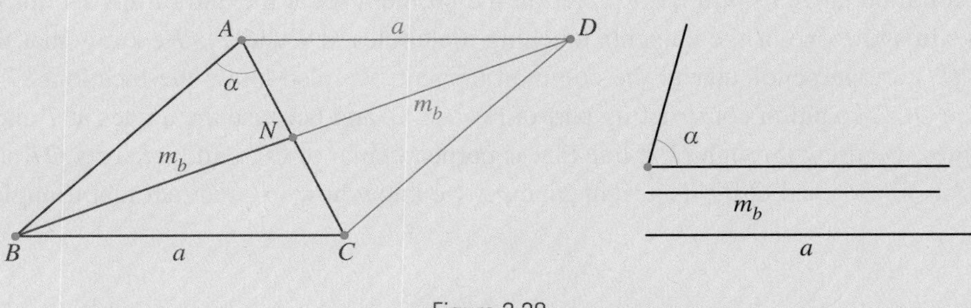

Figure 2.38

Construction 2.4

From a point P outside a circle, construct tangents to the circle with center O.

Investigation In Figure 2.39, imagine that $\overleftrightarrow{PT_1}$ is one of the required tangents. Connect O with T_1. Since $\overleftrightarrow{PT_1}$ is a tangent, it is perpendicular to the radius $\overline{OT_1}$. In addition to T_1 being on the given circle with center O, because $\angle OT_1P$ is 90°, we know that T_1 is also on the locus of all points from which $\overline{OP}$ is seen at a right angle. This locus (in one half plane) is a semicircle with center M, which is the midpoint of $\overline{OP}$ and radius $\overline{OM}$. The intersection of this semicircle with the given circle is T_1. The other semicircle intersects the given circle at T_2.

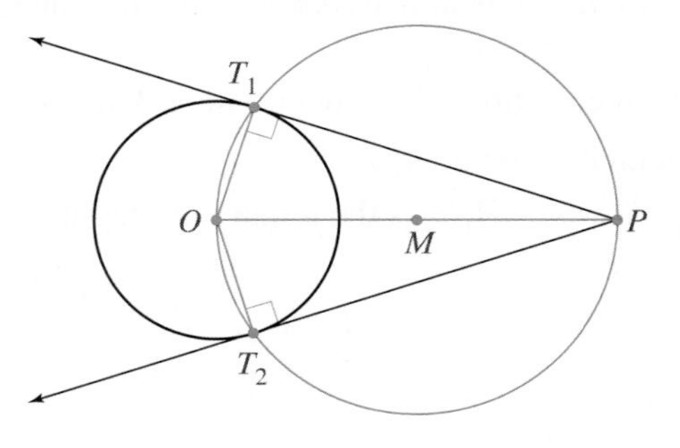

Figure 2.39

The construction and its proof follow directly from the investigation and are left to the reader.

Construction 2.5: CommonTangents to Two Circles

Given the circles with centers O and O_1, construct the circles' common tangents, as shown in Figure 2.40.

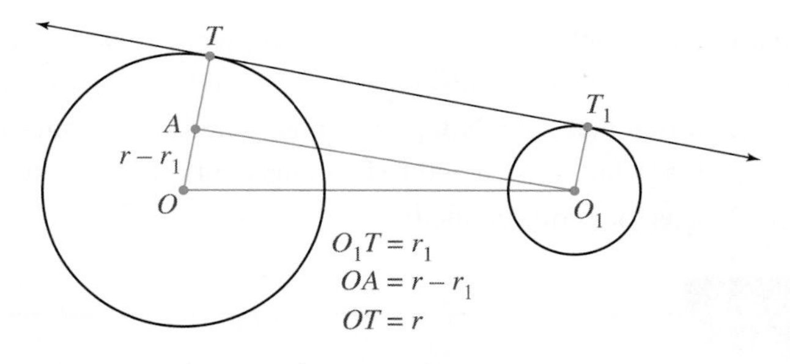

Figure 2.40

Investigation We focus only on the *external tangents* shown in Figure 2.40 (the construction of interior common tangents will be explored in the problem set at the end of this section). In the figure we imagine one of the tangents touching the circles at T and T_1. We know that the radii OT and OT_1 are perpendicular to the common tangent. We also know the locations O and O_1 and hence $\overline{OO_1}$. To aid in constructing part of Figure 2.40, and because the angles at T and T_1 are right angles, we draw through O_1 a line that is perpendicular to $\overline{OT}$ and intersects OT at A. Because $OA = r - r_1$, and OAO_1 is a right triangle, we know how to construct that triangle. After

$\triangle OAO_1$ is constructed, we extend $\overline{OA}$ to the point T, where the line intersects the circle. We notice that ATT_1O_1 is a rectangle.

Now we have several choices regarding how we construct the tangent. We could construct through T the line perpendicular to OT and claim that it is the common tangent, or we could construct through O_1 the line perpendicular to AO_1; where that line intersects the smaller circle is T_1. We would then claim that TT_1 is a common tangent.

Construction In Figure 2.41, we construct the right triangle OAO_1 using $r - r_1$ as one of the sides and $\overline{OO_1}$ as the hypotenuse. The steps of the construction are suggested in the preceding investigation and in Figure 2.41. We choose the second option for constructing the point of tangency T_1; that is, we construct through O_1 a line parallel to OT. That line intersects the smaller circle at T_1, which is the other point of tangency. $\overleftrightarrow{TT_1}$ is a common tangent.

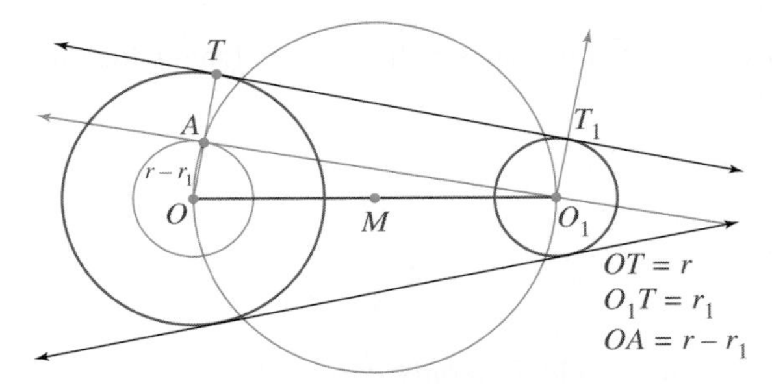

Figure 2.41

Proof

In Figure 2.41, by construction the angle at A is 90°. At O_1, we constructed the line parallel to $\overleftrightarrow{OA}$ and intersecting the smaller circle at T_1. We need to prove that $\overleftrightarrow{TT_1}$ is a common tangent—that is, that the angles at T and T_1 are 90°. For that purpose, we prove that ATT_1O_1 is a rectangle. By our construction, the angles of ATT_1O_1 at A and O_1 are 90°. We will show that ATT_1O_1 is a parallelogram. Notice that $AT = OT - OA = r - (r - r_1) = r_1$. Because $O_1T_1 = r_1$, we have $AT = O_1T_1$. Because $\overline{AT}$ and $\overline{OT}$ are parallel and congruent, ATT_1O_1 is a parallelogram. Because a parallelogram with a right angle is a rectangle, the proof is completed. □

Now Solve This 2.12

1. As mentioned in the investigation of Construction 2.5, the common tangent to the circles can be constructed by first finding the point of tangency T to the larger circle and then at T constructing the perpendicular to $\overline{OT}$. Prove that this perpendicular is also tangent to the smaller circle. (*Note:* You can't assume that the perpendicular intersects the smaller circle.)
2. Construct a figure like Figure 2.42, in which an internal tangent to two circles is shown. (Construct a line t and then two circles tangent to t.)
3. Given two circles like the one in Figure 2.42, construct their exterior tangents. (Here the two circles are given.) Write an investigation, describe the construction, and write a proof showing that the construction is valid. Now actually construct the two tangents.

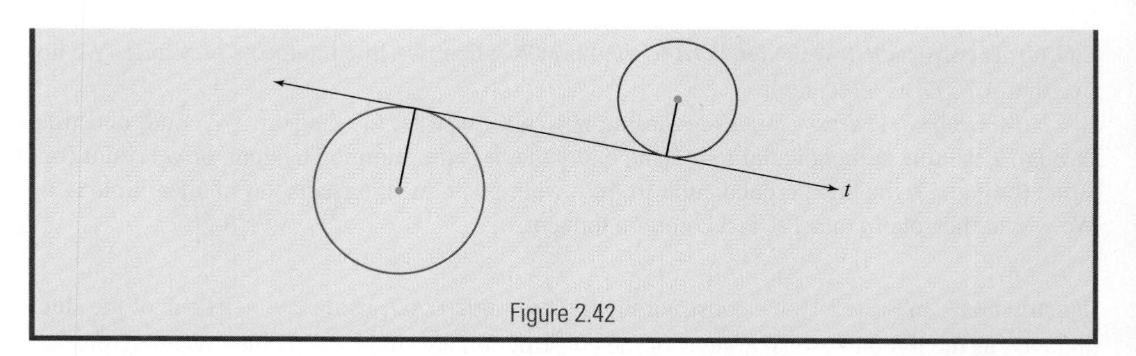

Figure 2.42

Problem Set 2.3

In all problems except Problem 1, include the three parts for each construction problem: investigation, construction, and proof.

1. Construct the locus of all points (in one half plane) from which a given segment (that you draw) is seen by
   a. 90°
   b. 60°
   c. 120°
   d. 105°
2. In $\triangle ABC$, α is an obtuse angle. Use only a, α, and m_b to reconstruct the triangle.
3. Draw a right triangle and then construct a triangle congruent to it using only a leg, the hypotenuse, and the fact that it is a right triangle.
4. Which of the following (taken from an actual triangle) determines a unique triangle (up to congruence; i.e., two congruent triangles are considered the "same"). Justify your answers.
   a. a, α, R
   b. a, h_a, h_b
   c. a, b, R
   •d. a, α, r
5. Construct a rhombus given a side and the radius of the inscribed circle.
•6. Construct a triangle given a, r, R. [*Hint:* Be sure to solve Problem 4(d) first.]
•7. Construct a square and four random points P, Q, S, and T on its sides, with one point on each side. Trace the four points on a semitransparent page without tracing the square. The problem is to construct the original square using only the four points P, Q, S, and T.

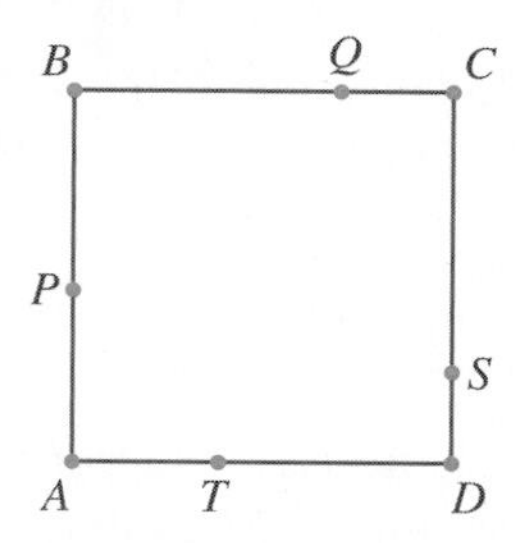

★ **8.** Given a line ℓ and points A and B on the same half plane determined by ℓ, construct point X on ℓ such that $\angle PXA$ is twice the size of $\angle QXB$ (the angles are marked by α and 2α, but neither is given). Is the answer unique?

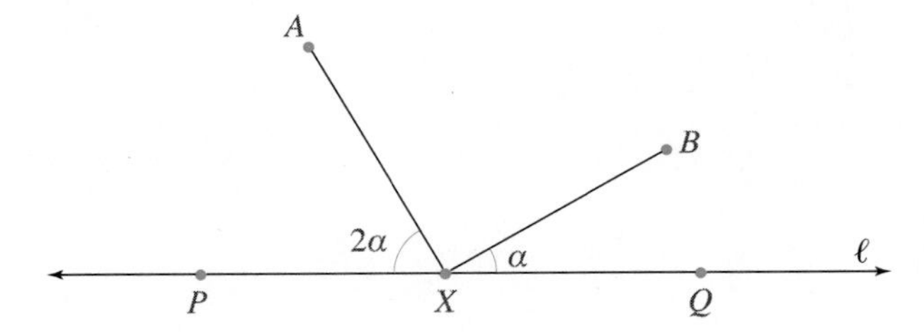

3 Area and the Pythagorean Theorem

I think it is said that Gauss had ten different proofs for the law of quadratic reciprocity. Any good theorem should have several proofs, the more the better. For two reasons: usually, different proofs have different strengths and weaknesses, and they generalise in different directions—they are not just repetitions of each other.

—Sir Michael Atiyah, interview in
European Mathematical Society Newsletter,
September 2004

Introduction

In this chapter we explore the concept of area, develop the formulas for the areas of various polygons, and solve related problems. We give several proofs of the Pythagorean Theorem using area, and apply the theorem to construction problems. In Section 3.3, we use the Pythagorean Theorem to derive the *distance formula* and from it the equation of a circle, a straight line, and, in the problem set, equations of other loci.

3.1 Areas of Polygons

To measure the length of a segment, we choose a unit segment and find how many of those unit segments cover the segment we want to measure. Similarly, to measure the area of a region, we choose a unit area and find how many of these units cover the region. Given a unit of length, the corresponding unit of area is the area of a square with sides of unit length. Thus area is measured in square units. In Figure 3.1a, the rectangle $ABCD$ has sides 3 units and 4 units long. In Figure 3.1b, 3 rows of 4 squares have been drawn. Because the rectangle contains $3 \cdot 4 = 12$ squares, the area of the rectangle is 12 square units. If the lengths of the sides of a rectangle are a units and b units, respectively, where a and b are whole numbers, we could divide the rectangle into a rows of b unit squares—that is, into ab unit squares.

If the lengths of the sides of a rectangle are rational numbers, we can show that its area is also the product of the lengths of its sides. For example, consider the rectangle $ABCD$ in Figure 3.2a, whose sides are $\frac{3}{2}$ and $\frac{5}{3}$ units long, respectively. We know how to find the

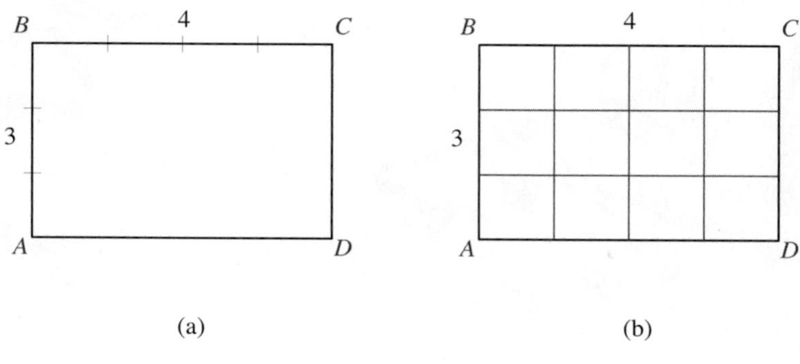

Figure 3.1

area of a rectangle if the rectangle is divided into a whole number of congruent squares. This can be accomplished if we could find a common measure for the sides. In other words, we could divide each of the sides AB and BC into a whole number of parts of the same length (using the same unit of length for each side). To find a common measure, we write the fractions with a common denominator:

$$\frac{3}{2} = \frac{3 \cdot 3}{2 \cdot 3} = \frac{9}{6} = 9 \cdot \frac{1}{6} \quad \text{and} \quad \frac{5}{3} = \frac{5 \cdot 2}{3 \cdot 2} = \frac{10}{6} = 10 \cdot \frac{1}{6}$$

Thus we can divide side AB, whose length is $\frac{3}{2}$ units, into 9 equal parts—each $\frac{1}{6}$ of the unit long—and side BC, whose length is $\frac{5}{3}$ units, into 10 equal parts—each also $\frac{1}{6}$ of the unit long ($\frac{1}{6}$ of the unit is the common measure for the two sides).

In Figure 3.2b, there are $9 \cdot 10$ squares created in this way. Notice that square $AGFE$ has sides that are one unit long and contains $6 \cdot 6$ of the smaller squares (each $\frac{1}{6}$ of a unit on a side). Because $6 \cdot 6$ of these small squares make one unit of area, one small square is $\frac{1}{6 \cdot 6}$ units of area. Thus $9 \cdot 10$ small squares (which cover rectangle $ABCD$) are $9 \cdot 10 \cdot \frac{1}{6 \cdot 6}$ units of area. Because $9 \cdot 10 \cdot \frac{1}{6 \cdot 6} = \frac{9}{6} \cdot \frac{10}{6} = \frac{3}{2} \cdot \frac{5}{3}$, we see that the area of the rectangle $ABCD$ in Figure 3.2 is the product of the lengths of its two adjacent sides. In an analogous way, it can be shown that if the lengths of adjacent sides of a rectangle are rational numbers, then the area of the rectangle is the product of their lengths (see Problem Set 3.1). What happens if the lengths of one or both of the sides are irrational numbers? In that case a common measure for the adjacent sides cannot always be found (see Problem Set 3.1). We can, however, approximate the lengths using rational numbers. For example, suppose that one of the sides is $\sqrt{2}$ units long and the other is $\sqrt{3}$ units long. The calculator displays the first eight digits of the decimal expansion like so:

$$\sqrt{2} \approx 1.41142136 \quad \text{and} \quad \sqrt{3} \approx 1.7320508$$

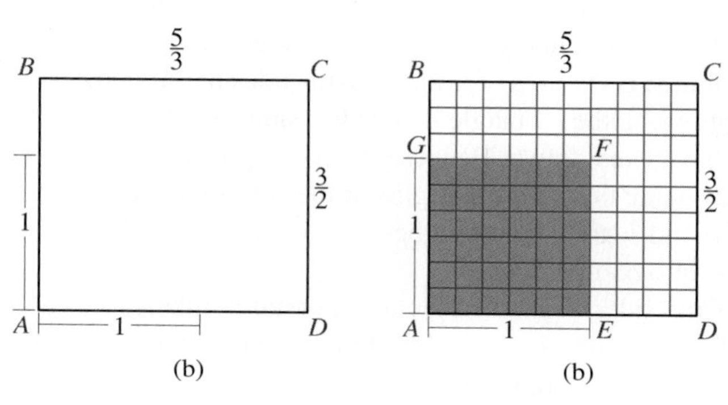

Figure 3.2

In Figure 3.3, $\varepsilon_1 = \sqrt{2} - 1.4142136$ and $\varepsilon_2 = \sqrt{3} - 1.7320508$. This suggests that the area of our rectangle whose sides are $\sqrt{2}$ and $\sqrt{3}$ should be very close to the area of the rectangle whose sides are rational numbers 1.4142136 and 1.7320508. The area of the nonshaded rectangle whose side lengths are rational numbers is $1.4142136 \cdot 1.7320508$, which is very close to $\sqrt{2} \cdot \sqrt{3}$. Because we can approximate $\sqrt{2}$ and $\sqrt{3}$ to any degree of accuracy using rational numbers, it seems that the exact area of the original rectangle is $\sqrt{2} \cdot \sqrt{3}$ square units. (A rigorous proof can be accomplished using the concept of a limit.) The preceding discussion suggests that the area of any rectangle is the product of the lengths of its two adjacent sides.

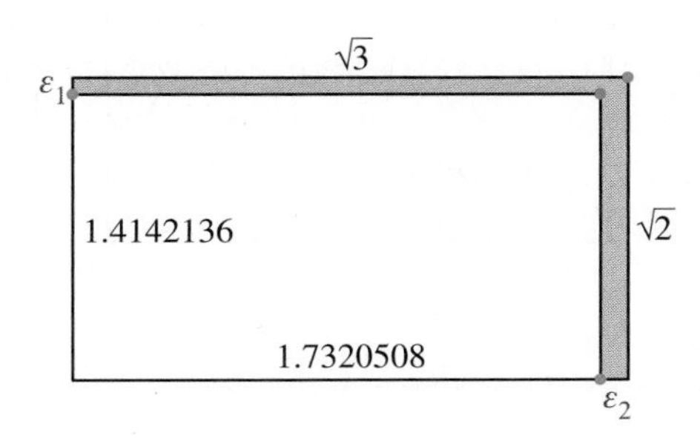

Figure 3.3

SPOTLIGHT on Teaching

When teaching the concept of area, intuitive "hands-on" approaches should precede the development of formulas. Such activities can be accomplished via a geoboard, which is constructed out of a piece of wood with equally spaced nails. Using rubber bands, one can create various polygons with vertices on the nails. Figure 3.4a shows a grid and a polygon created using a rubber band.

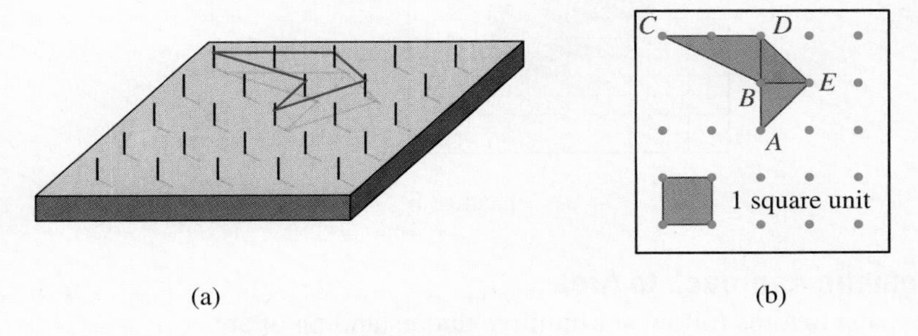

Figure 3.4

If the distance between two (horizontally or vertically) adjacent nails or dots is 1 unit, then the area of a square made of four adjacent dots is 1 square unit as shown in Figure 3.4b. Using that square unit, what is the area of the polygonal region *ABCDE* in Figure 3.4a? In Figure 3.4b, we have divided the polygon into three triangles *CDB*, *DBE*, and *ABE*. The area of $\triangle CDB$ is half the area of the rectangle *CDBF*. Because the area of the rectangle is 2 square units, the area of $\triangle CDB$ is $\frac{1}{2} \cdot 2$, or 1 square unit. Similarly, the areas of $\triangle DBE$ and $\triangle BEA$ are $\frac{1}{2}$ square unit each. Thus the area of the polygonal region is $1 + (\frac{1}{2}) + (\frac{1}{2})$, or 2 square units. Figure 3.5 and the corresponding computation suggest how to find the area of the polygon *ABCD* shown. Can you supply the details?

$$\text{Area } ABCD = (\text{Area } EFGD) - (\text{Area of 3 triangles})$$
$$= 4 \cdot 3 - \left(\frac{1}{2} \cdot 2 \cdot 1 + \frac{1}{2} \cdot 1 \cdot 3 + \frac{1}{2} \cdot 4 \cdot 1\right)$$
$$= 7\tfrac{1}{2} \text{ square units}$$

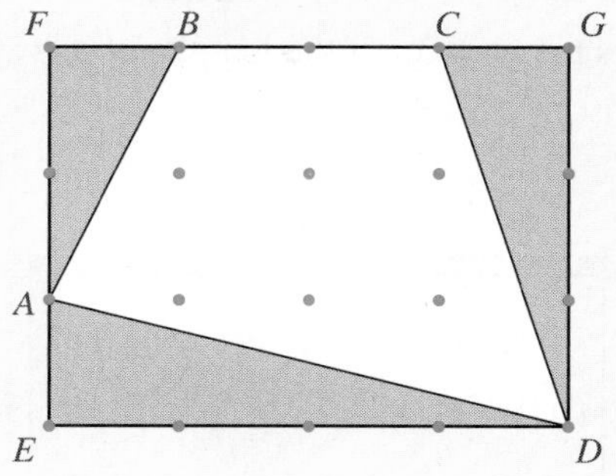

F B C G A E D

Figure 3.5

In Figure 3.6, the region is not a polygonal region and hence cannot be divided into triangles. It is possible, however, to estimate the area of the region by finding the areas of two polygonal regions: one contained in the region and another such that the curved region is contained in it. Can you show that the bounds in Figure 3.6 for the area A of the curved region are as shown? Can you find better bounds? This approach of "squeezing" the area of a region between two areas of polygons will be used later in this chapter to find the area of a circle.

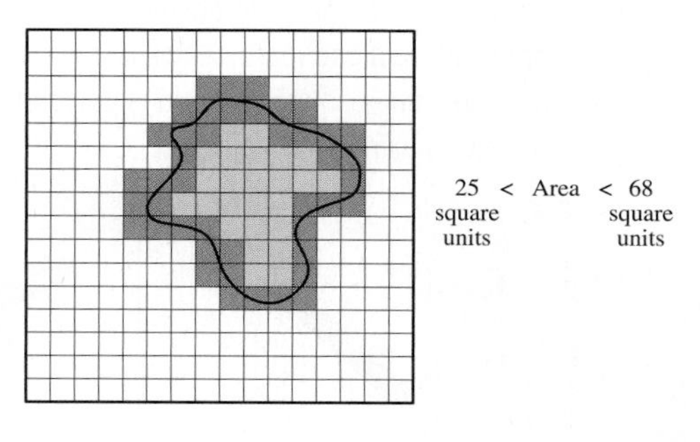

Figure 3.6

Axiomatic Approach to Area

The following axioms follow our intuitive understanding of area.

Axiom 3.1 *The area of any polygonal region (a polygon and its interior) is a unique positive real number.*

Axiom 3.2 *The area of any point or line segment is* 0.

Axiom 3.3 *Congruent polygonal regions have the same area.*

Axiom 3.4 *Area is additive. That is, if we can divide a figure into non-overlapping parts, or parts that share lines or line segments, then its area is the sum of the areas of the parts.*

Axiom 3.5 *A square of length a units has area of a^2 square units.*

Using some of these axioms, we will justify the well-known formulas for the areas of a rectangle, a parallelogram, a triangle, and a trapezoid.

Theorem 3.1

Area of a Rectangle

The area of a rectangle[1] *with sides of length a and b is ab.*

Proof

Consider a rectangle with sides of length a and b as in Figure 3.7a. Denote its area by A. To use Axiom 3.5, we construct a square as in Figure 3.7b. By Axiom 3.5, the area of the square with side $a + b$ is $(a + b)^2$. This square can also be regarded as the union of two rectangles with area A, and two squares with area a^2 and b^2, respectively. This is equivalent to each of the following:

$$(a + b)^2 = 2A + a^2 + b^2$$

$$a^2 + 2ab + b^2 = 2A + a^2 + b^2$$

$$2ab = 2A$$

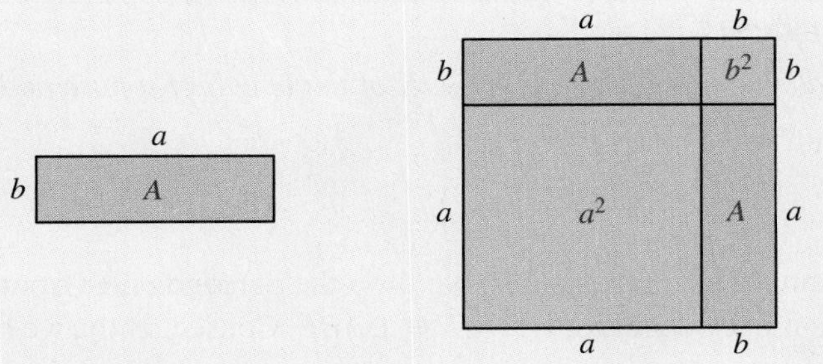

Figure 3.7

Therefore $A = ab$. □

Areas of a Parallelogram, a Triangle, and a Trapezoid

Knowing the formula for the area of a rectangle, we can use it to derive formulas for the area of a parallelogram, a triangle, and a trapezoid. We will use the following terminology. Any side

1. Strictly speaking, we should talk about the area of a rectangular region. For brevity's sake, however, we will adopt the convention of calling the area of any polygonal region the "area of the polygon."

of a triangle or a parallelogram will be referred to as a **base**. The **height of a triangle** is the distance from any vertex of the triangle to the opposite base—in other words, the length of an altitude. A triangle has three heights. A **height of a parallelogram** is the distance between two parallel sides. Thus a parallelogram has two heights.

There are many ways to derive formulas for the areas of a parallelogram, a triangle, and a trapezoid. Some approaches not developed in the text will be explored in the problem set at the end of this section. One important strategy is known as **dissection**. In dissection, we cut a figure with unknown area into a number of pieces. By reassembling these pieces, we can then obtain a figure whose area we know how to find.

Informally, to find the area of the parallelogram in Figure 3.8a, we can cut the shaded triangular piece of the parallelogram and move it to obtain the rectangle in Figure 3.8b. Because the shaded triangles are congruent, the area of the parallelogram in Figure 3.8a is the same as the area of the rectangle in Figure 3.8b. The area of that rectangle is bh.

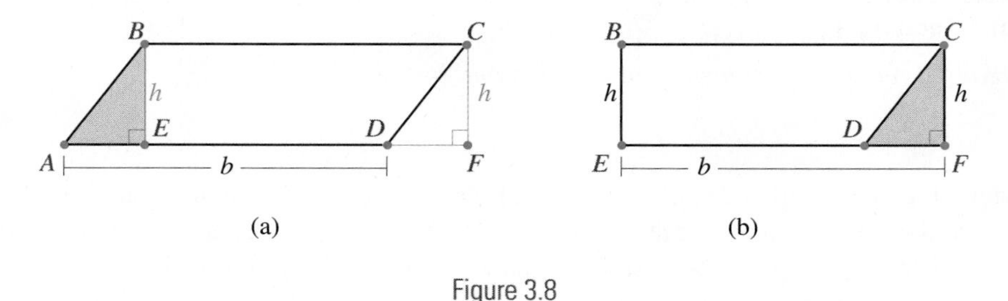

Figure 3.8

One way to make the preceding argument more rigorous is shown in the proof of the following theorem.

Theorem 3.2

Area of a Parallelogram

The area of a parallelogram equals the product of the length of a base and the corresponding height.

Proof

In the parallelogram $ABCD$ in Figure 3.8a, we drop the perpendicular from B to the side AD. From the H-L congruency condition, $\triangle DCF \cong \triangle ABE$. Consequently, we have

$$\begin{aligned}\text{Area}(ABCD) &= \text{Area}(ABE) + \text{Area}(EBCD)\\ &= \text{Area}(DCF) + \text{Area}(EBCD)\\ &= \text{Area}(EBCF)\\ &= bh\end{aligned}$$

since $EBCF$ is a rectangle. □

Remark The preceding proof of Theorem 3.2 is incomplete because it does not apply to a parallelogram like the one in Figure 3.9, where the perpendicular from B to $\overleftrightarrow{AD}$ does not intersect the side AD of the parallelogram. The proof in this case is explored in Now Solve This 3.1.

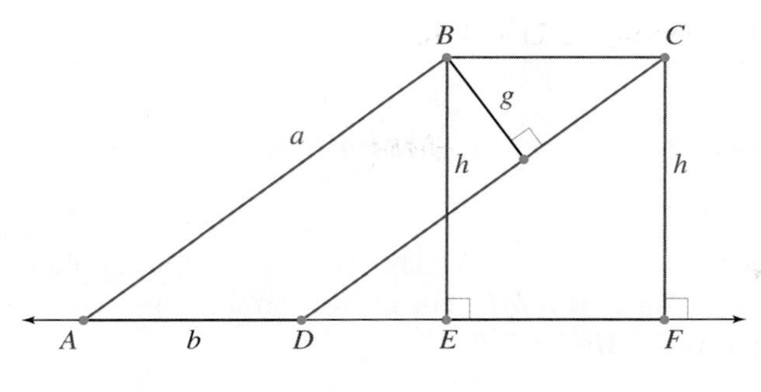

Figure 3.9

Now Solve This 3.1

Prove the formula given for the area of a parallelogram like the one in Figure 3.9 in two different ways:

1. Notice that

$$\text{Area}(ABCD) = \text{Area}(ABCF) - \text{Area}(\triangle DCF)$$
$$\text{Area}(BCFE) = \text{Area}(ABCF) - \text{Area}(\triangle ABE)$$

2. Enclose the parallelogram $ABCD$ in a rectangle with sides AF and CF, and subtract from the area of that rectangle the area of a rectangle with sides DF and CF.

Remark Axiom 3.1 tells us that the area of the parallelogram $ABCD$ in Figure 3.9 is unique. A proof analogous to the one given earlier, however, shows that the area of the parallelogram $ABCD$ is also ag, where a is the other base of the parallelogram and g is the corresponding height. Therefore, $bh = ag$.

To find a formula for the area of a triangle, we could also dissect the triangle and reassemble the pieces into a rectangle. This approach will be explored in the problem set at the end of this section. Here we use a somewhat different approach. We can trace or cut out the triangle ABC in Figure 3.10 and mark the traced triangle $A'B'C'$ on a separate patty paper (or any other paper). We retrace $\triangle A'B'C'$ next to the original triangle so that A falls on A', B falls on B', and C' is marked as shown in Figure 3.10. The resulting figure $ACBC'$ turns out to be a parallelogram. Consequently, the area of $\triangle ABC$ is half the area of $ACBC'$—that is, $\frac{1}{2}bh$. We state this result in the following theorem and outline a proof.

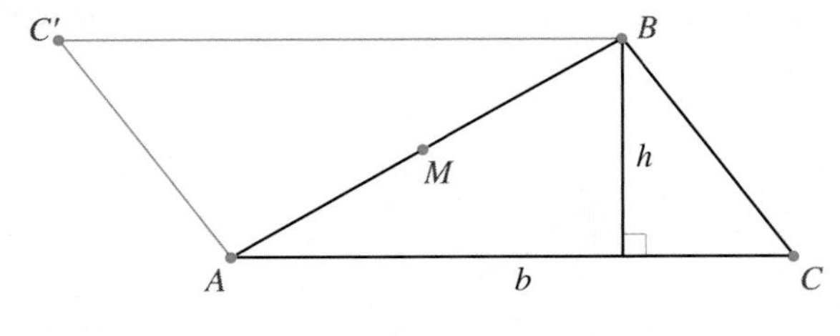

Figure 3.10

Theorem 3.3

Area of a Triangle

The area of a triangle is half the product of the length of a base and the corresponding height.

Proof

Let M be the midpoint of a side of the triangle (side AB in Figure 3.10) and C' the point collinear with M and C and such that $C'M = MC$. The quadrilateral $ACBC'$ is a parallelogram (Why?) whose area is bh. Because $\triangle ABC \cong \triangle BAC'$, we get

$$
\begin{aligned}
bh &= \text{Area}(ACBC') \\
&= \text{Area}(\triangle ABC) + \text{Area}(\triangle ABC') \\
&= 2[\text{Area}(\triangle ABC)]
\end{aligned}
$$

Hence $\text{Area}(ABC) = \frac{1}{2}bh$. □

Remark 1 Because any side of a triangle can be chosen as a base, the area of $\triangle ABC$ in Figure 3.11 is

$$\tfrac{1}{2}bh_b = \tfrac{1}{2}ah_a = \tfrac{1}{2}ch_c$$

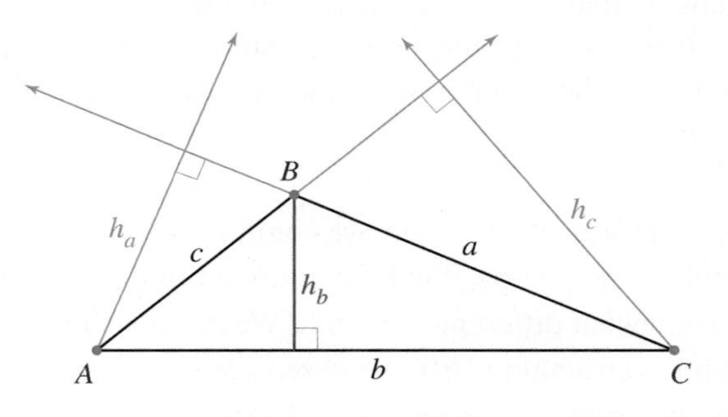

Figure 3.11

Remark 2 Theorem 3.3 implies that triangles with the same base and height have the same area. Figure 3.12 shows $\triangle ABC$ with base b and height h. Construct line k parallel to the side AC. Because the distance between the lines is h ($h_1 = h_2 = h_3 = h$), any triangle with vertex on k and base AC will have the area $(bh)/2$.

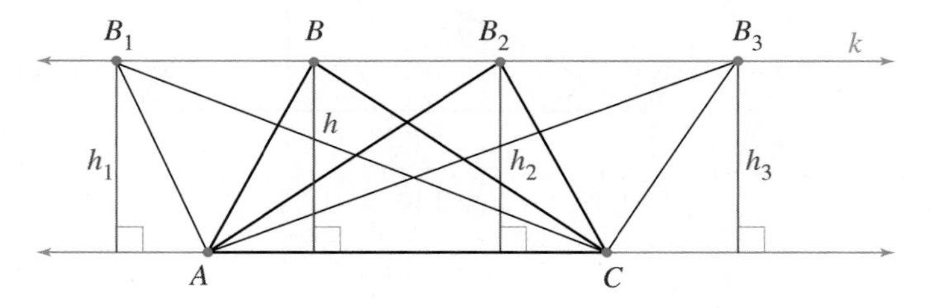

Figure 3.12

Example 3.1 **Redrawing a Border** Two farmers own large fields divided by a common border consisting of two sections of fence as shown in Figure 3.13. They want to replace the old crooked fence with a new straight one so that the areas of the new regions are the same as the old areas. In other words, each farmer should have the same amount of land as before the border was changed. Where should the new "straight" fence be placed?

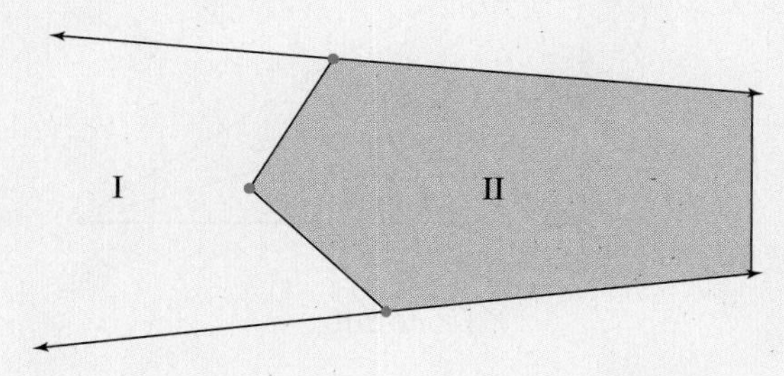

Figure 3.13

Solution

One of the farmers owns the region marked I in Figure 3.14, and the other owns the region marked II. The farmer who owns region II owns $\triangle ABC$, because region II includes $\triangle ABC$. The area of that triangle equals the area of any triangle with AC as a base and vertex on the line ED parallel to the base AC. Thus, if point B "moves" along the segment ED, we get a variety of triangles whose area is the same as the area of $\triangle ABC$. If we choose $\triangle ADC$, the area of region II does not change and the new border CD is straight. Making the border become AE is another possibility.

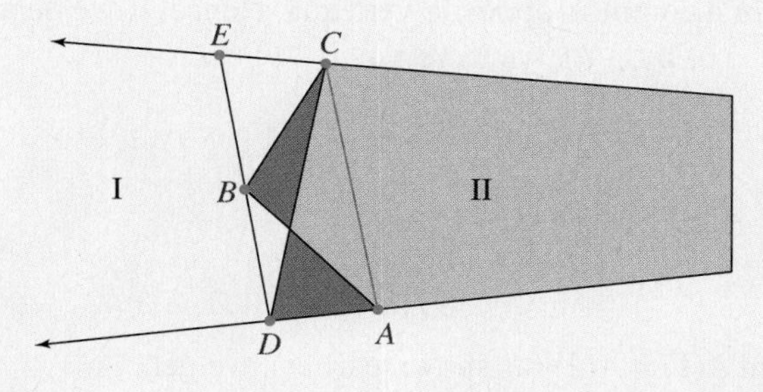

Figure 3.14

Now Solve This 3.2

Solve the "redrawing a border" problem in Example 3.1 for a border consisting of three segments like the one shown in Figure 3.15.

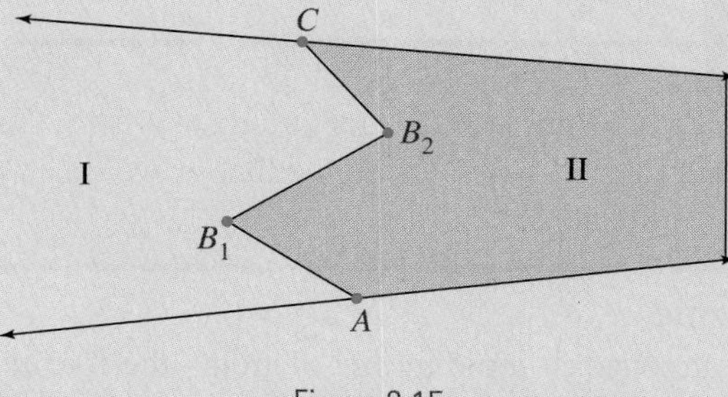

Figure 3.15

Example 3.2 **Triangles Created by Medians** Prove that the medians of a triangle "divide" the triangle into six non-overlapping smaller triangles of equal area, as shown in Figure 3.16.

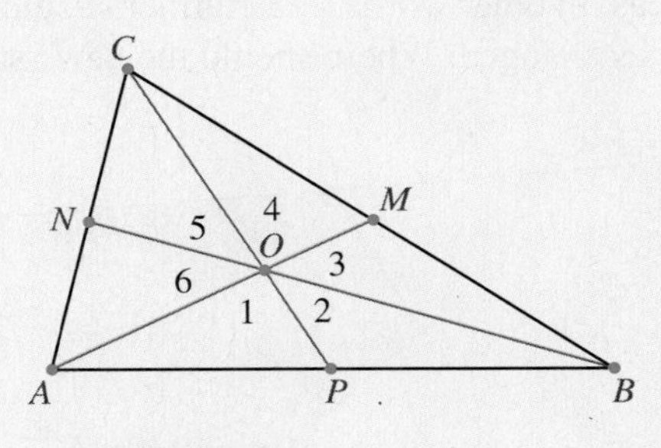

Figure 3.16

Solution

In Figure 3.16, M, N, and P are the midpoints of the sides of $\triangle ABC$. We need to show that the six triangles marked with numbers 1 through 6 have the same area.

We notice immediately that triangles 1 and 2 have the same area: Their bases AP and PB are congruent, and their corresponding heights (the distance from O to the line AB) are the same. Similarly, the pairs of triangles 3 and 4, as well as triangles 5 and 6, have equal areas.

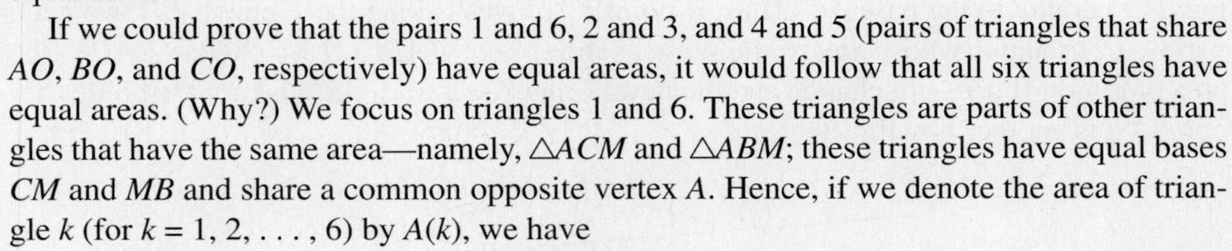

If we could prove that the pairs 1 and 6, 2 and 3, and 4 and 5 (pairs of triangles that share AO, BO, and CO, respectively) have equal areas, it would follow that all six triangles have equal areas. (Why?) We focus on triangles 1 and 6. These triangles are parts of other triangles that have the same area—namely, $\triangle ACM$ and $\triangle ABM$; these triangles have equal bases CM and MB and share a common opposite vertex A. Hence, if we denote the area of triangle k (for $k = 1, 2, \ldots, 6$) by $A(k)$, we have

$$A(6) + A(5) + A(4) = A(1) + A(2) + A(3)$$

Because $A(4) = A(3)$, it follows that

$$A(6) + A(5) = A(1) + A(2)$$

Since $A(5) = A(6)$ and $A(1) = A(2)$ (as shown earlier), we get

$$2A(6) = 2A(1)$$

$$A(6) = A(1)$$

In an analogous way, it follows that $A(2) = A(3)$ and that $A(4) = A(5)$. Consequently, the six triangles have the same area.

Now Solve This 3.3

Prove that the midpoints N and P in Figure 3.16 are equidistant from the median AM. State this property of midpoints and their distance to the median in words so that your statement includes any two of the three medians.

Area of a Trapezoid

We will now investigate our last topic in this section—the formula for the area of a trapezoid. Such a formula can be derived in a variety of ways. Two approaches are suggested in Now Solve This 3.4.

Now Solve This 3.4

1. (a) Trace trapezoid $ABCD$ on a separate sheet of paper and retrace it next to the original trapezoid so that the lower base is up and the upper base is down, as shown in Figure 3.17a. What kind of quadrilateral is $ABA'B'$? Find the area of that quadrilateral and then the area of the trapezoid.
   (b) Write a rigorous proof for finding the area of a trapezoid based on the ideas in part (a).
2. Derive a formula for the area of a trapezoid by an approach suggested in Figure 3.17b.

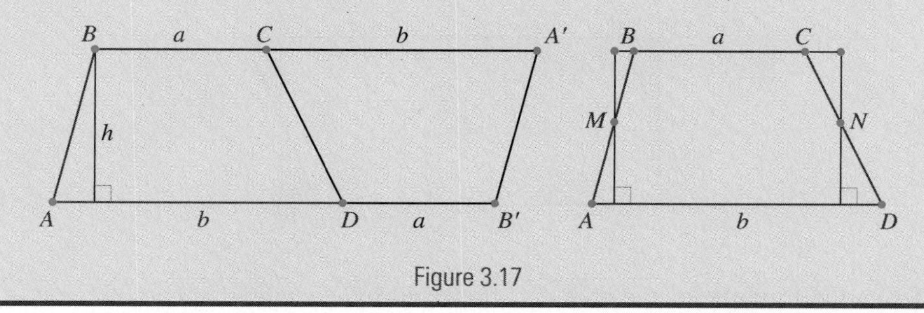

Figure 3.17

We now state the formula for the area of a trapezoid and prove it in yet another way.

Theorem 3.4

The area of a trapezoid whose bases have length a and b and whose height is h is given by $\frac{1}{2}(a + b)h$.

Proof

We dissect the trapezoid into two triangles. This can be accomplished by drawing either of the two diagonals. In Figure 3.18, the area of $\triangle BCD$ is $(ah)/2$ and the area of $\triangle ABD$ is $(bh)/2$. (Notice that $BE = FD = h$ because the lines BC and AD are parallel.) Consequently,

$$\text{Area}(ABCD) = \text{Area}(ABD) + \text{Area}(BCD) = \frac{ah}{2} + \frac{bh}{2} = \frac{(a+b)h}{2}$$

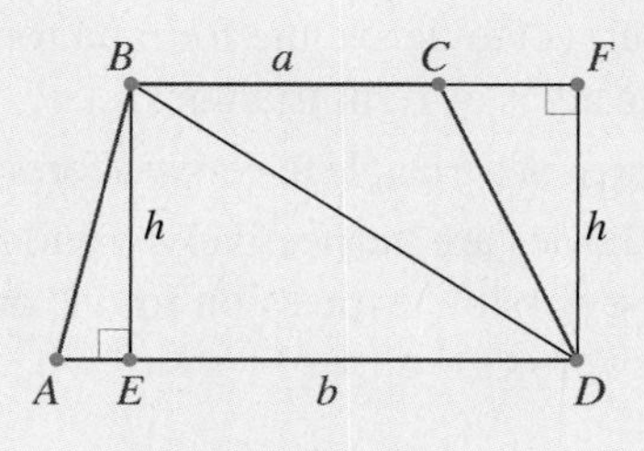

Figure 3.18 □

Problem Set 3.1

1. Find the areas of each of the following shaded figures if the distance between two adjacent dots in a row or column is one unit. In part (c), prove first that the shaded figure is a square.

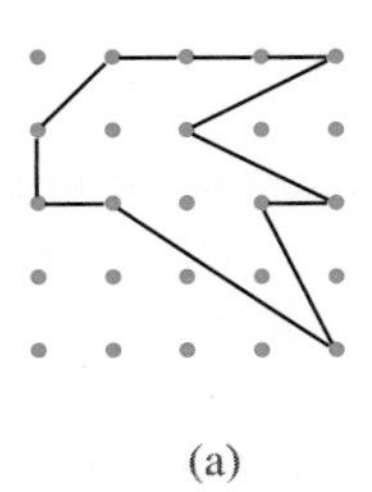

(a)

(b)

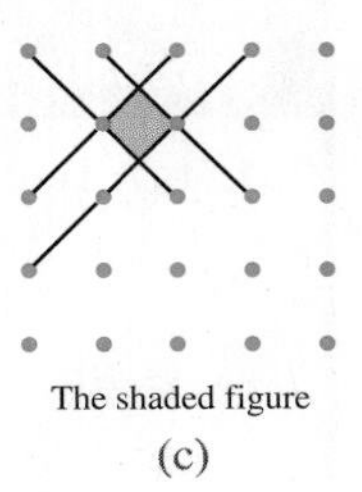

(c)

2. Prove the formula for the area of a parallelogram (Theorem 3.2) by dissecting it in a way suggested in the figure below. Which transformation will map $\triangle AME$ onto $\triangle BME'$?

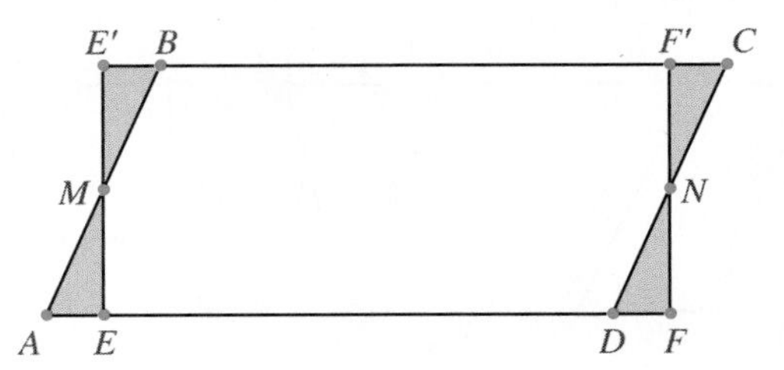

3. Derive the formula for the area of a trapezoid by extending $\overline{BM}$ to intersect $\overleftrightarrow{AD}$ at E (where M is the midpoint of $\overline{CD}$) and showing that its area equals the area of $\triangle ABE$.

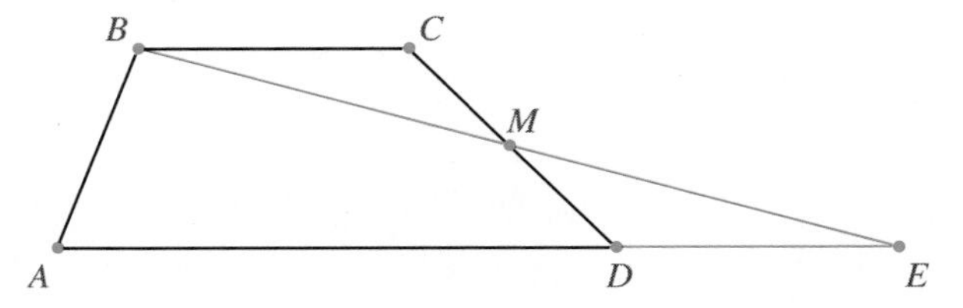

4. •a. Find the area of a rhombus in terms of the lengths of its diagonals d_1 and d_2.
   b. Will the same formula you derived in part (a) work for kites? Why or why not?
   c. For which other quadrilaterals will the formula you derived in part (a) work? (Find as many such quadrilaterals as you can.)
5. The formula for the area of a triangle can be derived before the formula for the area of a parallelogram is proved. To do so, use the suggestions in the following questions.
   a. Using only the formula for the area of a rectangle, find the area of a right triangle with legs a and b.
   b. Use your result from part (a) to derive the formula for the area of a triangle by using a sum or difference of the areas of right triangles.
   c. Use the formula for the area of a triangle to derive a formula for the area of a parallelogram.
•6. In a trapezoid, the parallel sides are, respectively, a and b units long. The base angles are 45° each. Find the simplest possible expression for the area of the trapezoid in terms of a and b (do not use trigonometry).

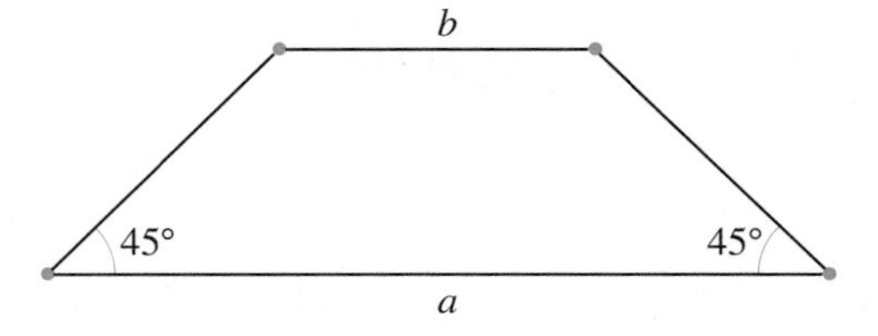

7. Suppose $ABCD$ is a parallelogram and P is a point on the diagonal AC. Prove that triangles ABP and APD have the same area.

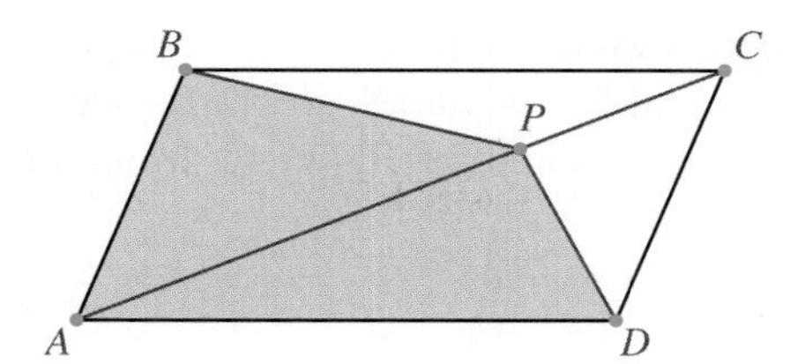

8. Let ABC be an equilateral triangle and P be any point in the interior of the triangle or on the triangle. Use the concept of area to prove that the sum of the distances from P to the three sides of the triangle is constant (the same for all the previously mentioned positions of P). That is, show that $PE + PD + PF$ is constant. How is that constant related to the height of the triangle?

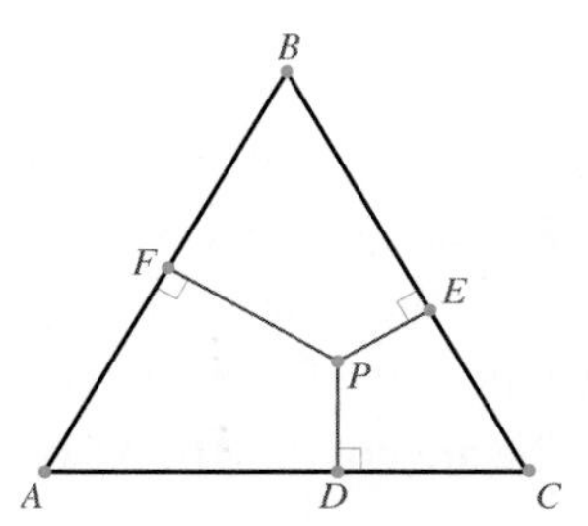

★ **9.** $ABCDEF$ is a regular hexagon. Let P be any point in the interior or on the hexagon.

a. Prove that the sum of the areas of the shaded triangles is constant.

b. State the property in part (a) in words (without reference to the letters in the figure).

c. How does the sum of the areas in part (a) relate to the area of the hexagon? Why?

d. What kind of property follows if the point P is one of the vertices? Sketch an appropriate figure.

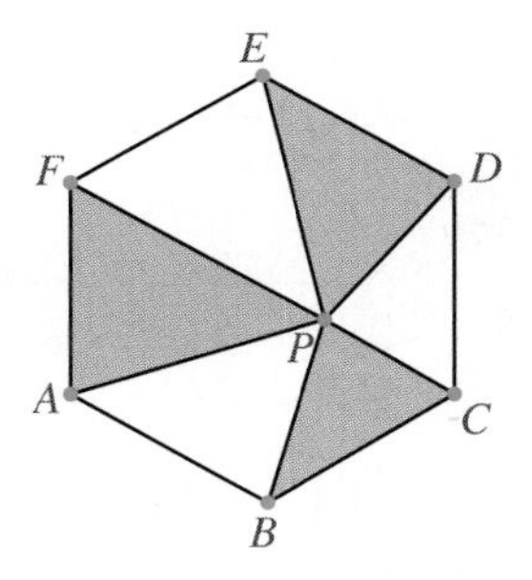

10. An arbitrary point P in the interior of a parallelogram is connected to the four vertices.

• **a.** If the areas of the triangles formed are A_1, A_2, A_3, and A_4 (as shown in the figure), how do $A_1 + A_3$ and $A_2 + A_4$ relate to the area of the parallelogram?

b. Does a relationship similar to the one you found in part (a) hold if instead of a parallelogram we have a trapezoid? Justify your answer.

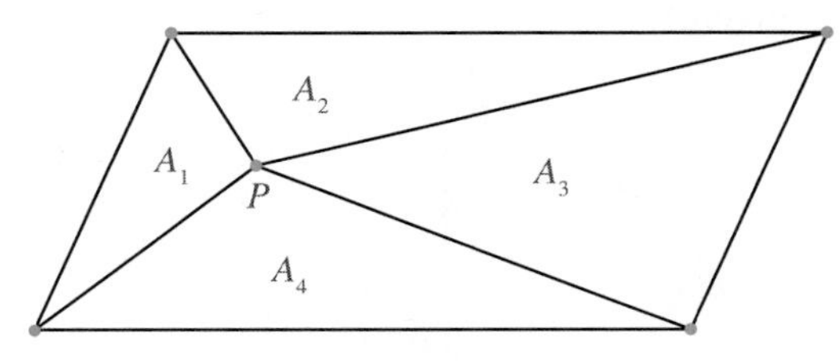

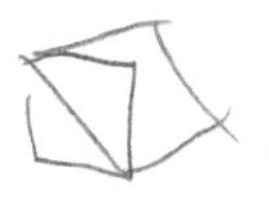

11. Construct any square, and then construct a square whose area is twice the area of the first square.

Homework

12. $\angle C$ is a right angle and P is a point in its interior. Through P we draw a line k intersecting the sides of the angle at A and B. If S_1 and S_2 are the respective areas of $\triangle APC$ and $\triangle BPC$, prove that $(1/S_1) + (1/S_2)$ is the same for all lines k through P.

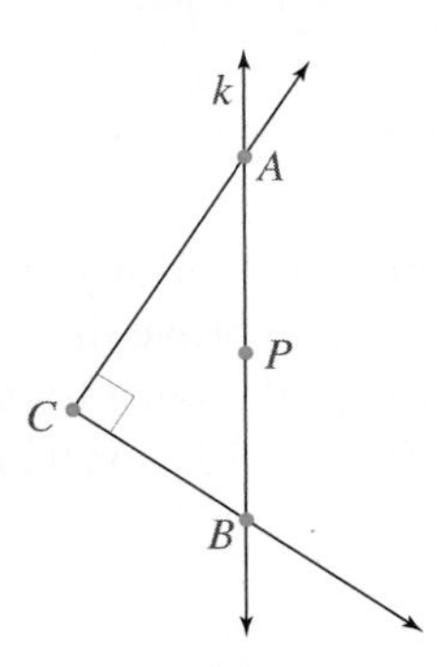

13. In $\triangle ABC$, $\angle C$ is a right angle. $ACDE$ and $ABFG$ are squares constructed on a side and the hypotenuse, respectively. $\overline{CH}$ is perpendicular to $\overline{AB}$. The segment CH has been extended to intersect $\overline{GF}$ at I. Prove the following:

a. $\triangle EAB \cong \triangle CAG$ and hence the triangles have the same area.

b. The area of $\triangle EAB$ equals the area of $\triangle EAC$. Also the area of $\triangle ACG$ equals the area of $\triangle AHG$.

c. The area of the square constructed on $\overline{AC}$ equals the area of the rectangle $AHIG$.

d. State and justify a result similar to the one in part (c) for the square constructed on $\overline{CB}$.

e. Using the results in parts (c) and (d), what can be said about the areas of squares constructed on the sides of the legs of a right triangle and the area of the square constructed on the hypotenuse? Justify your answer.

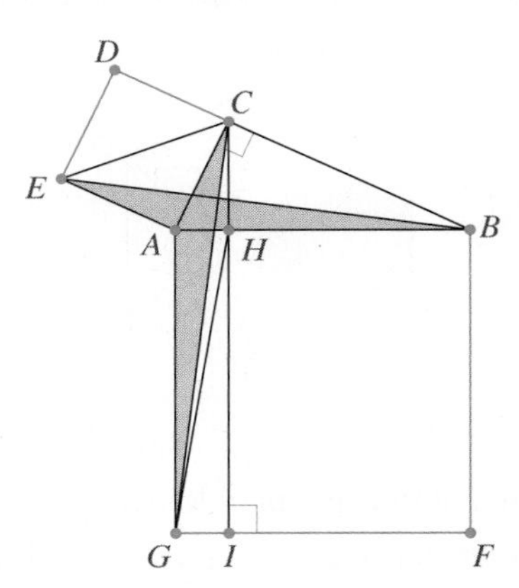

• **14.** Squares A and B are congruent. One vertex of B is at the center of A. What is the ratio of the shaded area to the area of square A? (*Note:* The lengths of the sides of the shaded quadrilateral are not given.)

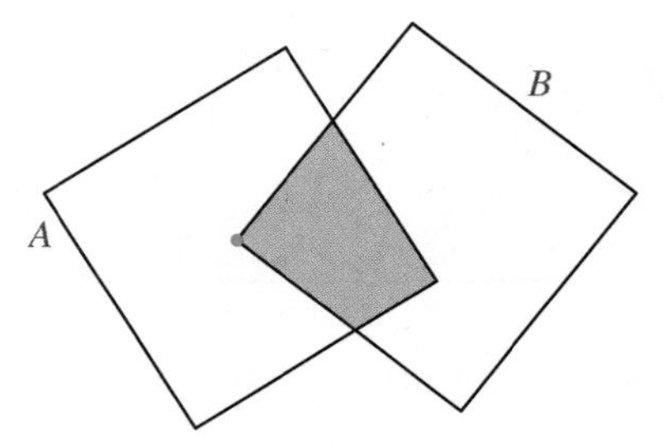

15. For each of the figures below, describe how to divide the region consisting of the interior of the squares in figure (a) and the interiors of the five congruent circles in figure (b) into two regions of equal area by a single straight line through the point P. Justify your answers. [*Note:* In figure (b), P is the center of the circle.] Show how to construct the required lines.

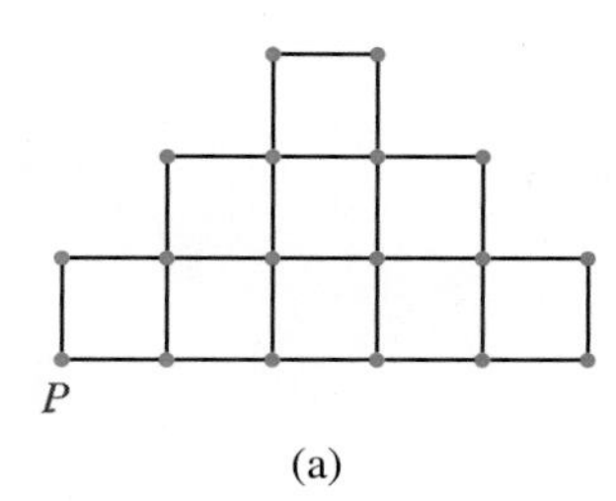

(a)

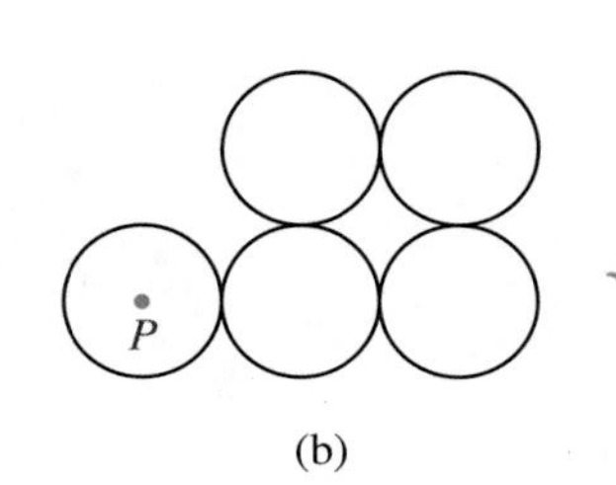

(b)

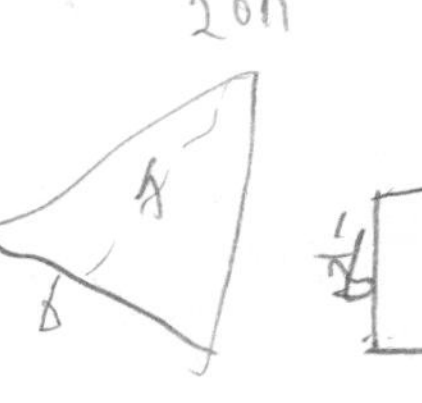

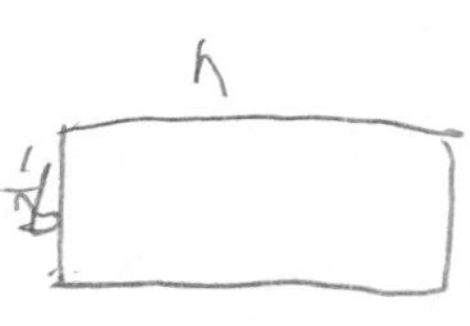

• **16.** Given a convex quadrilateral, construct a triangle whose area is the same as the area of the quadrilateral.

17. Construct a non-isosceles trapezoid, and then construct an isosceles trapezoid with the same area. Is the solution unique?

18. Use the concept of area to demonstrate geometrically each of the following properties for positive real numbers:

a. $a(b + c) = ab + ac$

b. $(a + b)^2 = a^2 + 2ab + b^2$

c. $a(b - c) = ab - ac$, where $b > c$

d. $(a + b)(a - b) = a^2 - b^2$, for $a > b$

19. The following figures suggest a geometric approach to finding the sum of the first n consecutive counting numbers, $1 + 2 + 3 + \cdots + n$. Explain this approach and find the sum.

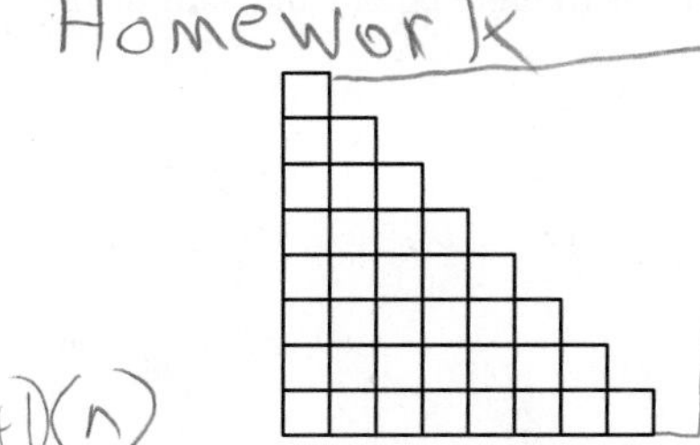

$8 + 7 + 6 + 5 + 4 + 3 + 2 + 1$

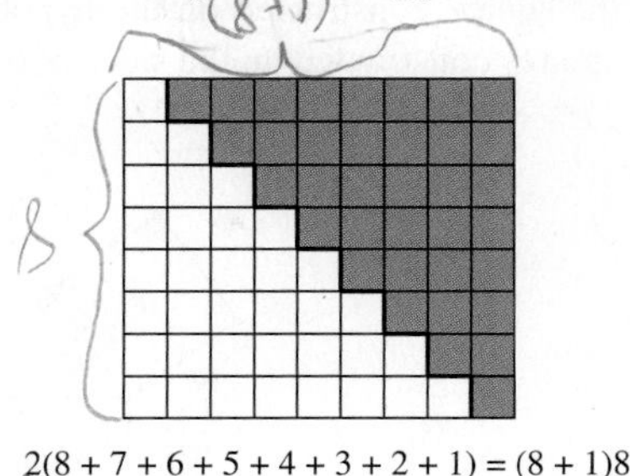

$2(8 + 7 + 6 + 5 + 4 + 3 + 2 + 1) = (8 + 1)8$

In Problems 20 and 21, assume that, as in Problem 1, the distance between two adjacent dots in a row or a column is one unit.

20. On the geoboard, find the area of the quadrilateral $ABCD$. Prove your answer.

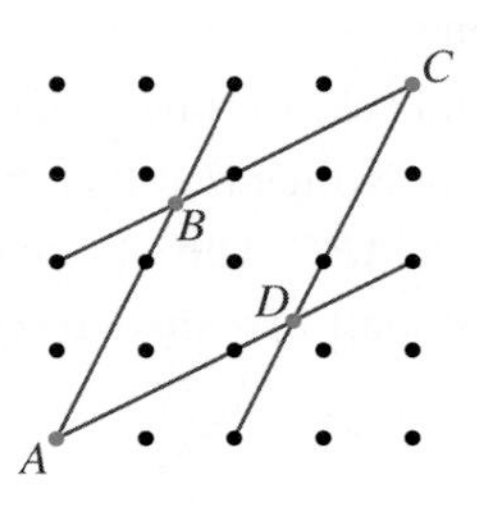

21. Consider $\triangle ABC$ and complete the following tasks:

a. Prove that it is an isosceles right triangle.

b. Find AC by finding the area of $\triangle ABC$ in two different ways (do not use the Pythagorean Theorem).

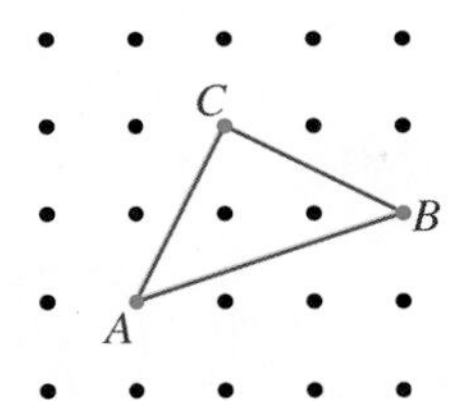

3.2 The Pythagorean Theorem

The Pythagorean Theorem is perhaps the most famous theorem in all of mathematics. It is simple, beautiful, and remarkably useful. William Dunham, in *The Mathematical Universe* (1994), says that "whether regarded algebraically or geometrically, the theorem is of supreme mathematical importance." Here it is:

Theorem 3.5

Pythagorean Theorem

In a right triangle with legs of length a and b and hypotenuse of length c, we have

$$a^2 + b^2 = c^2$$

Figure 3.19 shows the equivalent form of the Pythagorean Theorem in terms of area:

> The area of the square constructed on the hypotenuse of a right triangle equals the sum of the areas of the squares constructed on the sides of the triangle.

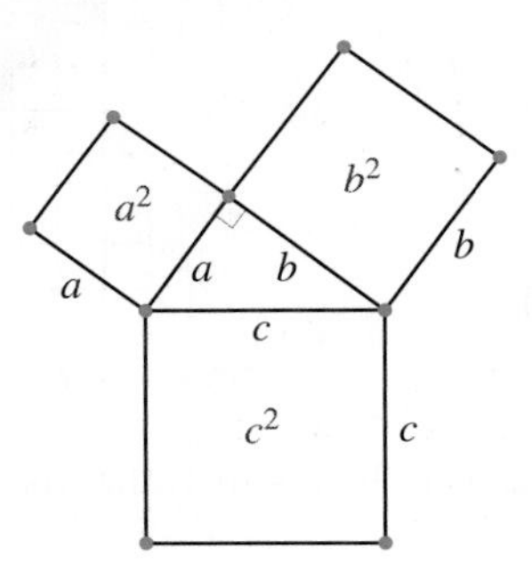

Figure 3.19

It is not clear if Pythagoras (circa 572–497 B.C.E.) had a proof of the theorem, but it is quite probable that he discovered a special case of it for isosceles triangles appearing in a floor tile pattern like the one shown in Figure 3.20. Notice that there are two congruent squares constructed on the legs of the isosceles right triangle ABC. The square on the hypotenuse $\overline{AB}$ consists of four triangles, each having an area equal to half of a shaded square; hence its area equals the area of the two shaded squares.

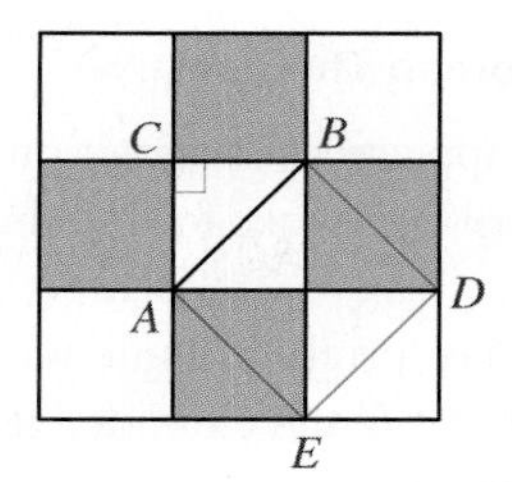

Figure 3.20

There are many known proofs of the Pythagorean Theorem. *The Pythagorean Proposition* by E. S. Loomis contains 370 different proofs, for example. We will discuss some of these proofs in this section; we will explore others in the problem set and in Chapter 4 when similarity of triangles is introduced. Perhaps one of the simplest proofs is the following:

Proof 1 of the Pythagorean Theorem

Consider $\triangle ABC$ in Figure 3.21a, whose legs have lengths a and b and whose hypotenuse has length c. In Figure 3.21b we see four triangles congruent to $\triangle ABC$ as part of a square with side $a + b$. Notice that $a^2 + b^2$—the sum of the areas of the two smaller squares in Figure 3.21b—is equal to the area of the large square (whose side is $a + b$) minus the area of the four congruent triangles. By contrast, in the large square with side $a + b$, the four triangles have been fitted as shown in Figure 3.21c. Here the inner figure seems to be a square. To prove this, first notice that all of this figure's sides are c. (Why?) Second, since $\alpha + \beta = 90°$, each angle of the inner quadrilateral is $180 - (\alpha + \beta)$ or $90°$. It follows now that c^2 (the area of the inner square) is equal to the area of the large square minus the area of the four congruent triangles, which as we showed earlier based on Figure 3.21b, is $a^2 + b^2$. Hence $a^2 + b^2 = c^2$. □

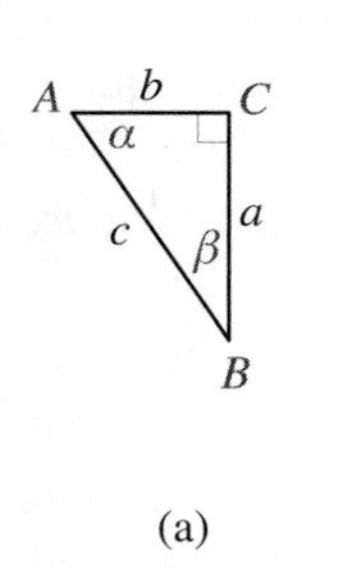

(a)

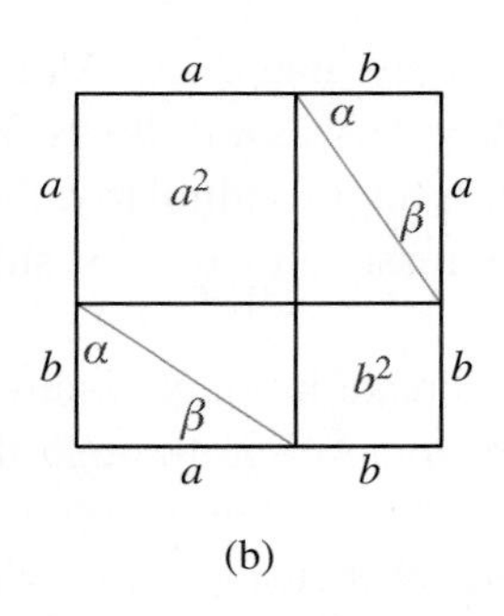

(b)

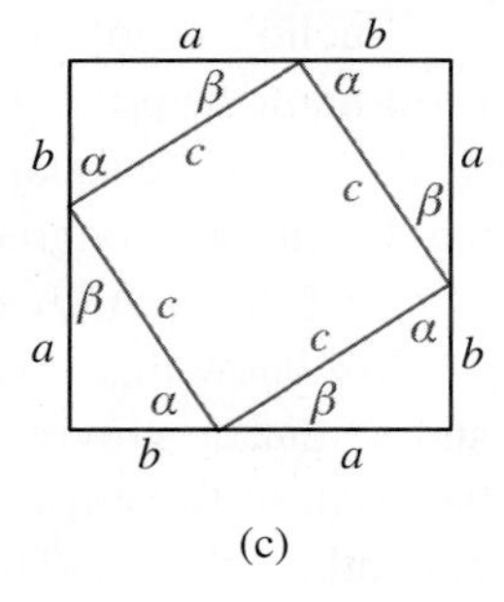

(c)

Figure 3.21

Our second proof of the Pythagorean Theorem is based on ideas that originated in China.

Proof 2 of the Pythagorean Theorem

We find the area of the larger square in Figure 3.21c in two different ways. Because the side of the square is $a + b$, its area is $(a + b)^2$. This larger square is made of four congruent triangles, each having area $\frac{1}{2}ab$ and the interior square having area c^2. Consequently,

$$(a + b)^2 = 4 \cdot \tfrac{1}{2}ab + c^2$$

$$a^2 + 2ab + b^2 = 2ab + c^2$$

$$a^2 + b^2 = c^2$$

□

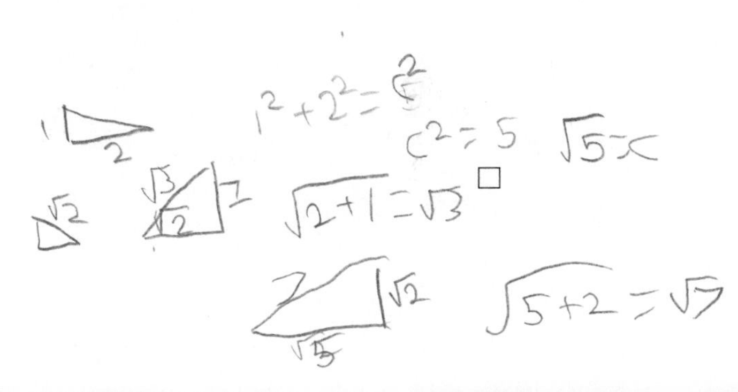

Proof 3: Euclid's Proof of the Pythagorean Theorem

A proof of the Pythagorean Theorem appears as Proposition 1.47 in Book I of Euclid's *Elements*. It was the standard proof in high school textbooks for hundreds of years. (See also Problem 13 in Problem Set 3.1.)

In this proof $\triangle ABC$ (in Figure 3.22) is a right triangle with the right angle at C. Euclid drew a perpendicular from C to the hypotenuse AB and extended it until it intersected the side HI of the large square at L. That perpendicular split the square on the hypotenuse into two rectangles, marked I′ and II′. Then Euclid proceeded to prove a remarkable property: The area of square $ACFG$ (marked I) is the same as the area of the rectangle $AKLH$ (marked I′). Similarly, the area of the square $BCED$ (marked II) is the same as the area of the rectangle $KBIL$ (marked II′). This immediately implied that I + II = I′ + II′ = area of $ABIH$, and hence the Pythagorean Theorem.

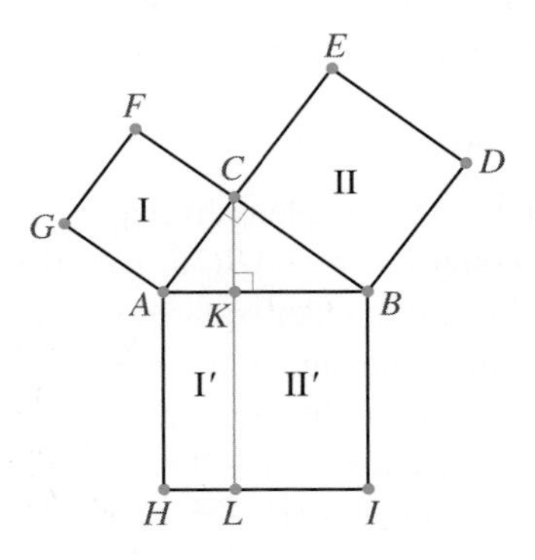

Figure 3.22

Euclid's proof that I = I′ and II = II′ is quite ingenious. Although we do not know how he came upon the proof, perhaps his thought process was as follows: The diagonal $\overline{GC}$ of the square $ACFG$ splits the square into two congruent triangles. Similarly, the diagonal $\overline{HK}$ splits the rectangle into two congruent triangles. For that reason, it would be sufficient to prove that the areas of $\triangle ACG$ and $\triangle AKH$ are equal.

We know that given a triangle, we can create infinitely many triangles with the same base and height by moving the opposite vertex along a line through that vertex and parallel to the base. All of these triangles will have the same area (see the second remark following Theorem 3.3 and Figure 3.12). This applies to our case by focusing on $\triangle AKH$ (Figures 3.23a and 3.23b) and moving point K along the line $\overleftrightarrow{CL}$ that is parallel to the base $\overline{AH}$ of $\triangle AKH$.

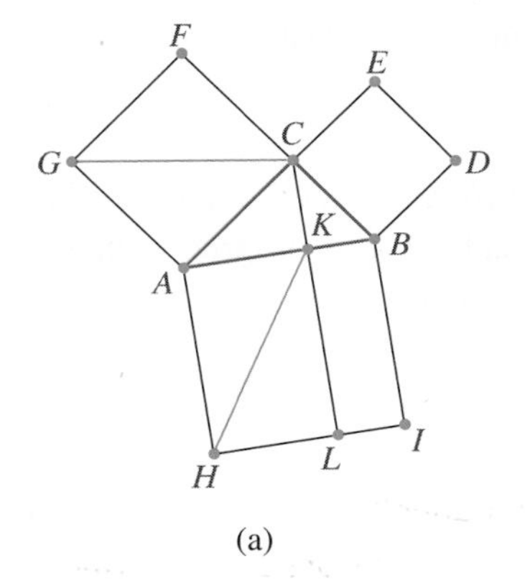

(a)

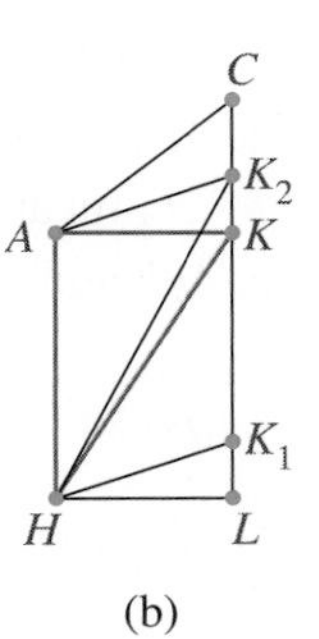

(b)

Figure 3.23

Triangles AKH, AK_1H, and AK_2H in Figure 3.23b are three representative triangles that have the same area. Let S_1 be the set of all the triangles with base $\overline{AH}$ and opposite vertex on $\overleftrightarrow{CL}$. (The set S_1 will be used in the next paragraph.)

A similar situation is obtained by focusing on $\triangle ACG$. We can keep the base $\overline{AG}$ fixed and move vertex C along the line through C parallel to $\overleftrightarrow{AG}$. Notice that the points F, C, and B are collinear (since $\angle ACF$ and $\angle ACB$ are right angles); hence C can be moved along $\overleftrightarrow{FB}$. Let S_2 be the set of all triangles with base $\overline{AG}$ and opposite vertex on $\overleftrightarrow{FB}$. Perhaps we could pick two triangles, one from S_1 and the other from S_2, that are congruent. These two triangles would have the same area, and hence every triangle in S_1 would have the same area as every triangle in S_2. Consequently, $\triangle ACG$ and $\triangle AHK$ would have the same area. This approach would be especially promising if the two chosen triangles have some of the sides of $\triangle ABC$ as their sides. (Try to choose such triangles before reading on.)

In Figure 3.24, we move point K all the way to C to get $\triangle ACH$. This triangle is a member of S_1 and its area is the same as the area of $\triangle AKH$ and hence half the area of rectangle $AKLH$. Next, we choose a triangle from S_2. Here the base is $\overline{AG}$ and the opposite vertex can move on $\overline{FB}$. We choose B to be the third vertex. $\triangle ABG$ will have two sides correspondingly congruent to two sides of $\triangle ABC$. The same is true of $\triangle AHC$. Moreover $\angle GAB \cong \angle CAH$ because the measure of each angle is $90° + m(\angle CAB)$. Thus $\triangle ABG \cong \triangle AHC$ by SAS. (Notice that the congruency of these triangles can be established by a transformational approach; when $\triangle AHC$ is rotated counterclockwise by a right angle about A, its image is $\triangle ABG$. Because the image of A is A itself, the image of H is B and the image of C is G.) As discussed earlier, the congruence of these triangles implies $\text{I} = \text{I}'$. Analogously $\text{II} = \text{II}'$ and hence $\text{I} + \text{II} = \text{I}' + \text{II}' = \text{Area}(ABIH)$. Consequently, the Pythagorean Theorem is proved. □

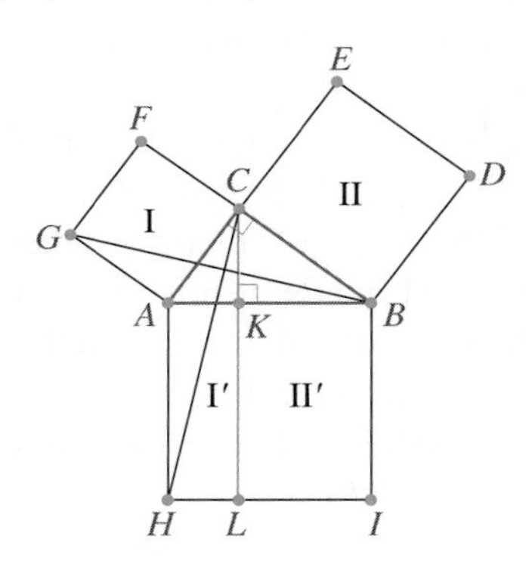

Figure 3.24

The preceding proof may seem long, mainly because we discussed possible motivations for taking the various steps in the proof. Next we give Euclid's proof without the motivation.

Euclid's Proof of the Pythagorean Theorem: A Short Version

Referring to Figure 3.24, we have $\overline{AH} \cong \overline{AB}$ and $\overline{AC} \cong \overline{AG}$. Also $\angle CAH \cong \angle GAB$ because the measure of each angle is $90° + m(\angle CAB)$. Consequently, by SAS

$$\triangle AHC \cong \triangle ABG \tag{3.1}$$

Because $\overleftrightarrow{CL} \parallel \overleftrightarrow{AH}$, it follows from the second remark after Theorem 3.3 that the areas of $\triangle ACH$ and $\triangle AKH$ are the same. Similarly, because points B, C, and F are collinear (the angles at C are right angles) and $\overleftrightarrow{CB} \parallel \overleftrightarrow{AG}$, it follows that the areas of $\triangle GAB$ and $\triangle GAC$ are equal. From

Equation 3.1, it follows that the areas of $\triangle AKH$ and $\triangle GAC$ are equal. Because the area of $\triangle AKH$ is half the area of the rectangle $AKLH$ and the area of $\triangle GAC$ is half the area of the square $ACFG$, it follows that the area of the rectangle is the same as the area of the square. In a completely analogous way, we find that the area of $KBIL$ is the same as the area of the square $BCED$. Consequently, the sum of the areas of the squares on the legs $\overline{AC}$ and $\overline{CB}$ equals the sum of the areas of the rectangles $AKLH$ and $KBIL$. This is the area of the square on the hypotenuse $\overline{AB}$. □

A Bonus

A closer look at Figure 3.24 and Euclid's proof of the Pythagorean Theorem suggests that a relationship exists between $\overline{HC}$ and $\overline{BG}$. In addition to being congruent, these segments are perpendicular to each other. (This can be easily verified if the rudiments of the transformational approach introduced in Chapter 5 are familiar to the reader. Rotating $\triangle AHC$ about point A by 90° (counterclockwise), the image of $\triangle AHC$ is $\triangle ABG$ because A is mapped onto itself, H onto B, and C onto G. Because the image of $\overline{HC}$ is $\overline{BG}$, these segments are congruent and perpendicular.)

Historical Note: Pythagoras (circa 580–500 B.C.E.) and U.S. President James Garfield (1831–1881)

Pythagoras of Samos was a Greek philosopher and mathematician. He founded a philosophical and religious school, known as the Pythagorean Brotherhood. This school formulated principles that influenced both Plato and Aristotle. Today Pythagoras is particularly remembered for the theorem that bears his name. In reality, this theorem—if not its proof—was known to the Babylonians and other cultures about 1000 years earlier.

The followers of Pythagoras were renowned for their ethical practices, unselfishness, honesty, and love of mathematics, especially geometry. The Pythagorean Theorem very likely has a greater number of different proofs than any other theorem in mathematics. The book *The Pythagorean Proposition* (Loomis, 1968) gives 367 different proofs of this theorem, among them a proof by James Garfield, the twentieth president of the United States. President Garfield studied mathematics at Williams College in Williamstown, Massachusetts, and after graduating taught math in public schools and at a college. He published his proof in the *Journal of Education* in 1876.

Garfield, like Pythagoras, was known for his character and honesty. Unfortunately, he served as president for only 200 days before becoming one of the four U.S. presidents who were assassinated while in office.

The Converse of the Pythagorean Theorem

If the sides of a triangle have lengths a, b, and c such that $c^2 = a^2 + b^2$, then the triangle is a right triangle with the right angle opposite the side of length c.

Proof

In $\triangle ABC$ in Figure 3.25a, $c^2 = a^2 + b^2$. We need to prove that $\angle C$ is a right angle. For that purpose, we construct in Figure 3.25b a right triangle $A_1B_1C_1$ with legs a and b and a right angle at C_1. We will show that the two triangles are congruent. Then it would follow that the angles at C and C_1 are congruent and hence that $\angle C$ is a right angle.

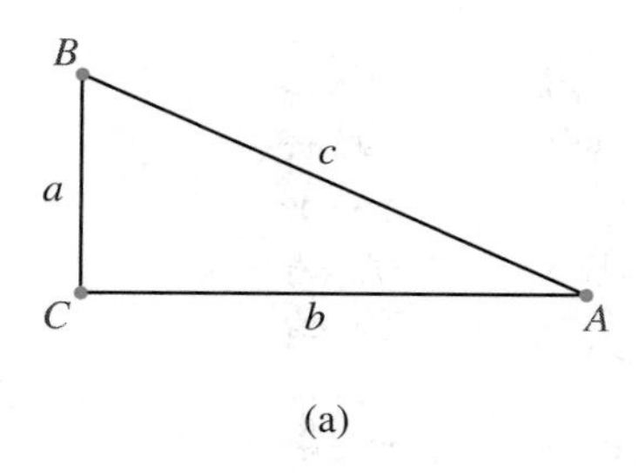

(a)

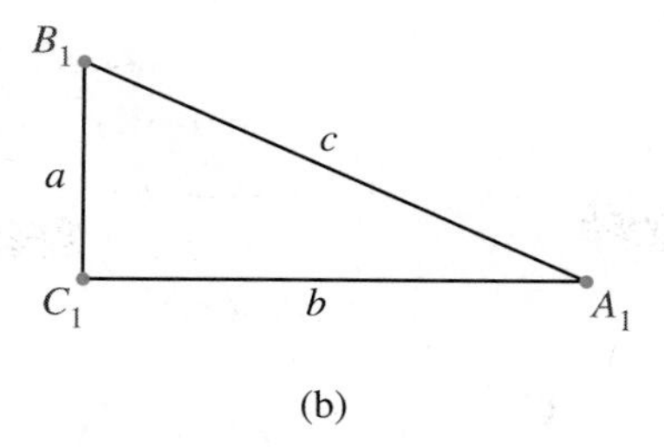

(b)

Figure 3.25

When applied to $\triangle A_1B_1C_1$, the Pythagorean Theorem gives $(A_1B_1)^2 = a^2 + b^2$. It is also given that $c^2 = a^2 + b^2$. Consequently, $c^2 = (A_1B_1)^2$ and hence $c = A_1B_1$. It follows now that $\triangle ABC \cong \triangle A_1B_1C_1$ by SSS. Thus $\angle C \cong \angle C_1$ and, therefore, $\angle C$ is a right angle. □

Remark The converse of the Pythagorean Theorem will be used later in this section in constructing $\sqrt{n}$, where n is an integer and a unit segment is given.

SPOTLIGHT on Teaching

Euclid's proof of the Pythagorean Theorem has disappeared from high school textbooks in the last several decades. Do you think it should be taught in high school? Why or why not? Should students be exposed to several proofs of the theorem or only one?

Now Solve This 3.5

Construct any segment to be one unit long, and then use the relationship $3^2 + 4^2 = 5^2$ to construct a segment whose length is $\sqrt{5}$.

Example 3.3 Find the area of an equilateral triangle with side a.

Solution

In Figure 3.26, ABC is an equilateral triangle. To find its area, we drop the perpendicular from a vertex to the opposite side. $\overline{BM}$ is such a perpendicular, and M is the midpoint of $\overline{AC}$. (Why?) In $\triangle ABM$, we have

$$h^2 + \left(\frac{a}{2}\right)^2 = a^2$$

$$h^2 = \frac{3}{4}a^2$$

$$h = \frac{a\sqrt{3}}{2}$$

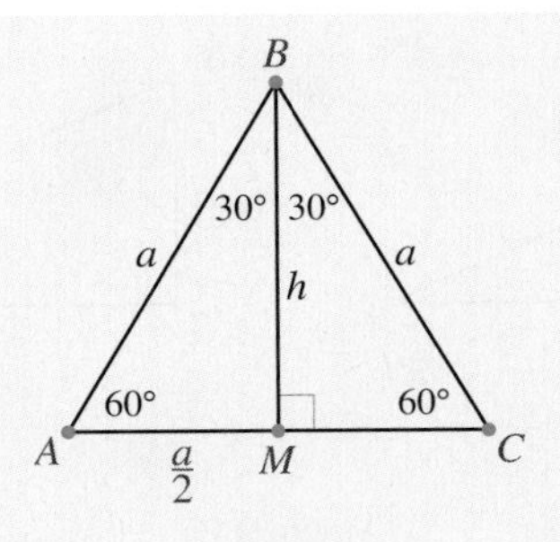

Figure 3.26

Consequently, the area of $\triangle ABC$ is $\frac{ah}{2} = \frac{a}{2} \cdot \frac{a\sqrt{3}}{2} = \frac{a^2\sqrt{3}}{4}$.

Now Solve This 3.6

1. Heron (first century C.E.) derived a remarkable formula for the area of a triangle given its sides a, b, and c. He denoted half the perimeter of the triangle by p—that is, $p = (a + b + c)/2$—and proved that the area A of the triangle is $A = \sqrt{p(p - a)(p - a)(p - b)}$. What is even more remarkable is that Heron was able to derive the formula without the use of algebra, which was not known to mathematicians at his time. By answering the following questions, you will prove Heron's formula using the Pythagorean Theorem and elementary algebra.
   (a) The area A of $\triangle ABC$ is $A = \frac{1}{2}ch$, where h is the height to side $\overline{BA}$ intersecting $\overline{AB}$ at D (draw your own triangle with the greatest angle at C, $BA = c$). Let $BD = x$; then $AD = c - x$. Show that $x = \frac{a^2 - b^2 + c^2}{2c}$.
   (b) Apply the Pythagorean Theorem in $\triangle BCD$ to express h in terms of a and x. Substitute for x the expression you obtained in part (a). You should get

   $$h^2 = a^2 - \left(\frac{a^2 - b^2 + c^2}{2c}\right)^2$$

   (c) Apply the difference of squares formula $u^2 - v^2 = (u - v)(u + v)$ to the result in part (b) to obtain

   $$4h^2c^2 = [2ac - (a^2 - b^2 + c^2)]\ [2ac + (a^2 - b^2 + c^2)]$$

   (d) Write the expression in the first brackets on the right side of the equation in part (c) as $b^2 - (c - a)^2$, and a similar expression for the expression in the second brackets. Apply the difference of squares formula again to obtain a product of four expressions, one of which will be $a + b - c$.
   (e) One of the expressions $a + b - c$ in the product you obtained in part (d) can be written using $a + b + c = 2p$ as follows:

   $$a + b - c = (2p - c) - c = 2(p - c)$$

   Write the other three expressions in the product in a similar way using p, and thus derive Heron's formula.

2. The Indian mathematician Brahmagupta (seventh century C.E.) found that the area of a cyclic quadrilateral (a quadrilateral that can be inscribed in a circle) with sides a, b, c, and d is given by the formula

$$\sqrt{(p-a)(p-b)(p-c)(p-d)} \quad \text{where } p = \tfrac{1}{2}(a+b+c+d)$$

Show that Heron's formula in part 1 follows from Brahmagupta's result.

Historical Note: Heron (first century C.E.)

Heron (also called Hero) of Alexandria was a first-century Greek mathematician whose *Metrica* treatise on geometry was lost until 1896, when the complete manuscript was found in Constantinople. (Constantinople, which was renamed Istanbul after World War I, is a city in present-day Turkey of about 8 million people that is located partially in Europe and partially in Asia). Heron also invented a jet-propelled rotary steam engine, wrote a text on mechanics for engineers, described how to dig tunnels through mountains, and formed the following principle on light reflection: *The angle of incidence is congruent to the angle of reflection.* In the delightful book *Journey Through Genius*, William Dunham discusses Heron's proof for his formula for triangular area and refers to it as a "great theorem of mathematics."

Example 3.4 Prove that it is impossible to construct an equilateral triangle on a geoboard no matter how large the geoboard is. In other words, prove that an equilateral triangle with vertices that are lattice points (have integer coordinates) does not exist.

Proof

Suppose such a triangle exists. We then set up a coordinate system so that one of the vertices is at the origin and the x-axis contains a row of lattice points, each one unit from its immediate neighbors (Figure 3.27). Because of our assumption that O, A, and B are at the lattice points, the coordinates of A and B are integers. We compute the area of $\triangle OAB$ in two ways. From Example 3.3, the area is

$$\frac{(OA)^2\sqrt{3}}{4} = \frac{(x_1^2 + y_1^2)\sqrt{3}}{4} \tag{3.2}$$

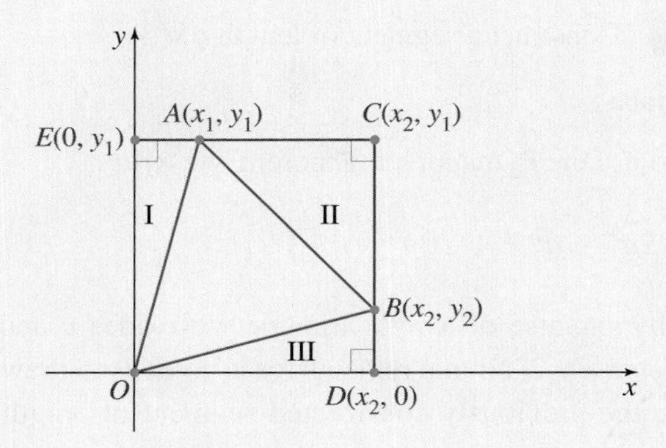

Figure 3.27

Because x_1 and y_1 are integers and $\sqrt{3}$ is irrational, each side of Equation 3.2 is irrational. If we compute the area of $\triangle OAB$ as the difference between the area of the rectangle $OECD$ and the sum of the areas of the three triangles marked I, II, and III, we get

$$(\text{Area of } \triangle OAB) = x_2y_1 - \tfrac{1}{2}(x_1y_1 + (x_2 - x_1)(y_1 - y_2) + x_2y_2)$$

This is clearly a rational number, which contradicts the fact that each side of Equation 3.2 is irrational. Consequently, it is impossible to have an equilateral triangle with all the vertices at lattice points. □

The Construction of $\sqrt{n}$

If each of the legs of a right isosceles triangle is a units long, and the hypotenuse is c units long (see Figure 3.28), then the Pythagorean Theorem tells us that

$$c^2 = a^2 + a^2$$

$$c^2 = 2a^2$$

$$c = a\sqrt{2}$$

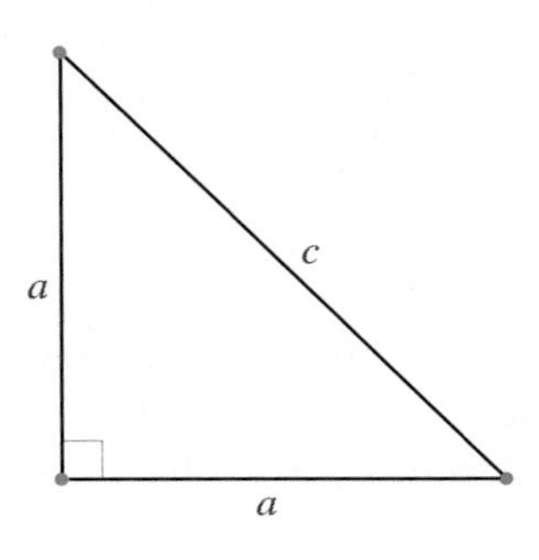

Figure 3.28

Thus, for a given segment of length a, we simply construct a right isosceles triangle with legs of length a. The hypotenuse will have the required length $a\sqrt{2}$. If a unit segment is given and $a = 1$, then the preceding approach gives us a segment of $\sqrt{2}$ units.

The Pythagorean Theorem can be used to construct $a\sqrt{n}$ for any positive integer n if n is not a perfect square. (If n is a perfect square, the construction does not require the Pythagorean Theorem!) Let's look at a few examples. (In all of these examples, a segment of length a is given.)

Example 3.5 Construct a segment of length $a\sqrt{3}$.

Solution: First Approach

Let $x = a\sqrt{3}$. Then to use the Pythagorean Theorem, we write

$$x^2 = 3a^2 = a^2 + 2a^2 = a^2 + (\sqrt{2}a)^2$$

Hence x is the hypotenuse of a right triangle with sides a and $a\sqrt{2}$. Because we have already constructed $a\sqrt{2}$, all that remains for us to do is to draw a segment of length a at a right angle to the previously constructed segment of length $a\sqrt{2}$, as shown in Figure 3.29.

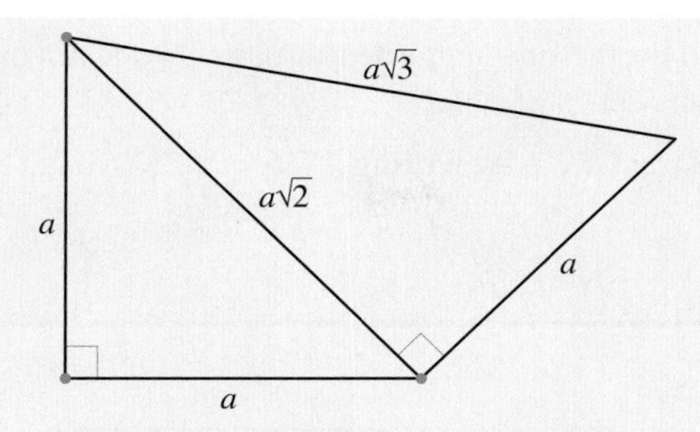

Figure 3.29

Solution: Second Approach

Notice that

$$x^2 = 3a^2 = 4a^2 - a^2 = (2a)^2 - a^2$$

and therefore

$$x^2 + a^2 = (2a)^2$$

Thus x is a leg of a right triangle whose hypotenuse is $2a$ and one leg is a. The two parts of Figure 3.30 suggest two ways to construct such a triangle and, therefore, two ways to construct $x = a\sqrt{3}$.

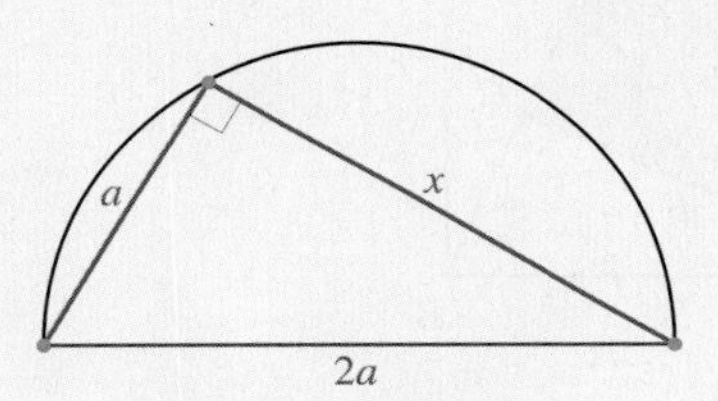

A Circle with diameter $2a$

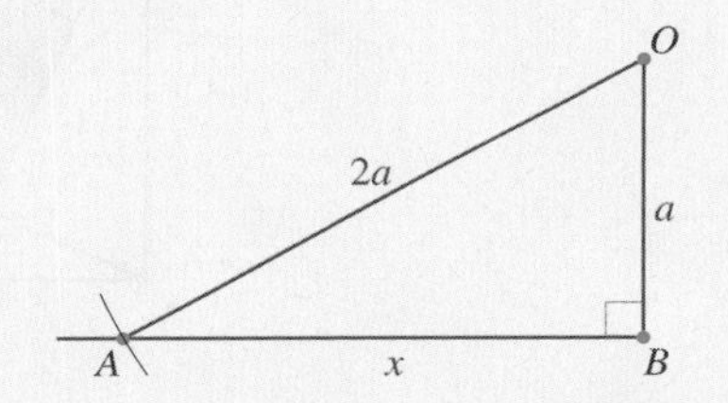

(1) Construct segment $\overline{BO}$. (2) Construct $\overline{BA}$ perpendicular to $\overline{BO}$. (3) Construct an arc with center O and radius $2a$

Figure 3.30

Example 3.6 Construct $a\sqrt{45}$.

Hints for a Solution

Because $45 = 3^2 \cdot 5$, $\sqrt{45} = 3\sqrt{5}$ and hence $\sqrt{45}a = 3\sqrt{5}a$. Thus we need construct only $\sqrt{5}a$. One way to accomplish this task is to notice that $5 = 2^2 + 1^2$. Another way is to write 5 as a difference of two squares. If $1 \cdot 5 = u^2 - v^2$, then $5 = (u - v)(u + v)$. Let $u - v = 1$ and $u + v = 5$; then $u = 3$ and $v = 2$. Hence $5 = 3^2 - 2^2$. Notice that any odd number can be expressed in this way as a difference of two squares.

You are asked to complete the solution of this problem and similar problems in Now Solve This 3.7.

Now Solve This 3.7

1. Complete the two approaches to the construction of $a\sqrt{5}$ as suggested in Example 3.6.
2. Show how to construct $a\sqrt{n}$ when n is an integer and square free (i.e., every prime in the prime factorization of n appears only once).

3. Can a method based on the one suggested in part 1 work to construct $a\sqrt{n}$ when n is even? Justify your answer.
4. Explain how to construct $a\sqrt{6}$ by writing $a\sqrt{6} = \sqrt{2}(\sqrt{3}a)$.
5. Construct $a\sqrt{\frac{3}{2}}$.

Problem Set 3.2

1. The figure below suggests yet another proof of the Pythagorean Theorem. $\triangle ABC$ is a right triangle with sides $AB = c$, $AC = b$, and $BC = a$. Four triangles congruent to $\triangle ABC$ have been "assembled" to form a rhombus of side length c.
   - **a.** Prove that the outer figure with side c (the rhombus) is a square.
   - **b.** Prove that the inner shaded figure is also a square.
   - **c.** Compute the area of the outer square in two different ways and obtain the Pythagorean Theorem.

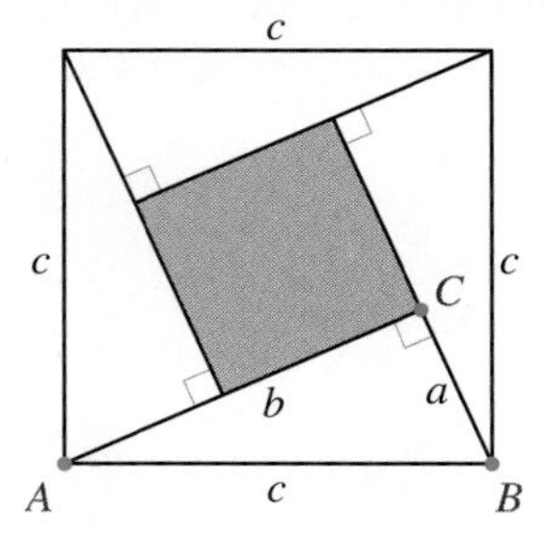

2. The following proof of the Pythagorean Theorem is attributed to President James Garfield. $\triangle ABC$ is a right triangle with sides a and b and hypotenuse c. $\triangle DEB \cong \triangle CBA$ and C, B, and D are collinear. Find the area of the trapezoid $ACDE$ in two different ways to obtain a proof of the Pythagorean Theorem.

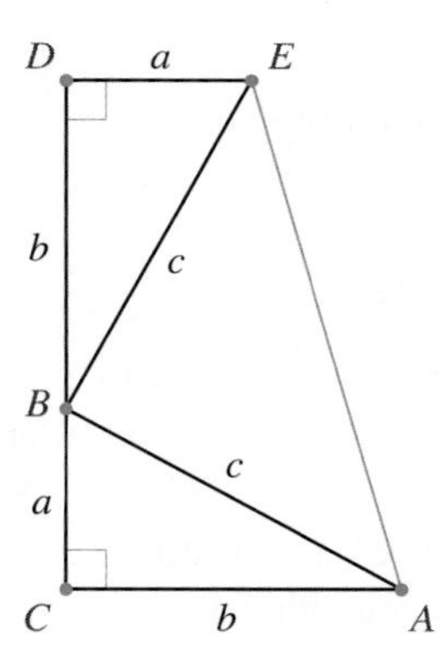

3. **a.** Prove that in a right triangle with a 30° angle, the side opposite that angle is half as long as the hypotenuse.
   b. Without using trigonometry, express the length of the legs of a 30°–60°–90° triangle in terms of the length of the hypotenuse c.

4. •a. $ABCD$ is a rectangle, and O is a point in its interior. The distances from O to the vertices are x, y, z, and w, as indicated in the figure. Prove that $x^2 + z^2 = y^2 + w^2$.

   b. Pose a problem whose solution will be easy if the result in part (a) is known.

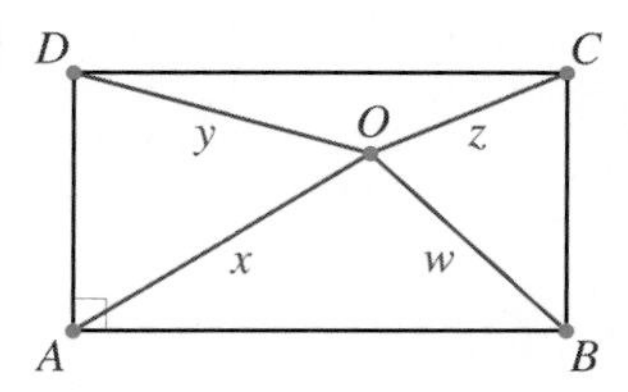

5. A boat starts at point A, moves 3 km due north, then 2 km due east, then 1 km due south, and then 4 km due east to point B. Find the distance AB.

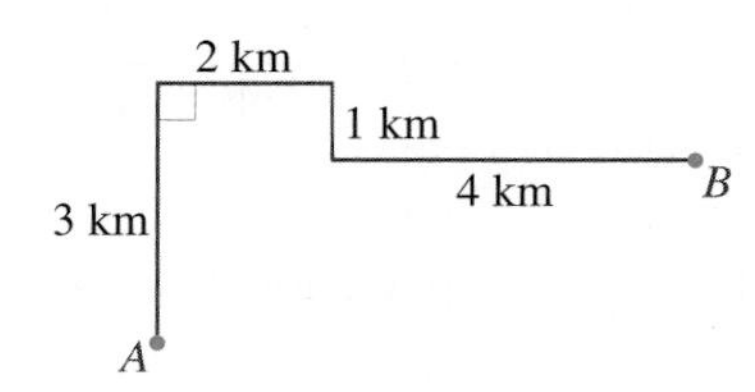

6. a. The distance from point A to the center of a circle with radius r is d. Express the length of the tangent segment $\overline{AP}$ in terms of r and d.

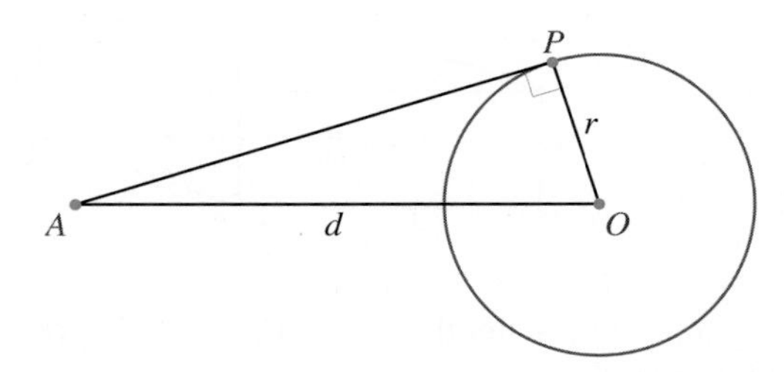

   b. The distance between the centers of two circles with radii r_1 and r_2 is d, where $O_1O_2 = d$. Line k is a common exterior tangent to the circles at P and Q. Find PQ in terms of r_1, r_2, and d. (Assume that $r_1 > r_2$.)

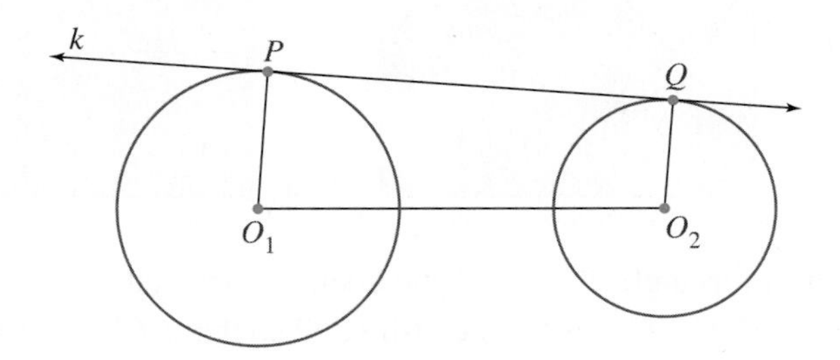

   c. Line ℓ is a common interior tangent to the circles with centers O_1 and O_2 at S and T. Find ST in terms of r_1, r_2, and d, where $O_1O_2 = d$.

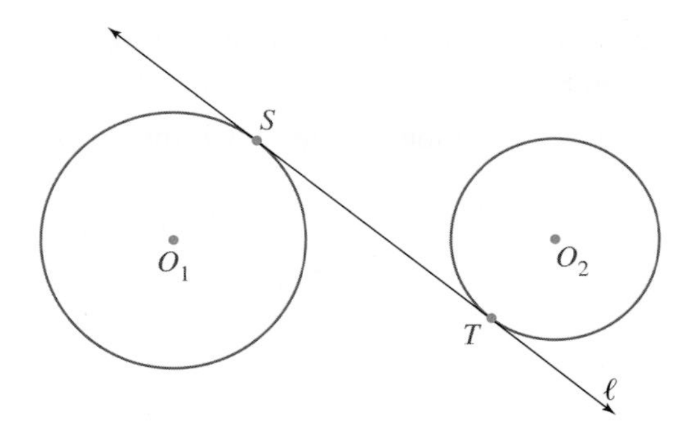

7. • **a.** Given a segment of length a, show two different ways for constructing a segment of length (i) $\sqrt{19}a$ and (ii) $a\sqrt{\frac{2}{3}}$.

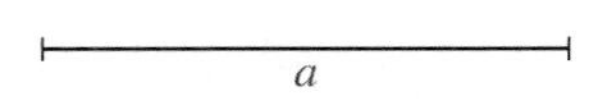

b. Given a segment of length a, explain how to construct $\sqrt{\frac{n}{m}}\,a$, where n and m are positive integers and assuming we know how to construct $\sqrt{k}\,a$ for any positive integer k.

8. $ABCDEFGH$ is a right rectangular prism (a box) of dimensions a, b, and c as shown in the figure. Find the length of the diagonal $\overline{DG}$ in terms of a, b, and c.

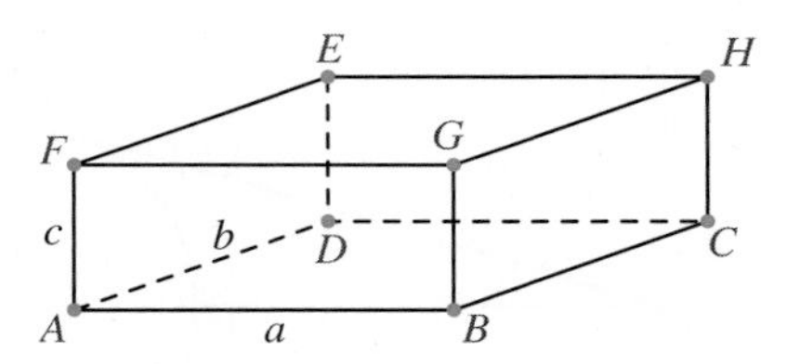

9. The following figures suggest another proof of the Pythagorean Theorem. Use the figures to write a complete proof.

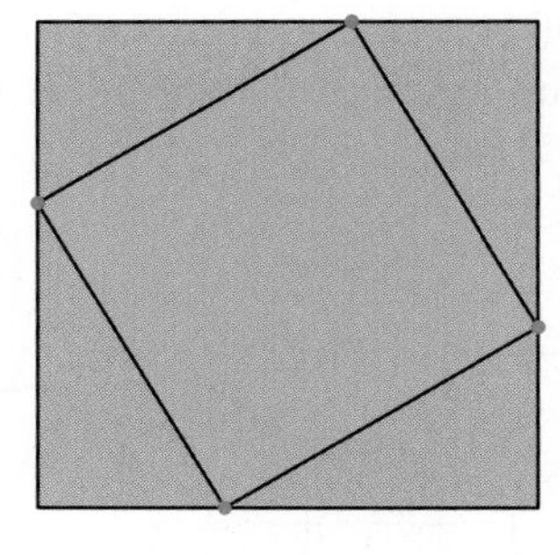

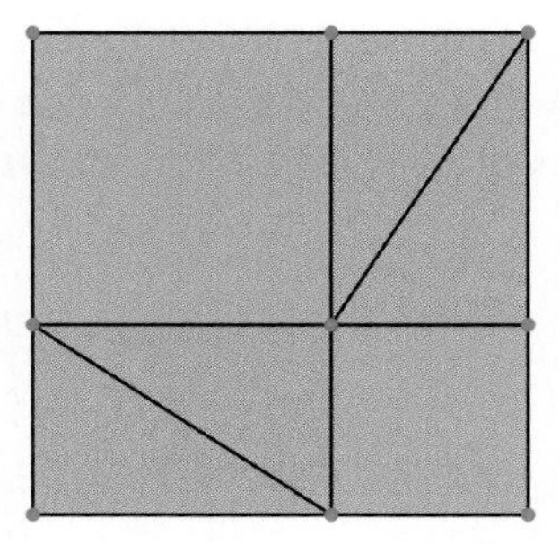

10. $\triangle ABC$ is a right triangle. Squares have been constructed on the hypotenuse and the legs of the triangle. $\triangle A'B'C'$ has been constructed so that $CC' = c$ and $\overline{CC'} \parallel \overline{AA'}$. Prove the following:

a. The area of the square $ABB'A'$ is equal to the sum of the areas of the parallelograms $ACC'A'$ and $CBB'C'$.

b. The height of the parallelogram $CBB'C'$ to side $\overline{CB}$ is a, and hence the area of the parallelogram is a^2. Similarly, show that the area of $ACC'A'$ is b^2.

c. Use your results from parts (a) and (b) to prove that $c^2 = a^2 + b^2$.

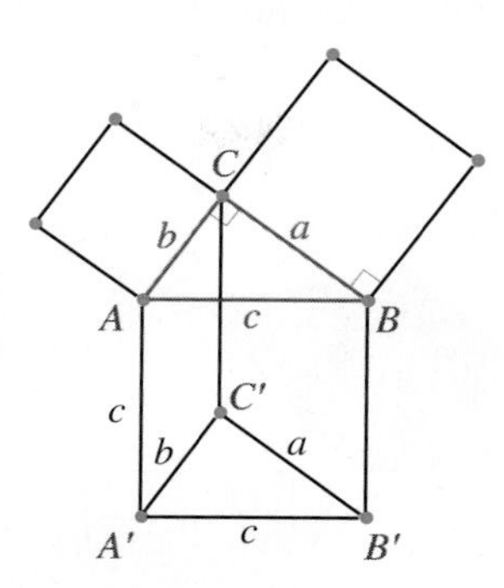

★**11.** Leonardo da Vinci (1452–1519) proved the Pythagorean Theorem with the help of the accompanying figure, in which $DC' = AC$ and $EC' = CB$. Use the figure to prove the Pythagorean Theorem.

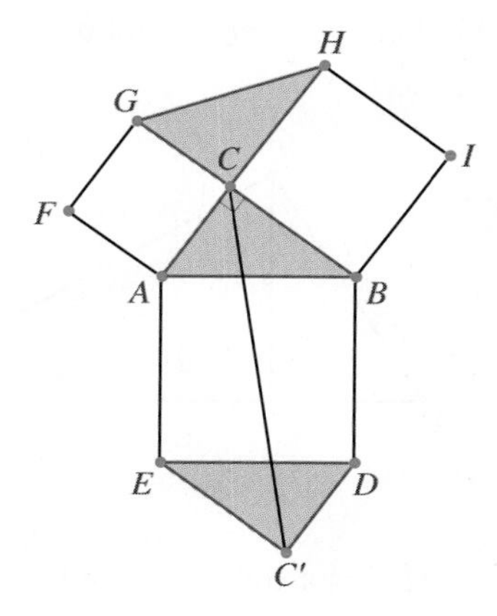

12. $ABCD$ is a rectangle with sides $AB = a$ and $AD = b$. $BEDF$ is a rhombus. Find EF in terms of a and b. (Simplify your answer.)

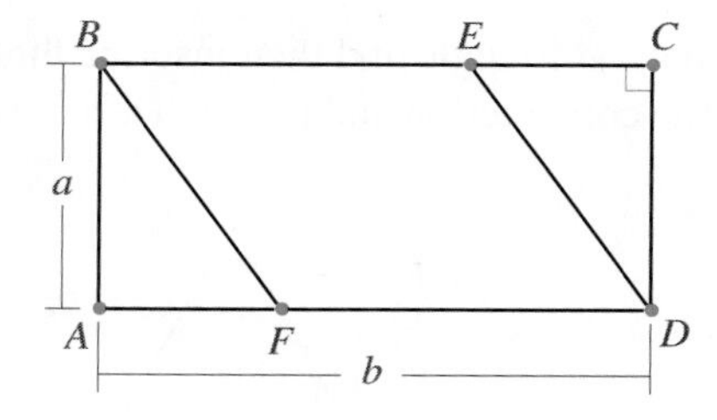

★**13.** The Greek geometer Pappus (circa 300 C.E.) proved the following generalization of the Pythagorean Theorem: $\triangle ABC$ is any triangle (not necessarily a right triangle). Arbitrary parallelograms are constructed on two sides of the triangle (sides $\overline{AC}$ and $\overline{BC}$ in the figure). D is the intersection of the lines containing the two sides of the parallelograms. A third parallelogram is constructed on side $\overline{AB}$ such that $AE = DC$ and $\overline{AE} \parallel \overline{DC}$. Pappus's Theorem asserts that

Area (I) + Area (II) = Area (III)

a. Prove Pappus's Theorem.

b. Show that the Pythagorean Theorem follows from Pappus's Theorem and therefore that Pappus's Theorem is a generalization of the Pythagorean Theorem.

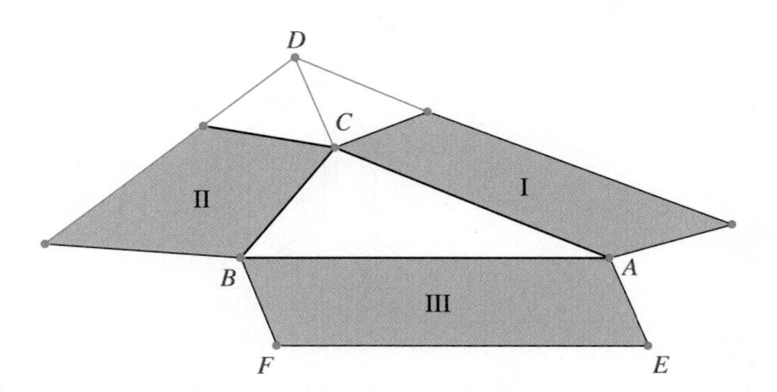

14. • **a.** $\triangle ABC$ is a right triangle with height h to the hypotenuse. The height divides the hypotenuse into segments a_1 and b_1. Prove that $h^2 = a_1b_1$. (This can be proved using similarity of Chapter 4, but here you should not use similar triangles.)

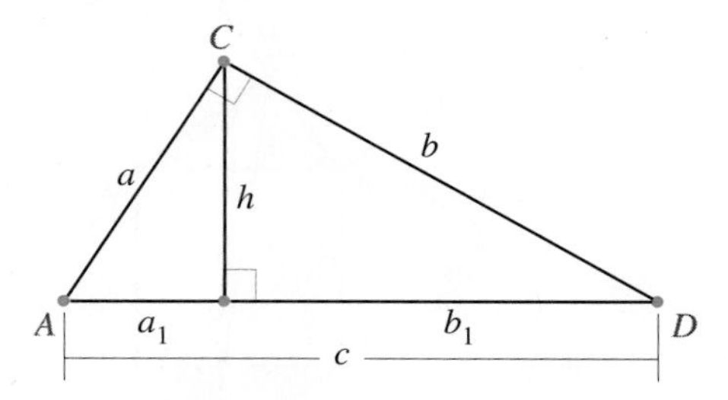

b. If it is not given that $\triangle ABC$ is a right triangle but it is known that $h^2 = a_1b_1$, prove that $\angle C$ must be a right angle.

15. $\triangle ABC$ is equilateral with side a. Three congruent circles are tangent to each other and to the sides of the triangle.

• **a.** Find the radius r of the circle in terms of a.

b. Construct an equilateral triangle and then inscribe three congruent circles as described above. Describe the construction and prove that it is valid.

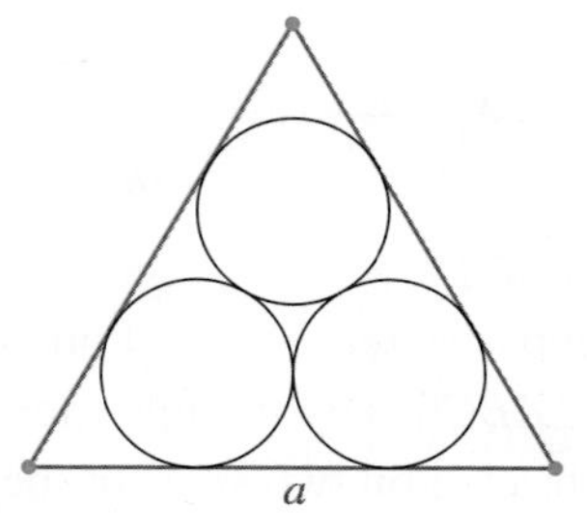

16. In the figure, $ABCD$ is a square with side measure a. $QSTP$ is also a square, and the four triangles are congruent. Congruent circles are inscribed in each triangle and in the inner square.

• **a.** Find the radius of the circle in terms of a.

b. Construct your own square, the triangles, the inner square, and the inscribed circles.

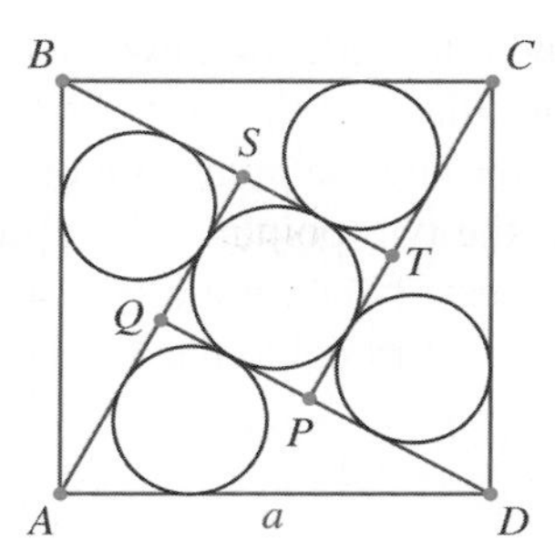

17. If a, b, and c are positive integers and $a^2 + b^2 = c^2$, then the triple a, b, c is called a **Pythagorean triple**. The triple 3, 4, 5 is perhaps the best-known example. Use it to complete the following tasks.

a. Prove that there are infinitely many Pythagorean triples.

b. Find all the Pythagorean triples that consist of consecutive positive integers. (Prove that you have found them all.)

c. Find all the Pythagorean triples that form an arithmetic sequence.

18. To find Pythagorean triples other than the ones in Problem 17, examine the following argument and answer the questions that follow.

Notice that

$$(u + v)^2 = (u - v)^2 + 4uv$$

Thus, if uv is a perfect square, this equation gives us a Pythagorean triple for all positive integers u and v with $u > v$. Let $u = n^2$ and $v = m^2$, where n and m are positive integers, and $n > m$. We get

$$u - v = n^2 - m^2$$
$$u + v = n^2 + m^2$$
$$4uv = 4n^2m^2 = (2nm)^2$$

Thus, for all positive integers $n > m$, $n^2 - m^2$, $2nm$, $n^2 + m^2$ is a Pythagorean triple.

a. Use the preceding expression to obtain the Pythagorean triples 5, 12, 13 and 7, 24, 25.

b. Does the argument in this problem prove that the preceding expression gives all the Pythagorean triples? Justify your answer.

3.3 The Distance Formula

Much of geometry and its applications hinges on being able to find the distance between two points. In this section we will develop the *distance formula*—a formula for the distance between two points. This formula is the equivalent form of the Pythagorean Theorem in a coordinate system, and is the basis for many topics in coordinate geometry. Here we will demonstrate its use in developing the equation of a circle. In the problem set at the end of this section, we explore its use in deriving the equations of the ellipse, the parabola, and the hyperbola.

Equations of straight lines are commonly developed using the concepts of slope and similarity. This approach will be taken in Chapter 4 on similarity; in addition, though not commonly done in textbooks, we will show in this section how the equation of a straight line can easily be developed using the distance formula.

If we are given the coordinates and the points are on the x-axis or on a line parallel to the x-axis, then the distance between the two points is the absolute value of the difference between their x-coordinates, as shown in Figure 3.31a. Similarly, Figure 3.31b shows the distance between the two points on the y-axis or on a line parallel to the y-axis.

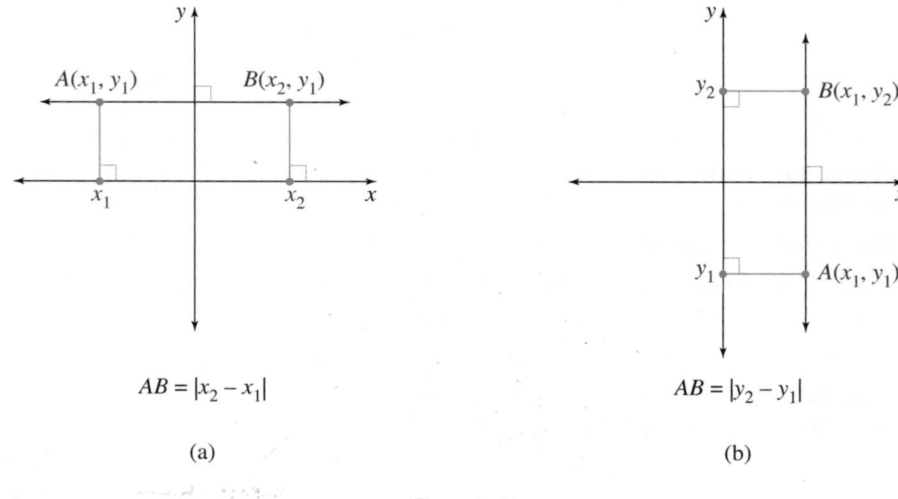

Figure 3.31

If the line through the two points is not parallel to either coordinate axis, we can find the distance between the points using the Pythagorean Theorem. We will show how this can be done in the following example:

Example 3.7

Given the coordinates of two points $A(x_1, y_1)$ and $B(x_2, y_2)$ such that the line through the two points is not parallel to either coordinate axis, find the distance AB.

Solution

Because we know how to find the distance between points on a horizontal or vertical line, we construct through A a line parallel to the x-axis and through B a line parallel to the y-axis, as shown in Figure 3.32. The lines intersect at C, which is the vertex of the right triangle ABC. We apply now the Pythagorean Theorem to $\triangle ABC$:

$$(AB)^2 = |x_x - x_1|^2 + |y_2 - y_1|^2$$

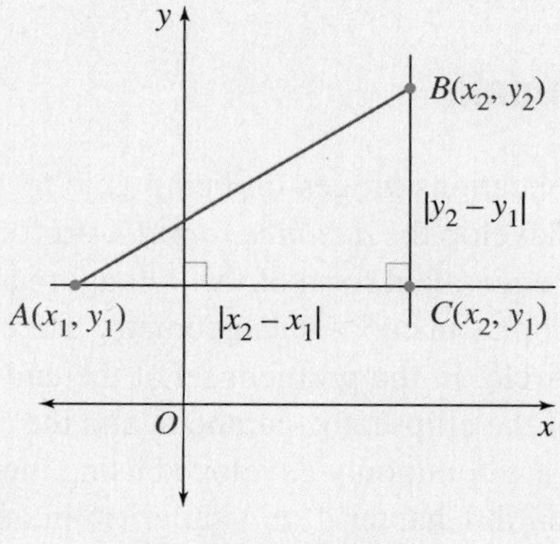

Figure 3.32

Because for all real numbers a, $|a|^2 = a^2$, we have $|x_2 - x_1|^2 = (x_2 - x_1)^2$ and $|y_2 - y_1|^2 = (y_2 - y_1)^2$. Thus

$$AB = \sqrt{(x_2 - x_1)^2 + (y_2 - y_1)^2}$$

Remark Because $x_2 - x_1 = -(x_1 - x_2)$, it does not make any difference in the preceding formula if we square $x_2 - x_1$ or $x_1 - x_2$, and similarly for the difference between the y-coordinates.

Let us also check whether this formula works for the case when the two points are on a horizontal or vertical line. If A and B are on a horizontal line, then $y_2 = y_1$ and we get $AB = \sqrt{(x_2 - x_1)^2 + (y_2 - y_1)^2} = \sqrt{(x_2 - x_1)^2} = |x_2 - x_1|$. This agrees with what was pointed out in Figure 3.31a. Similarly, we can show that the formula works in the case when the points are on a vertical line.

It is reasonable to wonder whether a similar formula can be developed for the distance between two points in space. This issue will be explored shortly, in Now Solve This 3.8. First, however, we state the solution of Example 3.7 in the following theorem:

Theorem 3.6

The Distance Formula

The distance between the points $A(x_1, y_1)$ and $B(x_2, y_2)$ is

$$AB = \sqrt{(x_2 - x_1)^2 + (y_2 - y_1)^2}$$

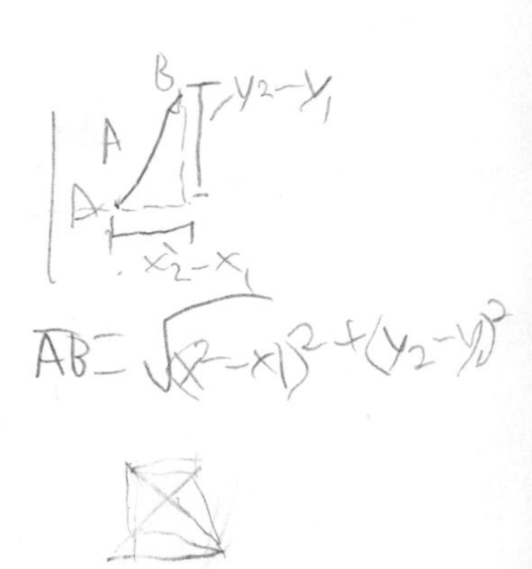

Now Solve This 3.8

1. Given the coordinates of two points in space, $A(x_1, y_1, z_1)$ and $B(x_2, y_2, z_2)$, use the ideas suggested in Figure 3.33 to derive a formula similar to the distance formula given in Theorem 3.6. (Show that $AC = A'B'$, and apply the Pythagorean Theorem to $\triangle ABC$.)

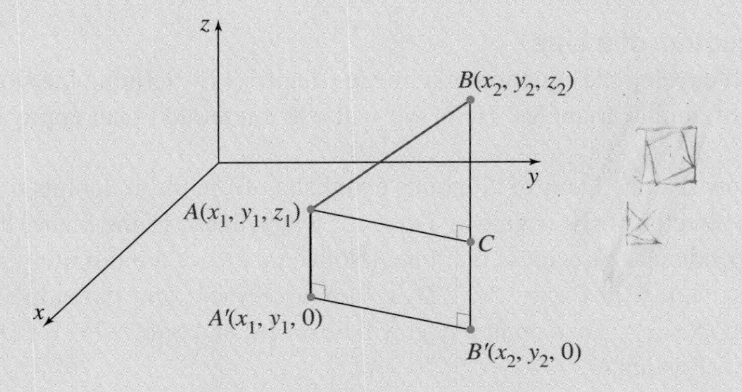

Figure 3.33

2. Show that the formula you derived in part 1 is a generalization of the distance formula in the plane; that is, show that the distance formula in the plane is a special case of the distance formula in space.

Equation of a Circle

A circle is the locus or set of all points in the plane at the same distance R (the radius of the circle) from a given point C (the center). Given $C(h, k)$ and the radius R as shown in Figure 3.34, a point $P(x,y)$ is on the circle if and only if $PC = R$. Using the distance formula, $PC = R$ if and only if

$$\sqrt{(x - h)^2 + (y - k)^2} = R \tag{3.3}$$

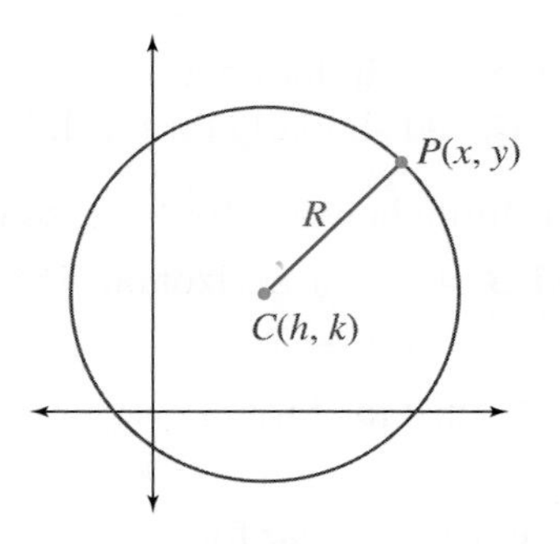

Figure 3.34

Whenever the variables x and y in Equation 3.3 are replaced by the coordinates of a point on the circle, the equation is satisfied (i.e., becomes a true statement). It is not satisfied (i.e., becomes a false statement) if the variables x and y are replaced by the coordinates of a point not on the circle. In this sense, Equation 3.3 is the equation of the circle. Because each side of Equation 3.3 is positive, the equation is equivalent to

$$(\sqrt{(x - h)^2 + (y - k)^2})^2 = R^2$$

and therefore to

$$(x - h)^2 + (y - k)^2 = R^2 \tag{3.4}$$

Equation 3.4 is more convenient to use and is commonly referred to as the equation of the circle.

Equation of a Line

We will develop the equation of a line in Chapter 4 by defining the slope of a line and using properties of similar triangles. Here, we will use a nontraditional approach using the distance formula.

We know that the locus of all points equidistant from the endpoints of a segment is the perpendicular bisector of the segment. Thus, for every line ℓ in the plane, there exists a segment whose perpendicular bisector is that line. (Notice that there are infinitely many such segments.) Suppose, as shown in Figure 3.35, $\overline{CD}$ is such a segment and its endpoints have coordinates $C(x_1,y_2)$ and $D(x_2,y_2)$. Then point $P(x,y)$ will be on ℓ if and only if $PC = PD$. This is equivalent to each of the following:

$$\sqrt{(x - x_1)^2 + (y - y_1)^2} = \sqrt{(x - x_2)^2 + (y - y_2)^2} \tag{3.5}$$

$$(x - x_1)^2 + (y - y_1)^2 = (x - x_2)^2 + (y - y_2)^2$$

$$x^2 - 2x_1x + x_1^2 + y^2 - 2y_1y + y_1^2 = x^2 - 2x_2x + x_2^2 + y^2 - 2y_2y + y_2^2$$

$$-2x_1x + x_1^2 - 2y_1y + y_1^2 = -2x_2x + x_2^2 - 2y_2y + y_2^2$$

$$2(y_2 - y_1)y = 2(x_1 - x_2)x + k$$

where $k = x_2^2 - x_1^2 + y_2^2 - y_1^2$ is a constant.

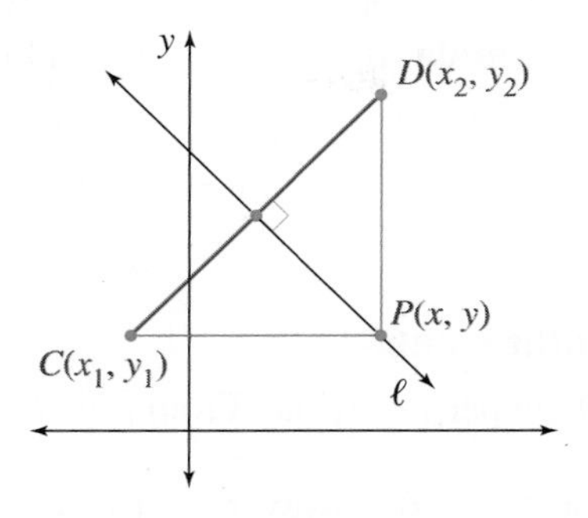

Figure 3.35

Before we divide both sides of Equation 3.5 by $2(y_2 - y_1)$ so that we can write y in terms of x, we need to consider the case when $y_2 - y_1 = 0$. Then $y_2 = y_1$ and so $\overline{CD}$ is parallel to the x-axis and therefore ℓ is a vertical line. We then get $0 = 2(x_1 - x_2)x + k$, which is equivalent to $x = \dfrac{-k}{2(x_1 - x_2)}$ (since C and D are different points $x_1 \neq x_2$). Because $x = \dfrac{-k}{2(x_1 - x_2)}$ is a constant, we obtained the equation of a vertical line as $x = c$, where c is a constant (of course, this should be evident without any calculations!). If $y_1 \neq y_2$ (i.e., $\overline{CD}$ is not horizontal and therefore ℓ is not vertical), we can divide both sides of Equation 3.5 by $2(y_1 - y_2)$ and obtain

$$y = \frac{(x_1 - x_2)x}{(y_2 - y_1)} + \frac{k}{2(y_2 - y_1)} \tag{3.6}$$

Thus we have proved that the equation of a nonvertical line has the form

$$y = mx + b \tag{3.7}$$

for some constants m and b; m is called the **slope** of the line and b is called the ***y*-intercept** (the y-coordinate of the point on the line for which $x = 0$). Notice that (x_1, y_1) and (x_2, y_2) are not on the line ℓ whose equation is given in Equation 3.6.

Finding the Equation of a Line Through Two Given Points

Given two points, there is a unique line through those points. Thus, if we are given the coordinates of the points, we should be able to find the equation of the line. Suppose that our two points are $A(x_A, y_A)$ and $B(x_B, y_B)$. We want to find the equation of the line AB. For that purpose, we substitute the coordinates of A and of B in the equation $y = mx + b$:

$$y_A = mx_A + b \tag{3.8}$$

$$y_B = mx_B + b \tag{3.9}$$

Subtracting, we obtain $y_B - y_A = m(x_B - x_A)$. If $x_B \neq x_A$ (i.e., if line AB is not vertical), we get

$$m = \frac{y_B - y_A}{x_B - x_A} \tag{3.10}$$

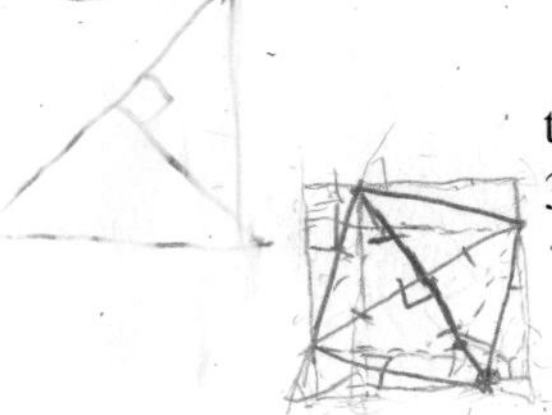

We can find the value of b by substituting in Equation 3.7 or 3.8 the value of m. The equation of the line can also be found by subtracting Equation 3.8 (or Equation 3.9) from Equation 3.7; we then get the **point-slope formula** for the equation of a line:

$$y - y_A = m(x - x_A)$$

Slopes of Perpendicular Lines

In Figure 3.35, lines ℓ and $\overleftrightarrow{CD}$ are perpendicular. From Equation 3.10, we find that the slope of line CD is $m_{CD} = \dfrac{y_2 - y_1}{x_2 - x_1}$. In Equation 3.6, however, we found that the slope m_ℓ of the perpendicular line is $m_\ell = \dfrac{x_1 - x_2}{y_2 - y_1}$. This implies that $m_{CD} \cdot m_\ell = \dfrac{y_2 - y_1}{x_2 - x_1} \cdot \dfrac{x_1 - x_2}{y_2 - y_1} = -1$. Thus we have proved the following theorem:

Theorem 3.7

If two lines (neither of which is vertical) are perpendicular, then the product of their slopes is -1.

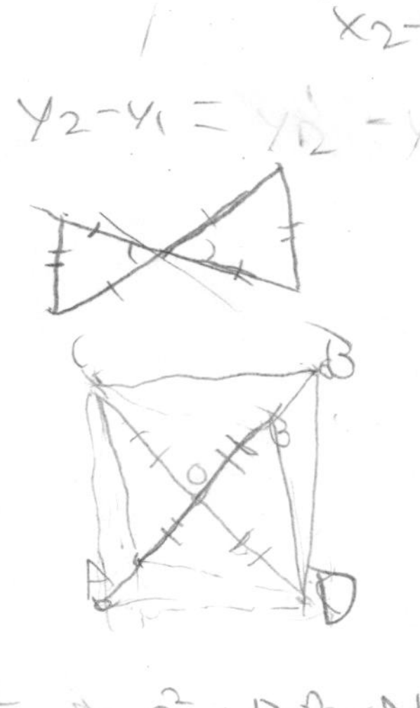

Now Solve This 3.9

1. Use Equation 3.9 to find a geometric interpretation of the slope.
2. We have proved that the equation of a nonvertical line has the form $y = mx + b$. Now prove the converse—that for every value of m and b, $y = mx + b$ is an equation of some line.
3. We have proved that if two nonvertical lines are perpendicular, then the product of their slopes is -1. Now prove the converse—that if two lines have slopes whose product is -1, then the lines are perpendicular.
4. Prove that two nonvertical lines are parallel if and only if they have the same slopes.

We can use the distance formula to find the equation of various loci, such as the parabola, the ellipse, and the hyperbola. These loci will be explored in the problem set at the end of this section. For now, we present an example whose solution is relatively easy if we use coordinate geometry and the distance formula, but fairly difficult otherwise.

Example 3.8 Given two points A and B in the plane, find the locus of all points P in the plane that are twice as far from A as from B.

Solution

For the sake of simplicity, we choose a coordinate system such that the origin is at A and the x-axis is on the line AB with the positive direction from A to B (Figure 3.36). Moreover, we let $AB = 1$. Then B is at $(1, 0)$.

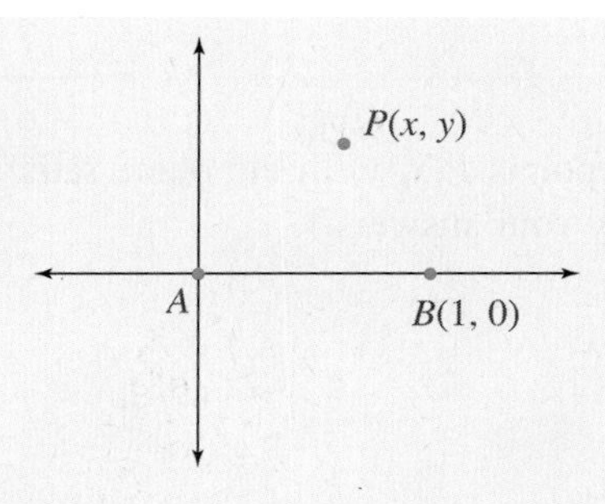

Figure 3.36

It is required that $PA = 2PB$. This is equivalent to each of the following:

$$\sqrt{(x-0)^2+(y-0)^2} = 2\sqrt{(x-1)^2+y^2}$$

$$x^2+y^2 = 4\left((x-1)^2+y^2\right)$$

$$3x^2+3y^2-8x+4=0 \tag{3.11}$$

$$x^2+y^2-\frac{8}{3}x+\frac{4}{3}=0$$

The last equation may remind you of the equation of the circle, $(x-h)^2+(y-k)^2=R^2$, when expanded and written as

$$x^2-2hx+h^2+y^2-2ky+k^2-R^2=0 \tag{3.12}$$

In fact, Equations 3.11 and 3.12 will be the same iff (if and only if) both the coefficients of x and y and the constant term in Equation 3.12 are equal to the corresponding coefficients in Equation 3.11. Thus

$$-2h = -\frac{8}{3} \quad -2k=0 \quad h^2+k^2-R^2=\frac{4}{3}$$

These equations give us $h=\frac{4}{3}$, $k=0$, $R^2=\frac{4}{9}$, or $R=\frac{2}{3}$. Hence Equation 3.11 represents a circle whose equation is $(x-\frac{4}{3})^2+y^2=(\frac{2}{3})^2$—that is, a circle with center at $(\frac{4}{3}, 0)$ and radius $\frac{2}{3}$. (Of course, the same result can be accomplished by completing the squares in Equation 3.11.)

Now Solve This 3.10

Generalize Example 3.8:

1. Given two points A and B in the plane, find the locus of all points P in the plane that are m times as far from A as from B.
2. (a) Substitute $m=2$ in your answer to part 1 to check that your answer is the same as the answer to Example 3.8.

   (b) Substitute $m=1$ in your answer to part 1. Do you get the same answer, as you would expect without any calculations?

   (c) Substitute $m=\frac{1}{2}$ in your answer to part 1. Do you get the same answer, as you would expect from purely geometric considerations? (*Hint:* Think about the symmetry of the locus for $m=2$ and $m=\frac{1}{2}$.)

Problem Set 3.3

1. Identify the locus of all points $P(x, y)$ in the plane satisfying each of the following expressions. (Briefly justify your answers.)

a. $x^2 + y^2 < R^2$

b. $(x - h)^2 + (y - k)^2 > R^2$

c. $\dfrac{x}{a} + \dfrac{y}{b} = 1$

d. $\left|\dfrac{x}{a}\right| + \left|\dfrac{y}{b}\right| = 1$

e. $|x| + |y| = 1$

f. $y = |x|$

g. $2x^2 - y^2 - xy + 3x + 3y = 2$

h. $x^2 + y^2 - 2x + 2y + 4 = 0$

2. Investigate the simultaneous solution of the equations $y = mx + b_1$ and $y = mx + b_2$ to prove that nonvertical lines are parallel if and only if their slopes are equal.

3. The following is a sketch of another proof that if two lines are perpendicular and neither line is horizontal, then the product of their slopes is -1. Complete the details of the proof.

Using Example 3.8, it is sufficient to prove the assertion for two perpendicular lines through the origin: $y = m_1x$ and $y = m_2x$. Through the point (0,1), draw the line parallel to the x-axis intersecting the two lines at A and B. Now find the coordinates of A and B (in terms of m_1 and m_2, respectively). Use the distance formula to find the sides of $\triangle ABO$. Next apply the Pythagorean Theorem to this triangle. Simplify and obtain that $m_1m_2 = -1$.

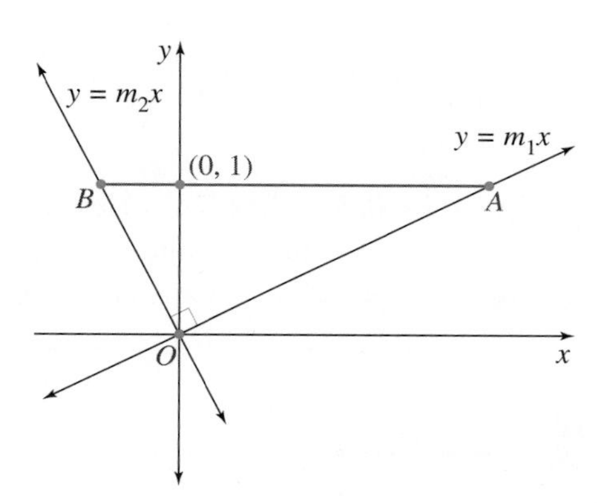

•**4.** Prove the converse of the statement that you proved in Problem 3: If the product of the slopes of the slopes of two lines is -1, then the lines are perpendicular.

5. A vertical line divides the triangle whose vertices are at (0,0), (1,1), and (9,1) into two parts; a triangle and a quadrilateral. If the areas of the triangle and the quadrilateral are equal, what is the equation of the vertical line?

6. $ABCD$ is a square where M is the midpoint of $\overline{AB}$ and $BN = \frac{1}{4}BC$.

a. Use the coordinate approach to prove that $\triangle DMN$ is a right triangle.

b. Prove that $\triangle DMN$ is a right triangle in another way.

7. The equations of two straight lines are $Ax + By + C = 0$ and $A_1x + B_1y + C_1 = 0$. Consider the lines

$$Ax + By + C + k(A_1x + B_1y + C_1) = 0$$

When k takes all possible real number values, this equation represents infinitely many lines. What do all these lines have in common? Distinguish two cases: (a) when the original lines intersect and (b) when the original lines are parallel.

•**8.** Find a necessary and sufficient condition on A, B, C, D, and E for $Ax^2 + By^2 + Cx + Dy + E = 0$ to be an equation of a circle.

9. Suppose $x^2 + y^2 + Ax + By + C = 0$ and $x^2 + y^2 + A_1x + B_1y + C_1 = 0$ are equations of two circles that intersect.

a. State and answer a question analogous to that posed in Problem 7.

b. Interpret the result in part (a) for $k = -1$.

10. A parabola is defined as the locus of all points equidistant from a given line called the **directrix** and a given point called the **focus**. Choose a coordinate system such that the x-axis is parallel to the directrix d and halfway between the focus F and the directrix. Let the distance between F and d be $2p$.

a. Prove that the equation of the parabola located as described above can be written as $y = \frac{1}{4p}x^2$.

b. Where are the focus and the directrix of the parabola $y = x^2$?

c. Is the graph of $y = \sqrt{x}$ a parabola? Justify your answer.

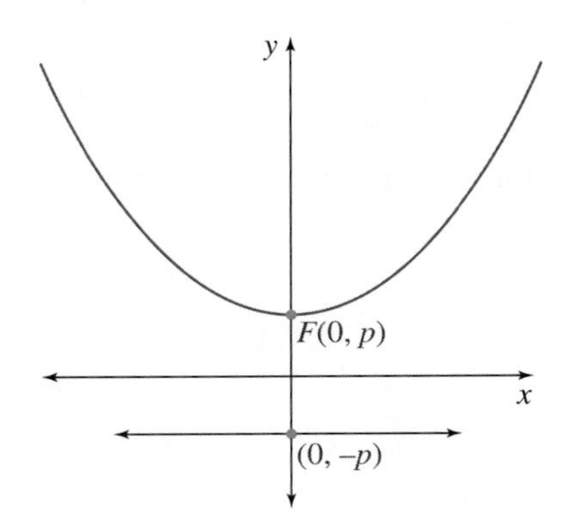

11. Two circles have the same radius R and respective centers at (a, b) and (b, a).

a. Find a simple necessary and sufficient condition on a, b, and R for the two circles to be tangent to each other (intersecting at exactly one point).

b. If the two circles intersect at exactly two points, find the length of their common chord in terms of a, b, and R. Simplify your answer.

•**12.** Given $A(x_1, y_1)$ and $B(x_2, y_2)$, derive the simplest formula possible for the coordinates of the midpoint M of $\overline{AB}$ in terms of x_1, y_1, x_2, y_2.

•**13.** The coordinates of three vertices of a parallelogram are $A(1, 1)$, $B(2, 4)$, $C(-2, 3)$. Find the coordinates of the fourth vertex. (*Note:* There is more than one answer!)

14. An **ellipse** is the locus of all points in the plane such that the sum of the distances from two fixed points is constant. The two fixed points are called the **foci** of the ellipse. To derive the equation of the ellipse, choose the x-axis to be the line through the foci and the y-axis to be the perpendicular bisector of the segment connecting the foci. Also, let the distance between the foci be $2c$ and the fixed sum of the distances from P to the foci be $2a$. (In the figure below, $PF_1 + PF_2 = 2a$.)

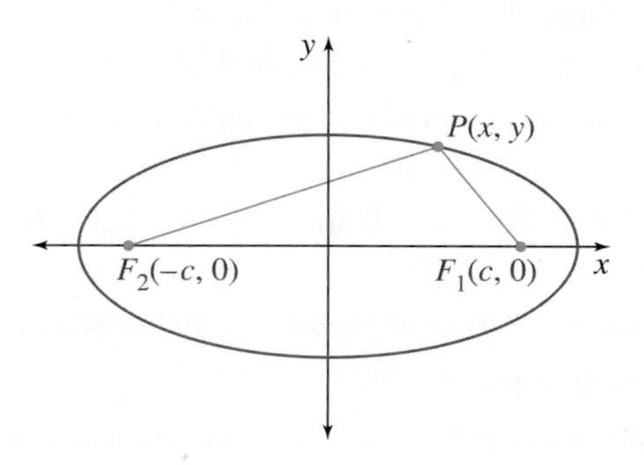

a. Use the distance formula to find an equation of the ellipse.

b. Eliminate the square roots in your answer to part (a). (You will need to square twice.) Show that the equation of the ellipse can be written as $(a^2 - c^2)x^2 + a^2y^2 = a^2(a^2 - c^2)$. Why, in spite of squaring twice, is this equation equivalent (has the same graph) to the equation you obtained in part (a)?

c. Explain why $a > c$. (Show that $2a > 2c$.)

d. Denote $a^2 - c^2$ by b^2, and show that the equation of the ellipse can be written as $\dfrac{x^2}{a^2} + \dfrac{y^2}{b^2} = 1$.

e. How can b be interpreted geometrically?

f. Show that the equation of the circle with center at (0, 0) follows from the equation of the ellipse.

Similarity 4

We could use up two Eternities in learning all that is to be learned about our own world and the thousands of nations that have arisen and flourished and vanished from it. Mathematics alone would occupy me eight million years.

—Mark Twain, Notebook #22, Spring 1883–September 1884

Introduction

An undistorted photograph of an object, what we call its *image,* has the same shape as the original object. We say that the image is similar to the object. If the ratio between the height and width of a window is 2, for example, the ratio will remain 2 in the photo of the window. This suggests the following definition:

Definition of Similar Figures Two figures are similar if and only if there exists a one-to-one correspondence between the figures such that the ratio between any two distances within one figure is the same as the ratio between the corresponding pair of distances in the other figure.

In Chapter 5, we will introduce the concept of size transformation and give a useful definition of similarity using transformations that is equivalent to the preceding definition.

4.1 Ratio, Proportion, and Similar Polygons

A ratio is a quotient of two real numbers. If $r \neq 0$, the number $\frac{s}{r}$ is the ratio of s to r. A proportion is an equation stating that two or more ratios are equal. We state some basic properties of ratios and proportion, most of which will already be familiar to you. (The proofs are left as exercises.)

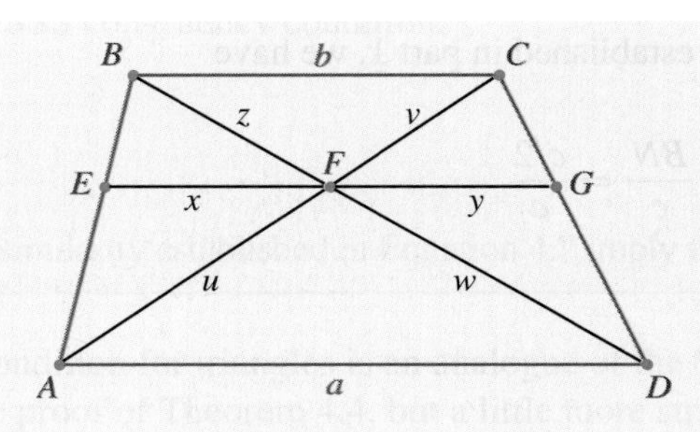

Figure 4.11

Solution

We find EF first. For convenience, we denote $EF = x$, $FG = y$, $AF = u$, $FC = v$, $BF = z$, and $FD = w$. To find x, we look for two similar triangles, one of which has a side of length x. One such pair of triangles is $\triangle AEF \sim \triangle ABC$. Consequently,

$$\frac{EF}{BC} = \frac{AF}{AC} \quad \text{or} \quad \frac{x}{b} = \frac{u}{u+v} \tag{4.10}$$

To find x in terms of a and b, it would be desirable to express u and v in terms of a and b, Notice that u, v, z, and w are lengths of sides of the similar triangles $\triangle BFC$ and $\triangle DFA$. From this similarity, we get

$$\frac{u}{v} = \frac{a}{b} \tag{4.11}$$

We could now substitute $u = \frac{a}{b}v$ in Equation 4.10 or write Equation 4.10 in an equivalent form containing $\frac{u}{v}$. We will demonstrate the latter approach. We divide the numerator and the denominator of the right side of Equation 4.10 by v and then substitute $\frac{a}{b}$ for $\frac{u}{v}$:

$$\frac{x}{b} = \frac{u}{u+v} = \frac{u/v}{u/v + v/v} = \frac{a/b}{a/b + 1} = \frac{a}{a+b}$$

Hence

$$x = \frac{ab}{a+b} \tag{4.12}$$

Analogous calculations result in $y = \dfrac{ab}{a+b}$ and, therefore,

$$EG = x + y = \frac{2ab}{a+b}$$

The quantity $\frac{2ab}{a+b}$ in Example 4.2 is referred to as the **harmonic mean** of a and b. To see a connection between the harmonic mean and the arithmetic mean, let $h = \frac{2ab}{a+b}$. Then $\frac{1}{h} = \frac{1}{2}(\frac{1}{a} + \frac{1}{b})$ and hence $\frac{1}{h}$ is the arithmetic mean of $\frac{1}{a}$ and $\frac{1}{b}$. Consequently, $\frac{1}{a}, \frac{1}{h}, \frac{1}{b}$ are in an arithmetic sequence.

Thus $\frac{1}{3}, \frac{1}{4}, \frac{1}{5}$ are in a harmonic sequence because the reciprocals 3, 4, 5 are in an arithmetic sequence.

Remark We encountered the arithmetic mean in Chapter 1 (the midsegment in a trapezoid). The harmonic mean will appear in several problems in this chapter. We will also introduce the geometric mean in this chapter. These means are used in various areas of mathematics, in physics, and in statistics.

Now Solve This 4.5

Example 4.2 can be solved in several different ways, two of which are suggested below.

1. Complete the argument and the derivation. Which solution do you like better, the one in Example 4.2 or the one below? Why?

   Through F, construct a line parallel to $\overline{AB}$ intersecting the bases of the trapezoid in I and H, respectively (see Figure 4.12). Then $BI = AH = x$, $IC = b - x$, and $HD = a - x$. We have

$$\frac{b-x}{x} = \frac{IF}{FH} = \frac{x}{a-x}$$

   Now solve the equation for x.

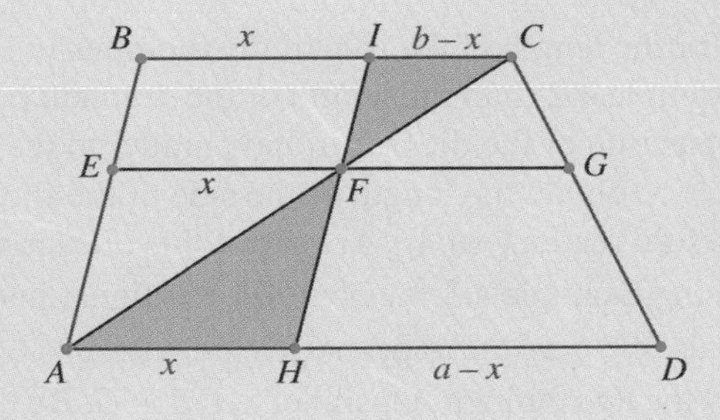

Figure 4.12

2. Solve the problem in a way similar to the one in part 1 but by drawing the line through F parallel to $\overline{CD}$.

Example 4.3 A rhombus $DEFC$ is inscribed in $\triangle ABC$ as shown in Figure 4.13. Find the length of a side of the rhombus in terms of a, b, or c, the lengths of the sides of $\triangle ABC$.

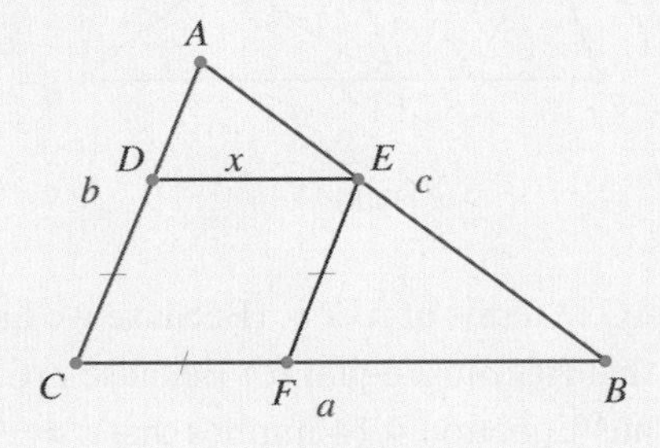

Figure 4.13

Solution

Let the length of a side of the rhombus be x. Because $\triangle DAE \sim \triangle FEB$, we have

$$\frac{DA}{EF} = \frac{DE}{FB}$$

Because $DA = b - x$ and $FB = a - x$, the last proportion is equivalent to each of the following:

$$\frac{b - x}{x} = \frac{x}{a - x} \tag{4.13}$$

$$x^2 = (a - x)(b - x)$$

$$x^2 = ab - ax - bx + x^2$$

$$x(a + b) = ab$$

$$x = \frac{ab}{a + b}$$

Reflections on Example 4.3

Notice that the expression for the length of the side of the rhombus in Example 4.3 is the same as the expression for EF in Example 4.2 (half the length of the segment obtained by the intersection of the line through the intersection of the diagonals and parallel to the bases of the trapezoid). It seems that we should be able to obtain the length of the side of the rhombus from the solution to Example 4.2. But how? (Before reading on, try to answer this question on your own.)

To find x in Figure 4.13 using Example 4.2, we complete the figure into a trapezoid with base $\overline{CB}$ and $\overline{AB}$ as one of the diagonals. We want the trapezoid to contain points C and E. Thus we extend $\overline{CE}$ in Figure 4.14 until it intersects the line through A, parallel to $\overline{CB}$ at G. By the result of Example 4.2,

$$x = \frac{a \cdot AG}{a + AG} \tag{4.14}$$

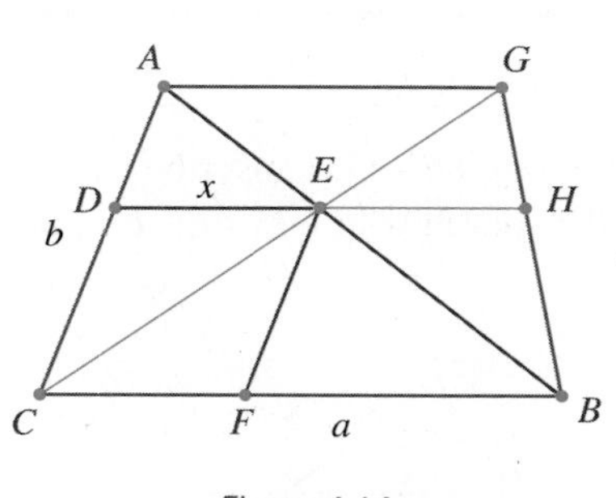

Figure 4.14

It remains now to express AG in terms of a or b. (Because we have solved the problem earlier, we know from Equation 4.13 that AG must equal b.) Because $\triangle CAG \sim \triangle CDE$, and $DE = DC$, we must have $AG = AC = b$. Thus Equation 4.14 implies that $x = \frac{ab}{a + b}$.

Problem Set 4.1

•**1.** In the figure below, if $\angle ABC \cong \angle ADE$, $AB = 8$, $BE = 14$, $AC = 12$, $CD = x$, and $ED = y$, find x and y if possible.

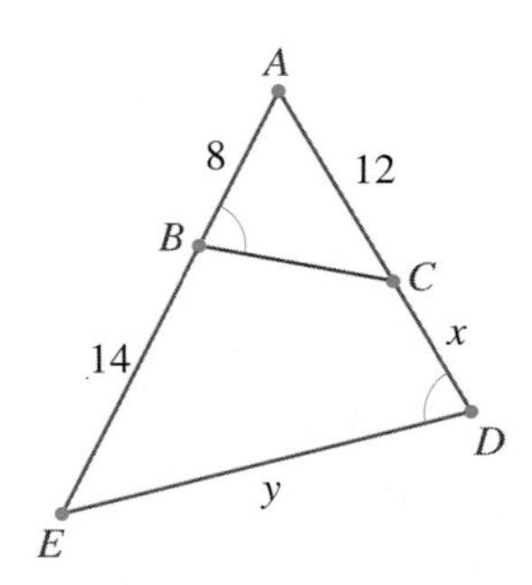

2. BD and AE are heights in $\triangle ABC$. List as many pairs of similar triangles as possible. Justify your similarity statements.

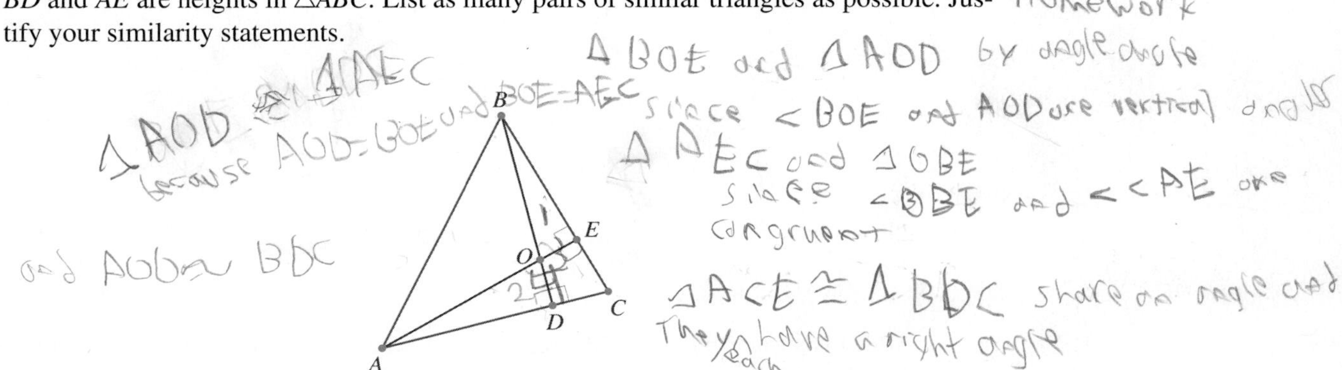

3. Let $ABCD$ be a trapezoid with parallel bases AD and BC. The diagonals meet at O.

a. Prove that $\dfrac{AO}{CO} = \dfrac{OD}{BO}$.

b. Based on the proportion in part (a) and the congruence of the vertical angles $\angle BOA$ and $\angle COD$, does it follow that $\triangle BOA$ and $\triangle COD$ are similar by the SAS similarity condition (Theorem 4.6)? Explain why or why not.

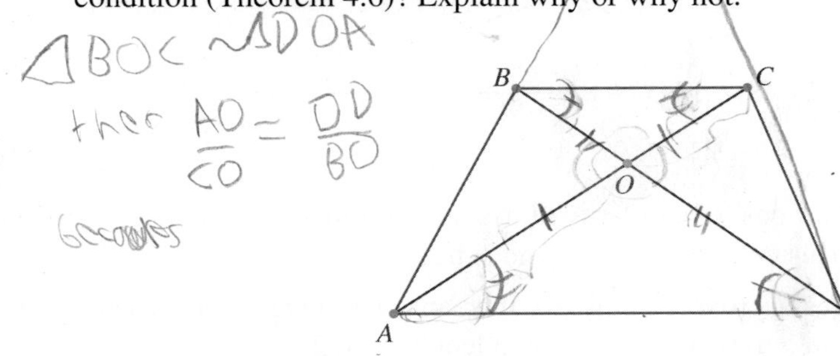

4. In the figure, $\overline{BC} \parallel \overline{B_1C_1}$ and $\overline{AB} \parallel \overline{A_1B_1}$. Prove that

a. $\triangle ABC \sim \triangle A_1B_1C_1$

b. $\overline{AC} \parallel \overline{A_1C_1}$

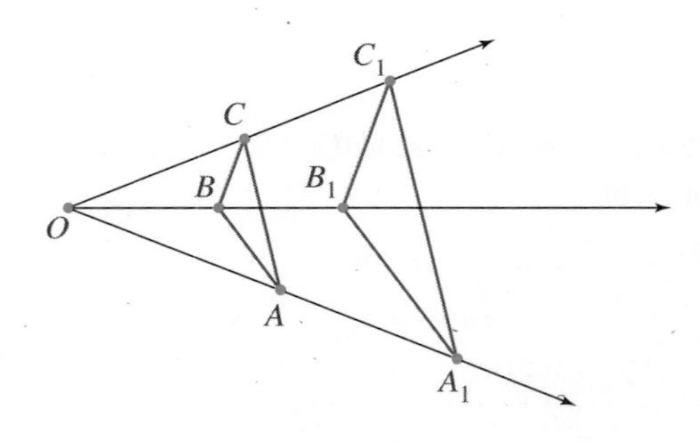

•**5.** In the figure, $\overline{PQ} \parallel \overline{BC}$.

•**a.** Prove that $\dfrac{PS}{SQ} = \dfrac{BT}{TC}$.

•**b.** If $\overline{AT}$ is a median in $\triangle ABC$, what can be concluded from the proof in part (a)? Explain.

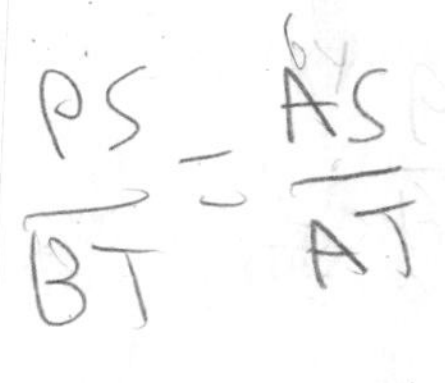

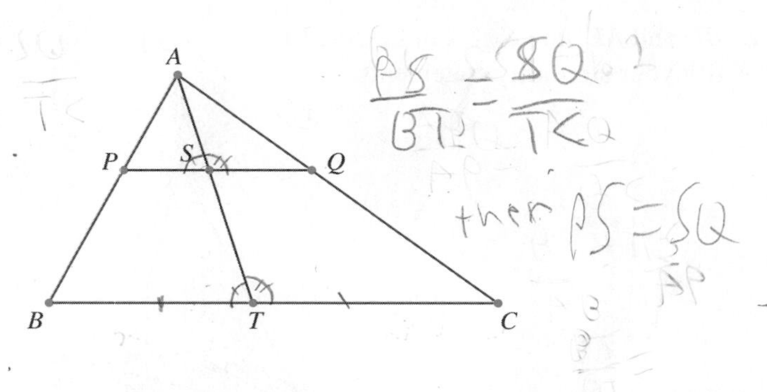

•**6.** *ABCD* is a trapezoid. The sides *AB* and *CD*, when extended, meet at *P*. The line through *P* and *O* (the intersection of the diagonals) intersects the bases of the trapezoid at *M* and *N*. Prove that *M* and *N* are the midpoints of the bases.

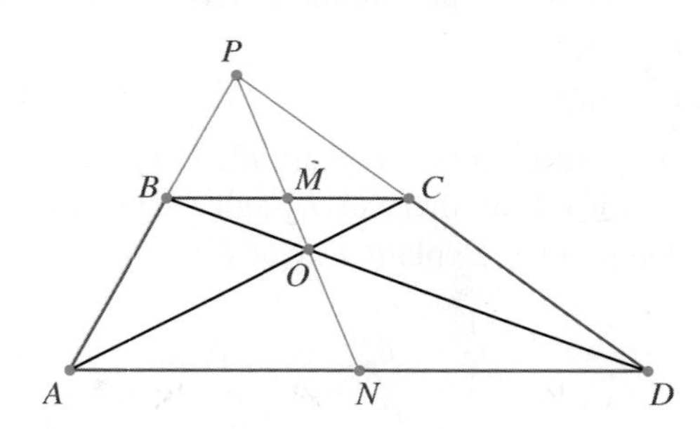

7. a. Use one of the conditions for similarity of triangles to prove that a midsegment of a triangle is parallel to a side of the triangle and is half the length of that side.

b. Use similar triangles to prove that any two medians of a triangle intersect at a point that divides each median into two segments with length ratio 2:1.

c. Use your result from part (b) to prove that the three medians of a triangle are concurrent.

8. *ABCD* is a trapezoid whose diagonals intersect at *O*. When extended, the sides *AB* and *CD* of the trapezoid meet at *E*. A line parallel to the bases *AD* and *BC* has been drawn through *E*. That line intersects lines *BD* and *AC* at points *F* and *G*, respectively. If $AD = a$ and $BC = b$, do the following:

a. Find FE and EG in terms of a and b.

b. In part (a), after computing just FE, how can you argue that $EG = FE$ without performing additional computations?

c. Through a point P on the side AB, draw a line parallel to the bases. That line intersects $\overline{AC}$, $\overline{BD}$, and $\overline{CD}$ at points Q, R, and S, respectively. Use Sketchpad, other software, or mechanical devices to conjecture a relationship between PQ and RS. (Check for different trapezoids and different positions of P.)

d. Prove your conjecture in part (c).

e. Does a result similar to the one in parts (c) and (d) hold if P is on $\overline{FB}$?

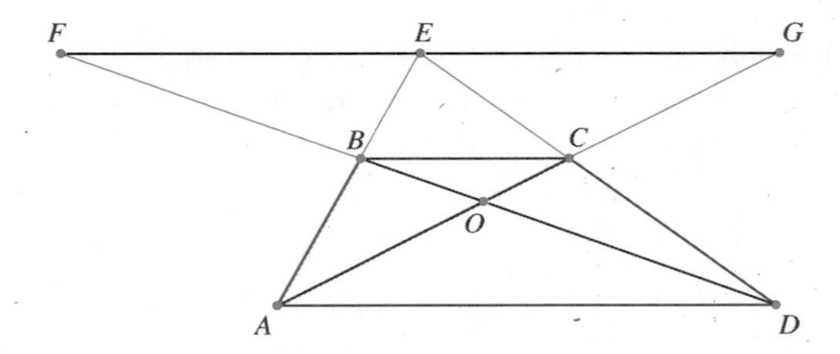

• **9.** $ABCD$ is a rectangle. The segments DF, AC, and BE are perpendicular to line FE.

• **a.** Prove that $AF = AE$.

• **b.** If $DF = a$ and $BE = b$, find FE in terms of a and b.

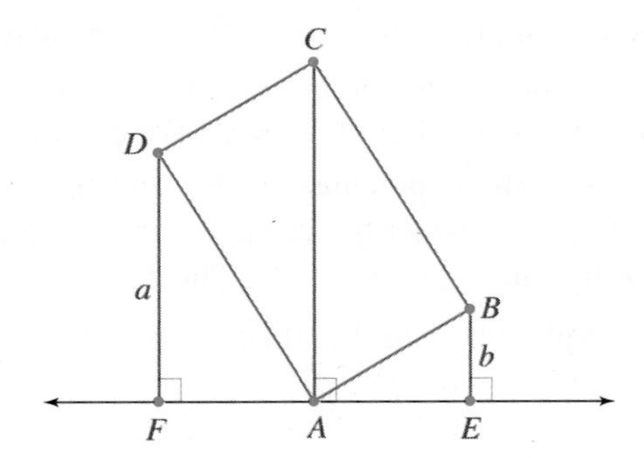

10. Suppose $\triangle ABC \sim \triangle A_1B_1C_1$ with scale factor λ; that is, $\dfrac{AB}{A_1B_1} = \dfrac{BC}{B_1C_1} = \dfrac{AC}{A_1C_1} = \lambda$. Prove that

a. If h and h_1 are the heights to corresponding sides of the triangles, then $\dfrac{h}{h_1} = \lambda$.

b. If r and r_1 are the radii of the corresponding inscribed circles, then $\dfrac{r}{r_1} = \lambda$.

c. If R and R_1 are the radii of the corresponding circumscribing circles, then $\dfrac{R}{R_1} = \lambda$.

d. If P and P_1 are the perimeters of the corresponding triangles, then $\dfrac{P}{P_1} = \lambda$.

• **11.** A square $EFGH$ is inscribed in triangle ABC. Express the side of the square in terms of measurable parts of $\triangle ABC$ (such as sides or heights).

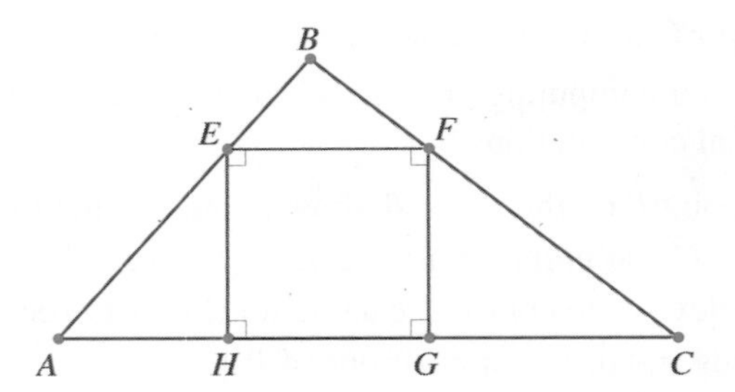

12. *MNOP* is a square with vertices *M* and *N* on side *AC* of $\triangle ABC$ and vertex *O* on side *AB*. There are infinitely many such squares. (This problem will also be solved in Chapter 5 using size transformation.)

- **a.** Prove that the vertices *P* of the squares are collinear.
- **b.** Construct a triangle and then use your result from part (a) to construct a square inscribed in the triangle.

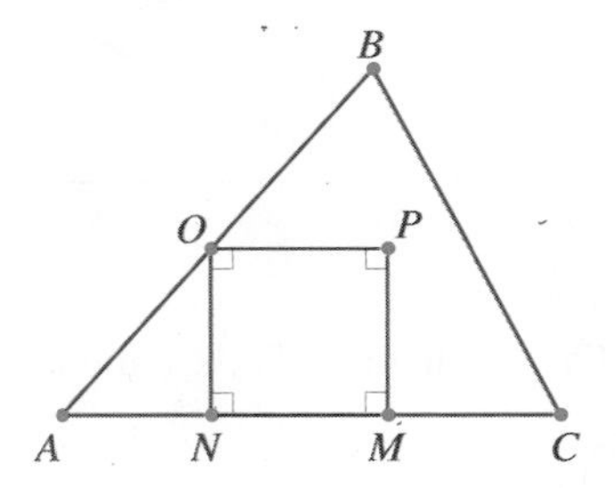

13. The right triangle in the figure has legs of length *a* and *b* and hypotenuse of length *c*. Let $d_1 = a$, d_2 be the length of the perpendicular from *C* to the hypotenuse, and D_2 be the point where that perpendicular intersects the hypotenuse. From D_2, a perpendicular to $\overline{AC}$ is drawn; it intersects $\overline{AC}$ at D_3. Let $D_2D_3 = d_3$. We continue in this way, drawing perpendiculars from points on the hypotenuse to $\overline{AC}$, and from points on $\overline{AC}$ perpendicular to $\overline{AB}$. Denote $C = D_1$, and consider the infinite path $D_1D_2D_3D_4\ldots$ made of segments perpendicular to the hypotenuse or to $\overline{AC}$. The length of the path is the infinite sum $S = \sum_{i=1}^{\infty} d_i$. Let's explore how to find this sum using a geometrical approach.

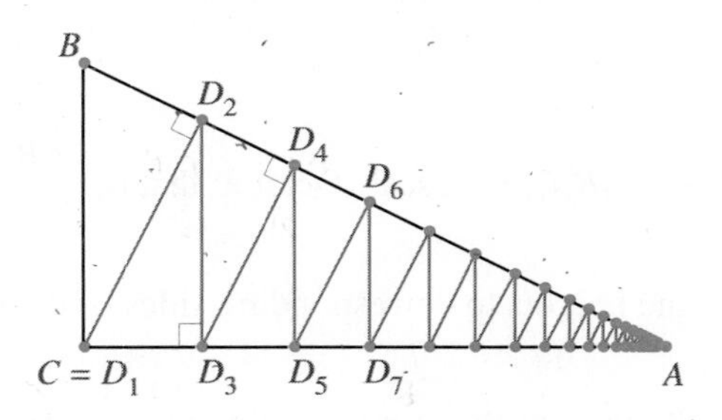

- **a.** Prove that $\dfrac{d_2}{d_1} = \dfrac{d_3}{d_2} = \dfrac{d_4}{d_3}$, and express this ratio in terms of *a*, *b*, or *c*.
- **b.** Explain why $\dfrac{d_k}{d_{k-1}}$ is constant for all $k \geq 2$. Express this constant in terms of *a*, *b*, or *c*.
- * **c.** Prove the assertion in part (b) by mathematical induction.
- **d.** Explain why the path associated with $\triangle CD_2A$ (that is, the path $D_2D_3D_4D_5\ldots$) is similar to the path associated with $\triangle ABC$ (that is, the path $D_1D_2D_3D_4D_5\ldots$).

e. Assume that $S = \sum_{i=1}^{\infty} d_i$ converges, and use your result from part (d) to show that if $\dfrac{d_k}{d_{k-1}} = r$ for $k \geq 2$, then

$$d_2 + d_3 + d_4 + \cdots = r(d_1 + d_2 + d_3 + \cdots)$$

$$S - d_1 = rS$$

$$S = \frac{d_1}{1 - r}$$

f. Using the ideas from part (e), find the sum S_n of a finite geometric sequence as suggested in what follows:

$$\begin{aligned} S_n &= 1 + r + r^2 + \cdots + r^{n-1} = 1 + r(1 + r + r^2 + \cdots + r^{n-2}) \\ &= 1 + r[(1 + r + r^2 + \cdots + r^{n-2} + r^{n-1}) - r^{n-1}] \\ &= 1 + r(S_n - r^{n-1}) \end{aligned}$$

Now solve for S_n.

• **14.** Consider the nested equilateral triangles in the figure, where the vertices of each triangle starting from the second are the midpoints of the sides of the previous triangle. A spiral path is created by connecting vertex A of $\triangle ABC$ with the midpoints $M_1, M_2, M_3, \ldots$ of sides in successive triangles.

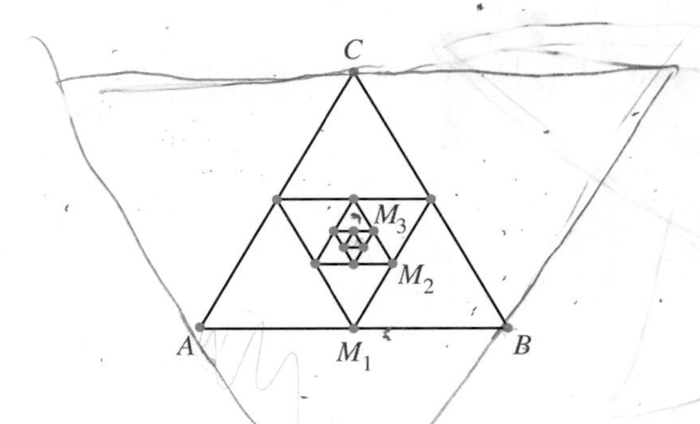

• **a.** Prove that the lengths of the successive segments of the path constitute a geometric sequence.

• **b.** Use the ideas from Problem 13 to find the length of the infinite path.

15. Create an infinite spiral using nested squares, rather than nested triangles as in Problem 14. Find the length of the path.

16. **a.** If the ratio of the corresponding sides of similar triangles is r, what is the ratio of the corresponding areas of the triangles? Prove your answer.

b. In $\triangle ABC$, M_1 is the midpoint of side AC. A line segment is drawn from M_1 parallel to side BC, intersecting side AB at M_2. A segment parallel to side AC is drawn from M_2, intersecting segment BM_1 at M_3. This process continues indefinitely, creating an infinite number of shaded triangles. If the area of $\triangle ABC$ is S, find the total shaded area in terms of S.

c. If $BC = a$ and $AC = b$, find the length of the infinite path $A{-}M_1{-}M_2{-}M_3{-}M_4{-}M_5{-}\cdots$. Why is the answer so simple?

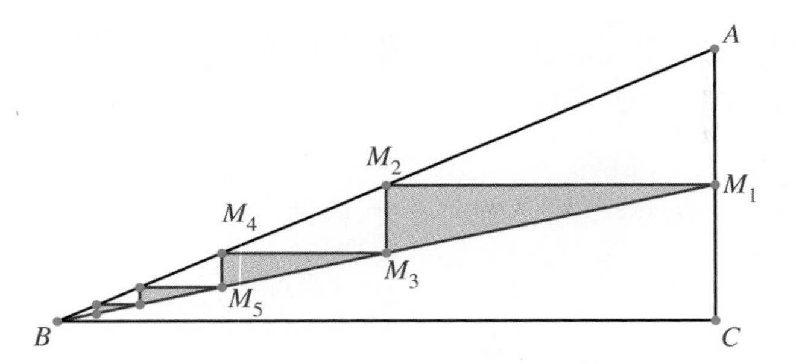

17. Construct an acute triangle ABC, and then inscribe in it a rectangle with one side on $\overline{AC}$ and the vertices of the opposite side on $\overline{AB}$ and $\overline{BC}$, respectively, such that the side of the rectangle on $\overline{AC}$ is twice as long as a side perpendicular to $\overline{AC}$. (See the figure.) Describe your construction and prove that it is valid.

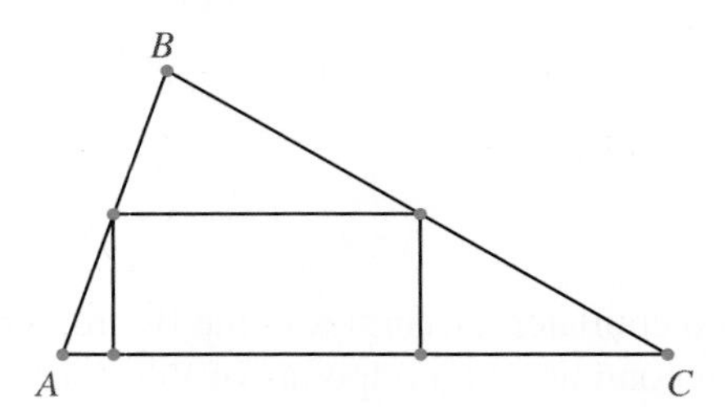

18. Given segments of length a, b, and c, the figure suggests how to construct a segment of length x such that $\frac{a}{b} = \frac{c}{x}$. Describe and justify the construction.

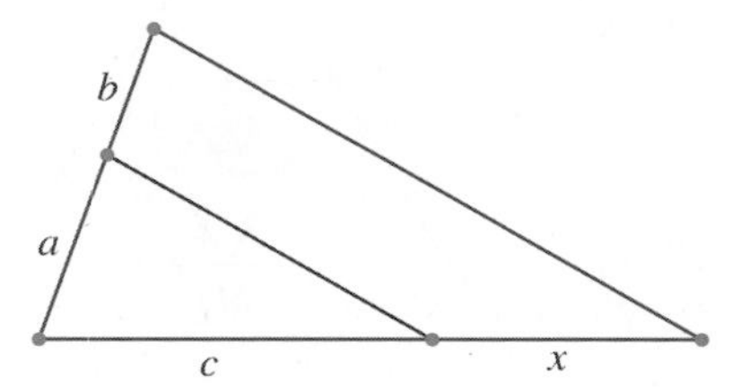

19. In Chapter 3, we developed the equation of a line using the distance formula. Now, after introducing the concept of similarity between triangles, we are ready to introduce the concept of slope and develop the equation of a line using slope.

a. Define the slope of a segment, and develop the equation of a line given two points on the line.

b. Prove that two lines, neither of which is vertical, are perpendicular if and only if the product of their slopes is -1.

4.2 Further Applications of the Side-Splitting Theorem and Similarity

In Example 3.8 (see page 150), we used the distance formula to prove that given two points A and B in the plane, the locus of all points P in the plane that are twice as far from A as from B is a circle. In this section, we will prove this fact without using coordinate geometry. For that purpose, we will need two theorems concerning the angle bisector of an exterior angle.

Theorem 4.7

If an interior angle of a triangle is bisected, the bisector divides the opposite side into segments whose lengths are in the same ratio as the lengths of the other sides of the triangle.

Restatement

Given Referring to Figure 4.15a, we have the fact that $\overrightarrow{AD}$ is the angle bisector of $\angle A$.

Prove

$$\frac{AC}{AB} = \frac{DC}{DB} \tag{4.15}$$

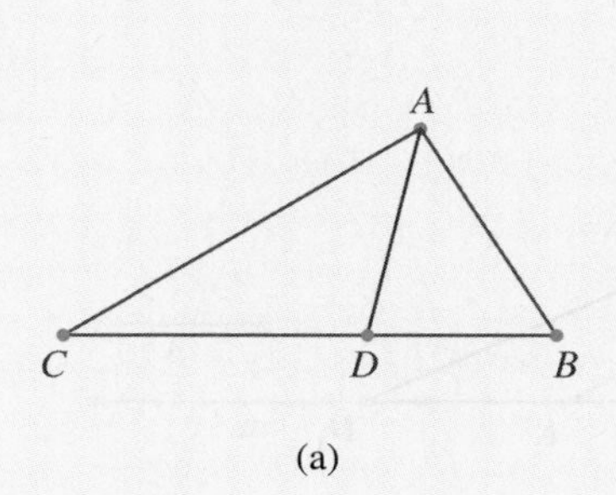

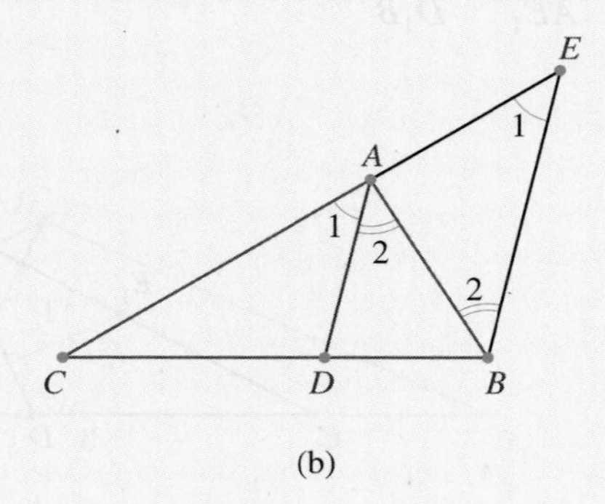

Figure 4.15

Proof

To use the Side-Splitting Theorem, we extend side CA and construct a line parallel to $\overline{AD}$ through B. That line intersects $\overleftrightarrow{CA}$ in E as shown in Figure 4.15b.

By the Side-Splitting Theorem (Theorem 4.2), we have

$$\frac{AC}{AE} = \frac{DC}{DB} \tag{4.16}$$

This proportion will become the one we are trying to prove in Equation 4.15 if $AE = AB$. Thus we need to prove that $\triangle BAE$ is isosceles or, equivalently, that $\angle E_1 \cong \angle B_2$.[2] Notice that

$\angle E_1 \cong \angle A_1$ (corresponding angles formed by the parallel lines EB and AD) (4.17)

We also have

$\angle B_2 \cong \angle A_2$ (alternate interior angles formed by the parallels EB and AD and the transversal AB) (4.18)

Because $\angle A_1 \cong \angle A_2$ (given), the relations in Equations 4.17 and 4.18 imply that $\angle E_1 \cong \angle B_2$. Consequently, $AE = AB$. Substituting AB for AE in Equation 4.16, we get

$$\frac{AC}{AB} = \frac{DC}{DB}$$ □

2. We are using here for the first time a different way to designate angles. There are several angles with vertex B; $\angle B_2$ is the one marked "2" in Figure 4.15b.

We have now proved the following theorem:

Theorem 4.11

If two chords intersect in the interior of a circle, then the product of the lengths of the segments of one chord equals the product of the lengths of the segments of the other chord. Each product equals $r^2 - d^2$, where r is the radius of the circle and d is the distance from the point of intersection of the chords to the center.

Now Solve This 4.9

1. Prove Theorem 4.11 on your own.
2. You can use Theorem 4.11 to find another proof of the Pythagorean Theorem. In Figure 4.22b, let $\overline{AB}$ be perpendicular to $\overline{A_1B_1}$. Then use Theorem 4.11 to prove that $(OP)^2 + (PB)^2 = (OB)^2$.

We now try to find a result analogous to Theorem 4.11 for a point P outside a circle. In Figure 4.23a, two secants are drawn through P outside the circle with center O: one secant intersects the circle at A and B, and the other secant intersects it at A_1 and B_1. We can also think about this situation as having chords $\overline{AB}$ and $\overline{A_1B_1}$ that, rather than intersecting in the interior of the circle, have their extensions intersect in the exterior of the circle.

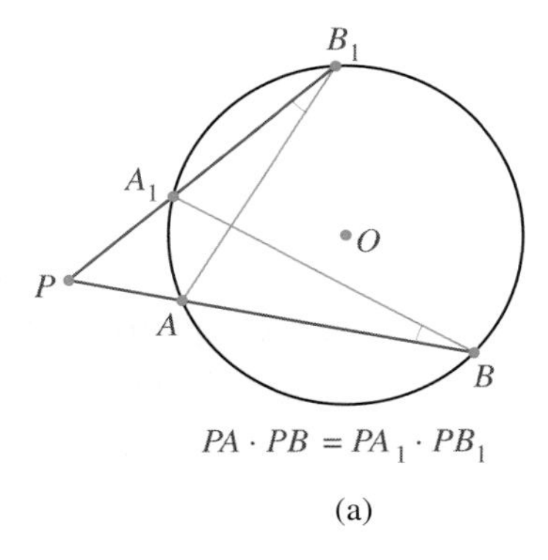

$PA \cdot PB = PA_1 \cdot PB_1$

(a)

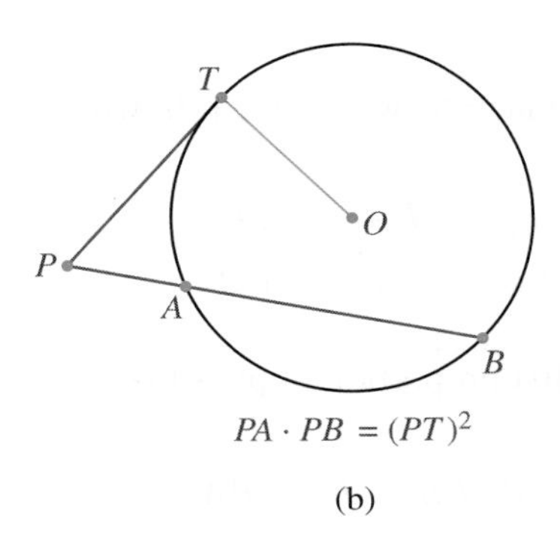

$PA \cdot PB = (PT)^2$

(b)

Figure 4.23

In analogy to Theorem 4.11, we look at $PA \cdot PB$—that is, the product of the distance from P to one endpoint of the chord AB and the distance from P to the other endpoint. We have the following theorem:

Theorem 4.12

If the lines containing two chords (AB and A_1B_1 in Figure 4.23a) in a circle intersect at point P in the exterior of a circle, then the product of the distance from P to one endpoint of a chord (PA) and the distance to the other endpoint (PB) is the same for each chord.

Restatement

Referring to Figure 4.23a, we have the following information:

Given Chords AB andA_1B_1 intersect at P in the exterior of the circle.

Prove $PA \cdot PB = PA_1 \cdot PB_1$.

Proof

To obtain a proportion, we connect some of the points in Figure 4.23a and obtain similar triangles. One option is to connect A_1 with B_1 and AA_1 with B (another option—connecting A with A_1 and B with B_1—is left for you to explore). Notice that $\angle B \cong \angle B_1$ because these inscribed angles intercept the same arc AA_1. Because $\triangle PB_1A$ and $\triangle PBA_1$ also share $\angle P$, they are similar by the AA similarity condition. We have

$$\triangle PB_1A \sim \triangle PBA_1$$

$$\frac{PB_1}{PB} = \frac{PA}{PA_1}$$

$$PB_1 \cdot PA_1 = PB \cdot PA$$

□

Using the concept of limits, we can obtain from Theorem 4.12 an immediate corollary concerning the length of the tangent segment from a point outside a circle. In Figure 4.23b, the secant PB intersects the circle at points A and B. Imagine that the secant PB is rotating about P in a counterclockwise direction. The points A and B get closer to each other and eventually coincide at T, where $\overleftrightarrow{PT}$ is tangent to the circle. In the limit, PA and PB become PT, and we get

$$\begin{aligned} PA \cdot PB &= PT \cdot PT \\ &= (PT)^2 \end{aligned}$$

We have proved the following corollary:

Corollary 4.1 *If the line containing a chord and a tangent intersect at point P outside a circle, then the product of the distance from P to one endpoint of the chord and the distance to the other endpoint equals the square of the length of the tangent segment.*

Problem Set 4.2

• **1.** $\overline{BM}$ and $\overline{NC}$ are medians in $\triangle ABC$. $\overrightarrow{BQ}$ is the angle bisector of $\angle B$, which intersects $\overline{NC}$ at P. Do the following:

• **a.** If $BC = 8$, $AB = 10$, and $NC = 9$, find OP.

• **b.** If $BC = a$, $AB = c$, and $NC = m$, express OP in terms of a, c, and m. Substitute the values of BC, AB, and NC from part (a) into the expression you obtained for OP and confirm that you get the same answer as in part (a).

c. Without using your answer to part (b), find OP if $a = c$. Does your answer agree with the one you can obtain by substituting $a = c$ in the expression for OP you found in part (b)?

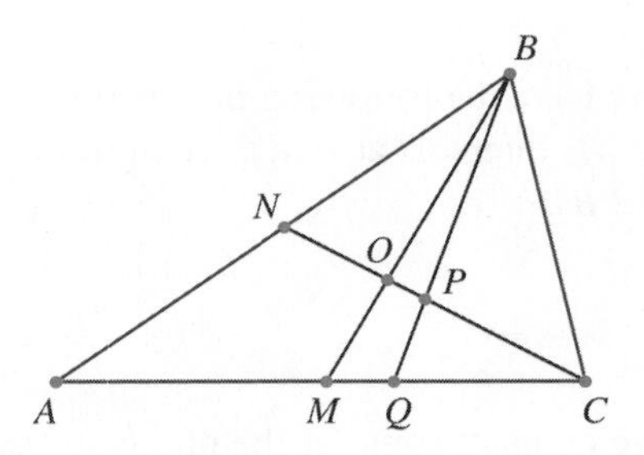

2. In the right triangle ABC, $\overrightarrow{AD}$ is the angle bisector of $\angle A$. Which is greater: BD or DC? Prove your answer.

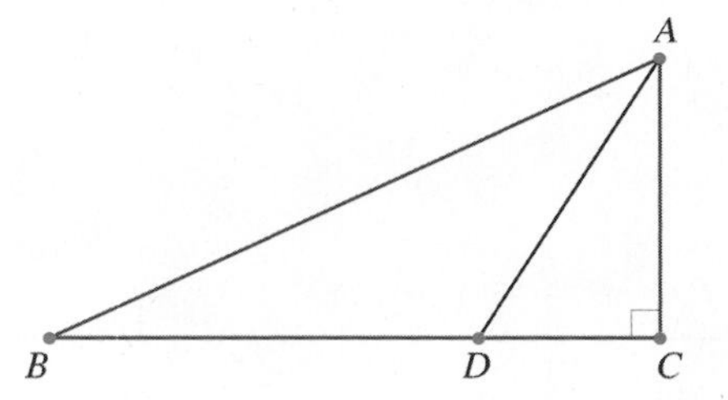

3. $\overline{BD}$ and $\overline{AC}$ are the diagonals of the convex quadrilateral $ABCD$. If $\overrightarrow{BE}$ and $\overrightarrow{BF}$ are the angle bisectors of $\angle ABD$ and $\angle DBC$, respectively, prove that $\overline{EF} \parallel \overline{AC}$. Is this assertion true if the quadrilateral $ABCD$ is not convex? Justify your answer.

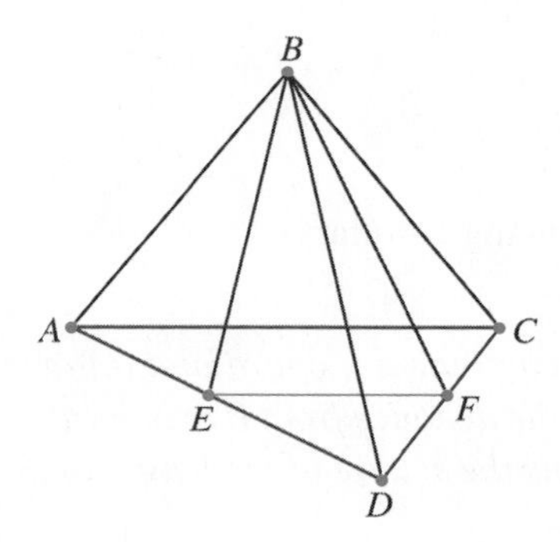

• **4.** In $\triangle ABC$, the angle bisectors of $\angle A$ and $\angle B$ intersect in O.

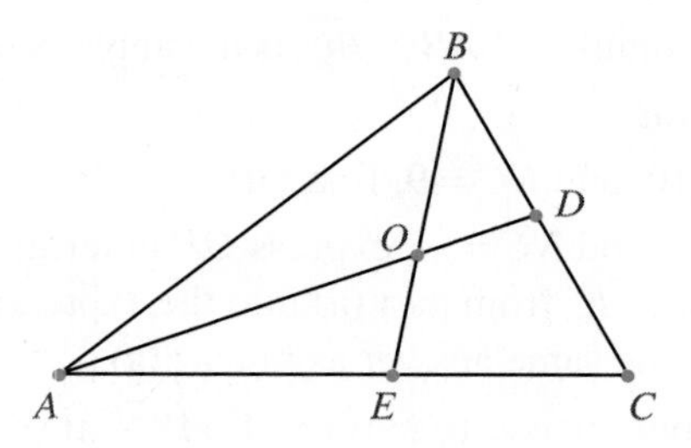

• **a.** If $AB = c$, $BC = a$, and $AC = b$, find $\frac{BO}{EO}$ in terms of a, b, and c.

b. Based on your answer to part (a), and without additional computations, find $\frac{AO}{OD}$. Justify your answer.

5. A, B, and C are points on the circle with center O. The diameter GF is perpendicular to $\overline{AB}$. Lines CF and AB intersect at D, while lines GC and AB intersect at E. Prove that D and E divide $\overline{AB}$ harmonically.

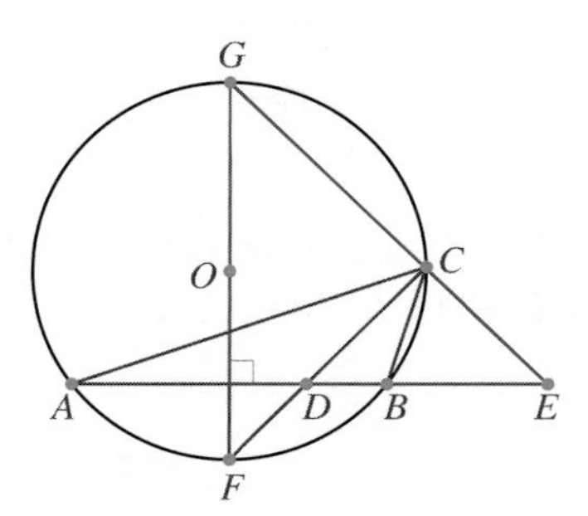

6. $ABCD$ is a trapezoid with two right angles at A and B. The diagonals intersect at P. The line through P parallel to the bases intersects the sides AB and CD at E and F, respectively. Prove that $\overrightarrow{EF}$ is the angle bisector of $\angle CED$.

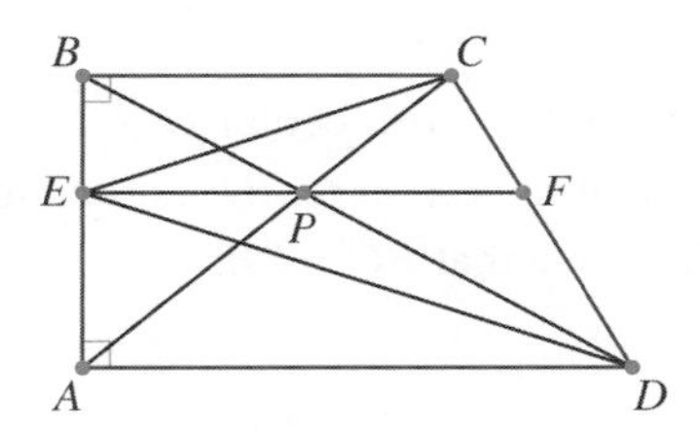

7. In $\triangle ABC$, the altitude CH and the median CM divide $\angle C$ into three congruent angles. Find $\frac{AB}{AH}$ and $m(\angle ACB)$.

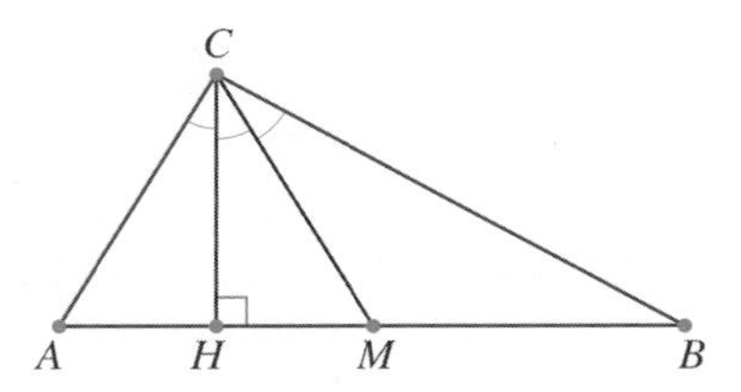

• **8.** M, B, and T are on the circle with center O. A tangent is drawn through T that intersects the line MB at A. Do the following:

• **a.** If $AM = MB = a$, find AT in terms of a.

b. If A is a point in the exterior of the circle, construct a secant through A that intersects the circle in two points M and B such that M is the midpoint of $\overline{AB}$. Investigate the construction, describe the steps in the construction, and prove that the construction is valid.

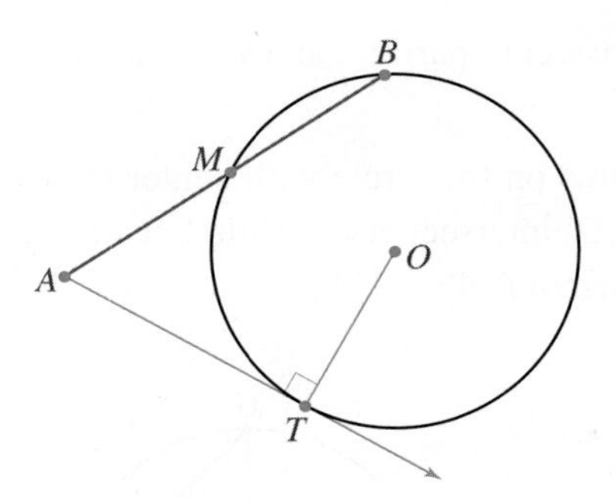

9. An isosceles triangle ABC is inscribed in a circle. $\overline{BD}$ is a chord that intersects $\overline{AC}$ in E. If $AB = BC = x$, $BE = a$, and $BD = b$, find x in terms of a and b.

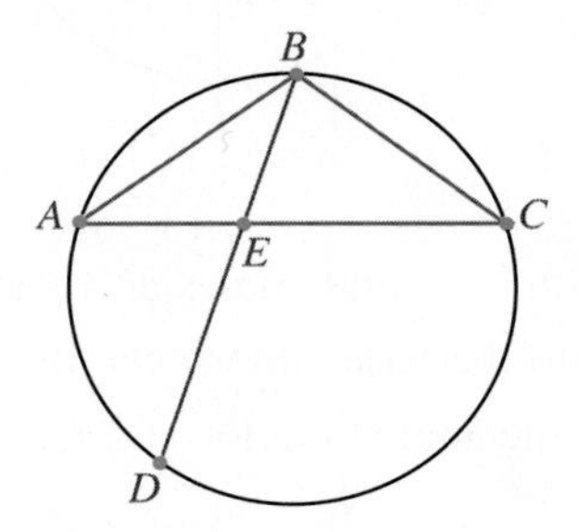

10. $\triangle ABC$ is inscribed in the circle with center O. $\overline{CD}$ is the altitude to the hypotenuse AB, which is also a diameter of the circle O. $\overline{BE}$ is the tangent segment to the smaller circle, whose diameter is $\overline{AD}$. Prove that $BC = BE$.

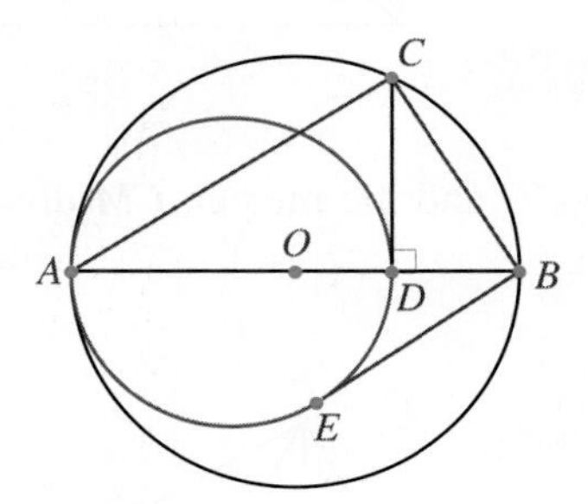

• 11. In the figure, $\angle A$ is a right angle, $\overline{AC}$ is a diameter, $AB = a$, and $AE = b$. Find BD in terms of a and b.

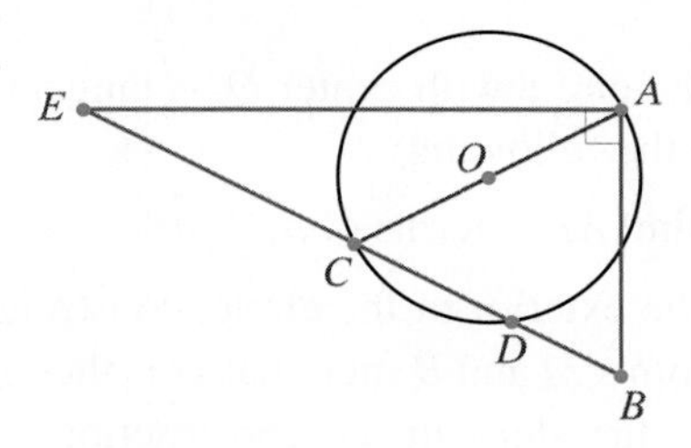

12. $\triangle ABC$ with $AB = BC$ is inscribed in a circle. Point D bisects $\overset{\frown}{BC}$. Lines $\overleftrightarrow{BD}$ and AC intersect at E.

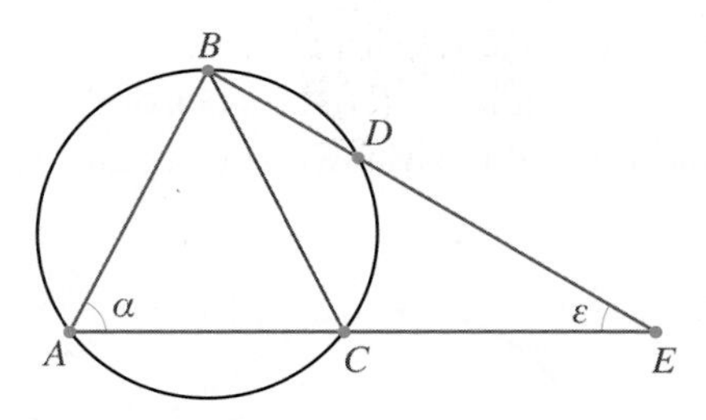

a. Find a relationship between α and ε.

b. If $BE = a$ and $BD = b$, express BC in terms of a and b.

13. $\triangle ABC$ is inscribed in a circle with center O. The tangent t to the circle is constructed at B. From the vertices A and C, perpendiculars to t intersect t at P and Q, respectively. If $AP = a$ and $CQ = b$, express BD in terms of a and b.

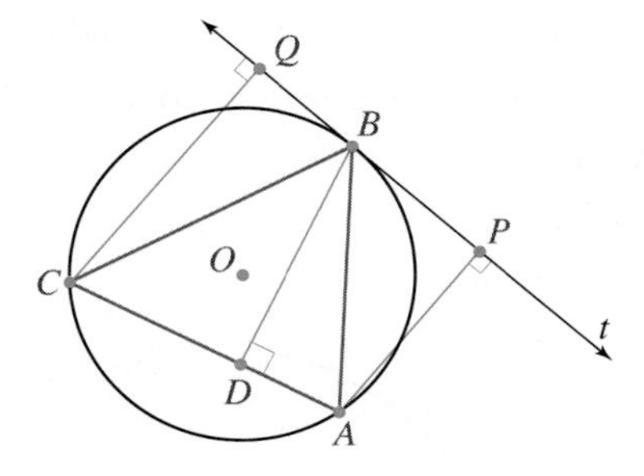

14. T is on one side of $\angle A$, while B and C are on the other side. If $(AT)^2 = AB \cdot BC$, prove that $\overleftrightarrow{AT}$ is tangent to the circle through B, C, and T.

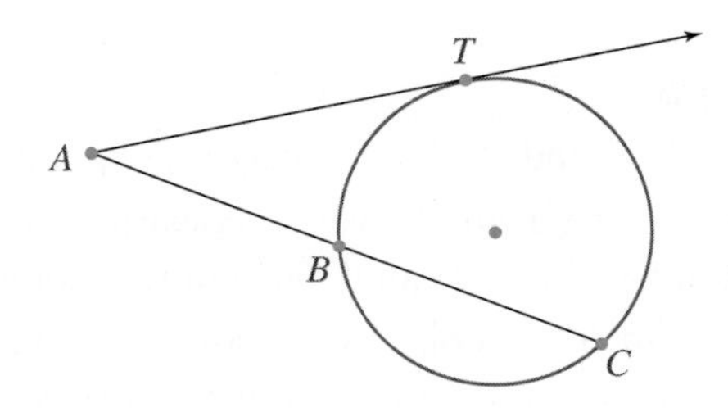

15. $\triangle ABC$ is inscribed in a circle. If the sides of the triangle have lengths a, b, and c, and if the radius of the circle is R, prove that the area of the triangle is $\frac{abc}{R}$.

• **16.** Two circles intersect in points A and B. If C is a point for which the tangent segments CD and CE are congruent, prove that A, B, and C are collinear.

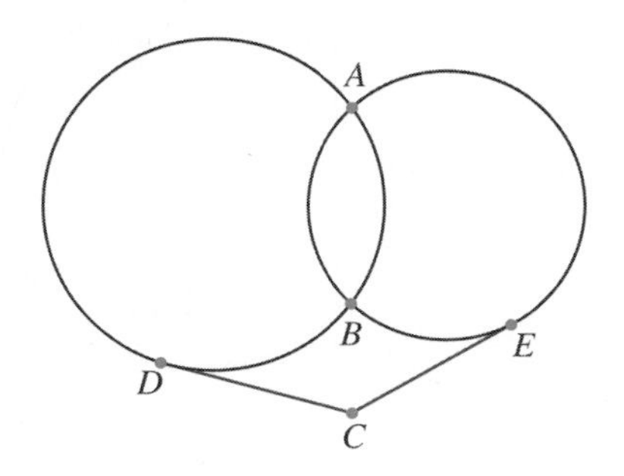

17. Two tangent segments CA and CB are drawn from C outside the circle O. From an arbitrary point P on $\overset{\frown}{AB}$, perpendiculars to the sides of $\triangle ABC$ are constructed. These perpendiculars intersect the sides of the triangle at Q, S, and T, respectively. Prove that $(PT)^2 = PS \cdot PQ$.

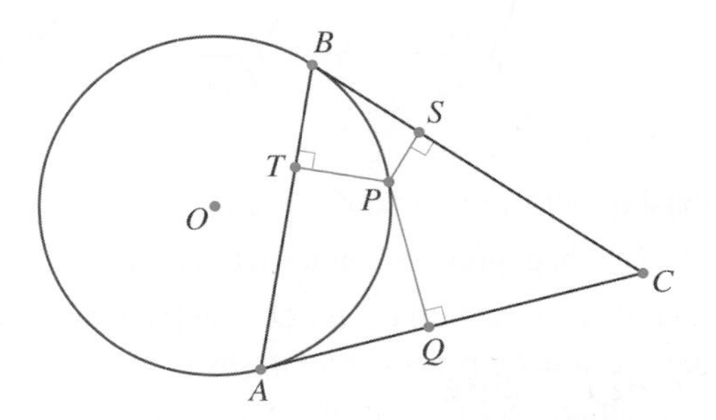

18. The figure shows a semicircle with diameter AC and two other semicircles with diameters AB and BC (B is on $\overline{AC}$). A common tangent to the smaller semicircles intersects one at X and the other at T. Also, $\overleftrightarrow{PB}$ is a common tangent to the smaller semicircles through their common point B.

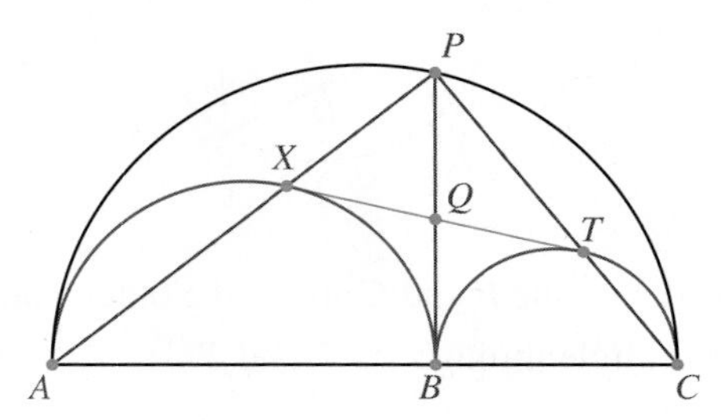

a. Prove that $XT = PB$.

b. Prove that points A, X, and P are collinear, as are points P, T, and C.

c. Parts (a) and (b) suggest a simple way to construct a common external tangent to two tangent circles (circles that have only one point in common). Explain how.

d. Use the ideas in part (c) to construct a common tangent to two intersecting (in two points) circles. Describe your construction and justify it.

• **19.** A square with side d is inscribed in a circle. A smaller square is then inscribed in a lune of the circle (bounded by an arch and a chord), as shown in the figure.

• **a.** Find the side of the smaller square in terms of d.

b. Use your answer to part (a) to construct a circle and the two squares.

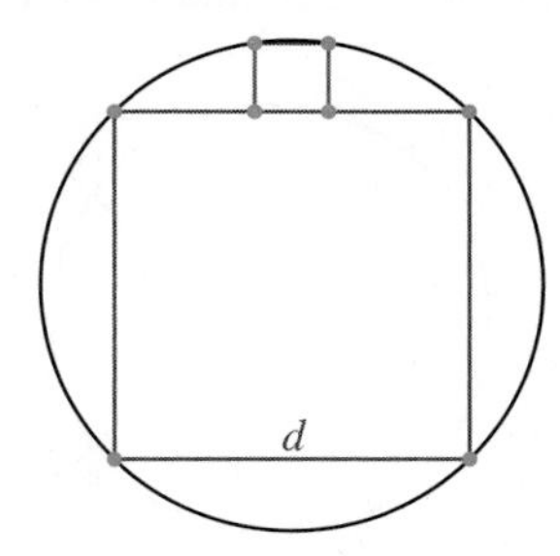

20. Construct a semicircle and two squares inscribed in the semicircle as shown. Investigate, construct, and prove that your construction is valid.

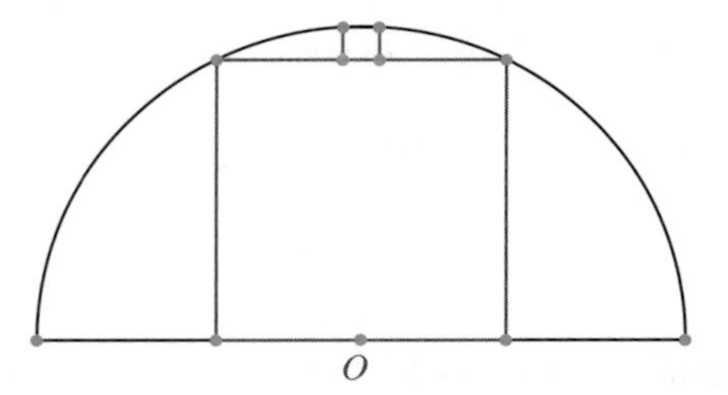

4.3 Areas of Similar Figures

The size of a rectangular television screen is commonly given as the length of the diagonal of the screen. Assuming that in two television screens the rectangular shapes are similar, how many times larger is the viewing area of a 35-inch screen than the viewing area of a 25-inch screen? To answer this question, we need to solve the following problem.

Problem 4.1 Given two similar triangles with similarity ratio r, what is the ratio of their areas?

Solution

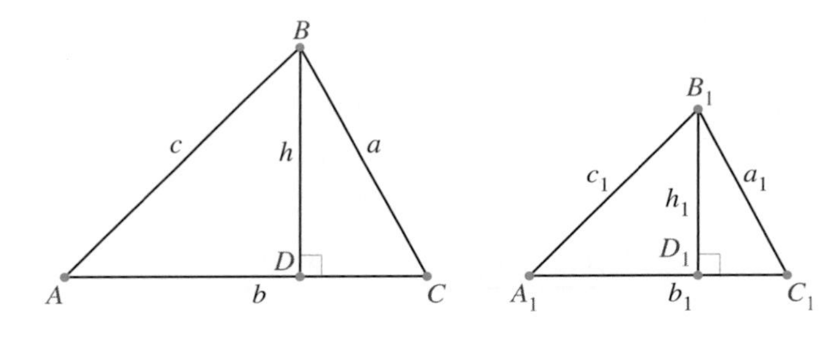

Figure 4.24

In Figure 4.24, $\triangle ABC \sim \triangle A_1B_1C_1$. If

$$\frac{a}{a_1} = \frac{b}{b_1} = \frac{c}{c_1} = r$$

we need to find the ratio between the areas of the triangles. If h and h_1 are the heights corresponding to the sides b and b_1, respectively, we get

$$\frac{\text{Area}(\triangle ABC)}{\text{Area}(\triangle A_1B_1C_1)} = \frac{bh/2}{b_1h_1/2} = \frac{b}{b_1} \cdot \frac{h}{h_1} \tag{4.30}$$

To find how the ratio $\frac{h}{h_1}$ relates to the given ratio r, we consider two triangles that have those heights as sides. One such pair is $\triangle ABD$ and $\triangle A_1B_1D_1$. These triangles are similar by AA (Why?) and therefore

$$\frac{h}{h_1} = \frac{c}{c_1} \tag{4.31}$$

Because $\frac{c}{c_1} = \frac{b}{b_1}$, we see that $\frac{h}{h_1} = \frac{b}{b_1}$. Substituting this last proportion in Equation 4.30, we get

$$\frac{\text{Area}(\triangle ABC)}{\text{Area}(\triangle A_1B_1C_1)} = \frac{b}{b_1} \cdot \frac{h}{h_1} = \left(\frac{b}{b_1}\right)^2 = r^2 \tag{4.32}$$

We have now proved the following theorem:

Theorem 4.13

The ratio of areas of similar triangles equals the square of the ratio of corresponding sides.

Now Solve This 4.10

Is it true that the viewing area of a 35-inch television screen is about twice the viewing area of a 25-inch television screen? Justify your answer.

Example 4.4 In Figure 4.25, M is any point on side AB of $\triangle ABC$. $\overline{MQ}$ and $\overline{MP}$ are parallel to the sides AC and CB, respectively. If the areas of $\triangle MBQ$ and $\triangle AMP$ are S_1 and S_2, respectively, find the area of $\triangle ABC$ in terms of S_1 and S_2.

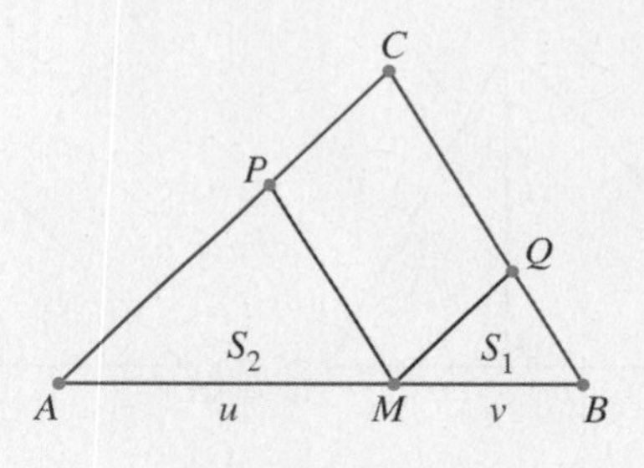

Figure 4.25

Solution

Because $\triangle APM \sim \triangle MQP$ (Why?), the ratio of their areas equals the square of the ratio of corresponding sides (Theorem 4.13). Using the notation in Figure 4.25, we have

$$\frac{S_1}{S_2} = \left(\frac{u}{v}\right)^2 \tag{4.33}$$

Let S denote the area of $\triangle ABC$. Then, because $\triangle ABC \sim \angle MBQ$, we have

$$\begin{aligned}\frac{S}{S_1} &= \left(\frac{AB}{MB}\right)^2 = \left(\frac{u+v}{u}\right)^2 \\ &= \left(1 + \frac{v}{u}\right)^2 \\ &= \left(1 + \sqrt{\frac{S_2}{S_1}}\right)^2 \quad \left(\text{substituting for } \frac{v}{u} \text{ from Equation 4.33}\right)\end{aligned}$$

Consequently,

$$S = S_1\left(1 + \frac{S_2}{S_1} + 2\sqrt{\frac{S_2}{S_1}}\right)$$
$$= S_1 + S_2 + 2\sqrt{S_1 S_2}$$

Therefore $S = S_1 + S_2 + 2\sqrt{S_1 S_2} = (\sqrt{S_1} + \sqrt{S_2})^2$, or in the attractive form $\sqrt{S} = \sqrt{S_1} + \sqrt{S_2}$.

Notice that our answer implies that the area of the parallelogram $PMQC$ is $2\sqrt{S_1 S_2}$.

Example 4.5 Given $\triangle ABC$, construct a line parallel to one of the sides of the triangle that divides the triangle into two regions: a triangle and a trapezoid of equal areas.

Solution

Figure 4.26 shows $\overline{DE}$, which halves the area of $\triangle ABC$. We need to find the location of D.

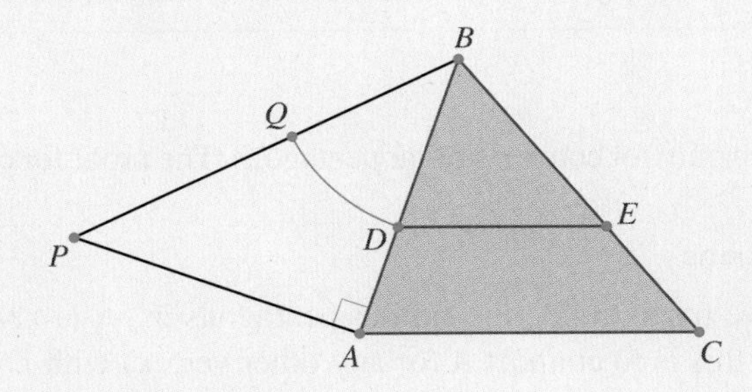

Figure 4.26

We need the area of $\triangle DEB$ to be half of the area of $\triangle ABC$. Because $\triangle DEB \sim \triangle ACB$, the ratio of the areas of the triangles is the square of the ratio of corresponding sides. Thus

$$\left(\frac{DB}{AB}\right)^2 = \frac{1}{2} \tag{4.34}$$

$$\frac{DB}{AB} = \sqrt{\frac{1}{2}} = \frac{\sqrt{2}}{2}$$

$$DB = \frac{AB\sqrt{2}}{2}$$

We know that $AB\sqrt{2}$ is the length of the hypotenuse of an isosceles right triangle whose legs have length AB. In Figure 4.26, we construct a right triangle PAB such that

$PA = AB$. Applying the Pythagorean Theorem, we get $PB = AB \cdot \sqrt{2}$. To obtain a segment congruent to $\overline{DB}$, Equation 4.34 suggests finding the midpoint Q of $\overline{PB}$. Then the arc with center at B and radius $\overline{BQ}$ intersects $\overline{AB}$ at the desired point D. A line through D parallel to $\overline{AC}$ intersects $\overline{BC}$ at E. Line DE is the required line.

We can now generalize Theorem 4.13 to the ratios of areas of similar polygons. Because similar polygons can be divided into corresponding similar triangles, it seems that the ratio of their areas is still the square of the ratio of corresponding sides. This is, in fact, true for both convex and non-convex similar polygons, but we will prove the relevant theorem for convex polygons only.

Ratio of Areas of Similar Polygons

Theorem 4.13, which deals with the ratios of areas of similar triangles, has an analog for similar polygons.

Theorem 4.14

The ratio of areas of similar polygons equals the square of the ratio of corresponding sides.

We prove the theorem only for convex similar pentagons. The proof for convex n-gons is similar.

Proof for Convex Pentagons

To use Theorem 4.13, we try to divide the similar pentagons in Figure 4.27 into similar triangles. One way to accomplish this is to connect A (or any other vertex) with D and C, and A' with D' and C'. Notice that $\triangle 1$ and $\triangle 1'$ are similar by the SAS similarity condition because

$$\frac{AB}{A'B'} = \frac{BC}{B'C'} \quad \text{and} \quad \angle B \cong \angle B'$$

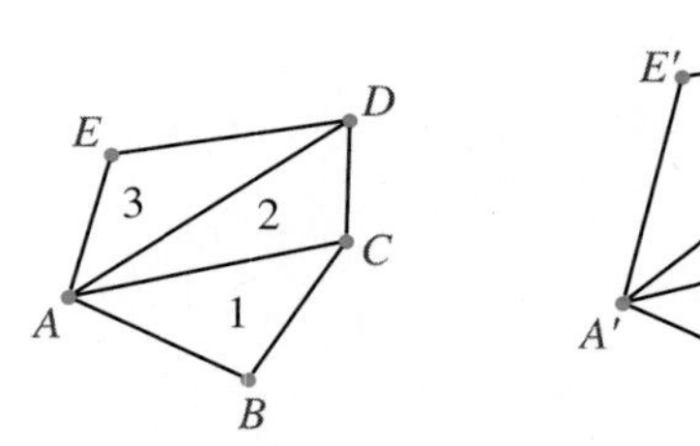

Figure 4.27

In the same way, $\triangle 3$ and $\triangle 3'$ are similar. Next, $\triangle 2$ and $\triangle 2'$ are similar by the AA similarity condition because $\angle ACD \cong \angle A'C'D'$ and $\angle ADC \cong \angle A'D'C'$ (this follows from the just-proved fact that triangles 1 and $1'$ are similar, triangles 3 and $3'$ are similar, and the angles at D and C of the first pentagon correspond to the angles at D' and C' of the second pentagon). If we designate the areas of triangles 1, 2, and 3 by S_1, S_2, and S_3, respectively, and the areas of triangles $1'$, $2'$, and $3'$ by $S_{1'}$, $S_{2'}$, and $S_{3'}$, respectively, then

$$\frac{S_1}{S_{1'}} = \left(\frac{BC}{B'C'}\right)^2 = \lambda^2$$

$$\frac{S_2}{S_{2'}} = \left(\frac{DC}{D'C'}\right)^2 = \lambda^2$$

$$\frac{S_3}{S_{3'}} = \left(\frac{DE}{D'E'}\right)^2 = \lambda^2$$

From the ratio of the sum of numerators to the sum of denominators property introduced in Section 4.1, we have

$$\frac{S_1 + S_2 + S_3}{S_{1'} + S_{2'} + S_{3'}} = \lambda^2$$

and thus

$$\frac{S}{S'} = \lambda^2$$

where S and S' are the areas of $AEDCB$ and $A'E'D'C'B'$, respectively. □

Now Solve This 4.11

Construct a convex quadrilateral and then a quadrilateral similar to it whose area is three-fourths of the area of the original quadrilateral. Investigate the construction, describe the construction, and prove that it is valid.

A Generalization of the Pythagorean Theorem

The Pythagorean Theorem states that the area of the square on the hypotenuse of a right triangle equals the sum of the areas of the squares on the legs. We might wonder if the assertion about the sum of the areas will hold for other figures constructed on the sides of a right triangle. In Figure 4.28, notice that we can divide each square in half by its midsegment. The resulting rectangles marked I, II, and III are figures constructed on the sides of $\triangle ABC$. Because each is half of the corresponding square, the area of the rectangle on the hypotenuse equals the sum of the areas of the rectangles on the sides of the triangle. We could now use the diagonals to divide each rectangle into two congruent triangles. The shaded triangles again satisfy the additive area property.

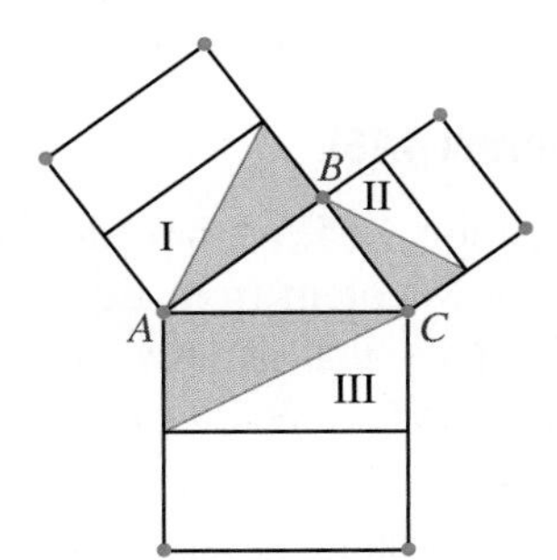

Figure 4.28

What do the squares, the rectangles, and the triangles have in common? The shapes are similar, which suggests the following theorem:

Theorem 4.15

If three similar polygons are constructed on the three sides of a right triangle, then the area of the polygon on the hypotenuse equals the sum of the areas of the polygons on the other sides.

Proof

We denote the areas of the polygons marked I, II, and III in Figure 4.29 by S_1, S_2, and S_3, respectively. From Theorem 4.14, we have

$$\frac{S_1}{S_3} = \frac{b^2}{c^2}$$

$$\frac{S_2}{S_3} = \frac{a^2}{c^2}$$

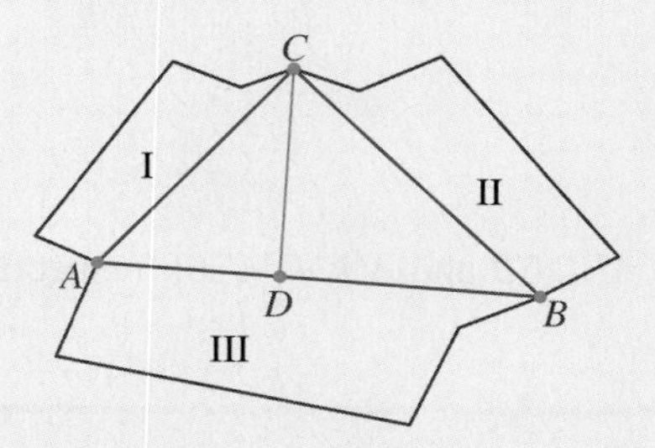

Figure 4.29

To obtain $S_1 + S_2$, we add the equations side by side:

$$\frac{S_1 + S_2}{S_3} = \frac{a^2 + b^2}{c^2} \tag{4.35}$$

Because $\triangle ABC$ is a right triangle, $a^2 + b^2 = c^2$ and Equation 4.35 becomes

$$\frac{S_1 + S_2}{S_3} = \frac{c^2}{c^2} = 1$$

$$S_1 + S_2 = S_3$$

□

Theorem 4.15 is a generalization of the Pythagorean Theorem because the latter is a special case of Theorem 4.15 (when the similar polygons are squares).

Looking Back (at the Proof of Theorem 4.15)

Notice that up to and including Equation 4.35, we have not used the Pythagorean Theorem. In fact, we could prove the Pythagorean Theorem from Equation 4.35 if we knew that for some similar polygons their areas are additive—that is, they satisfy the equation $S_1 + S_2 = S_3$. Such polygons are the shaded triangles in Figure 4.30. We saw earlier that the altitude to the hypotenuse ($\overline{CD}$ in Figure 4.30) divides a right triangle into two similar triangles, each of which is similar to the original triangle. These triangles satisfy the additive area property because the area of $\triangle ABC$ equals the sum of the areas of $\triangle ACD$ and $\triangle CDB$. To better see that these triangles

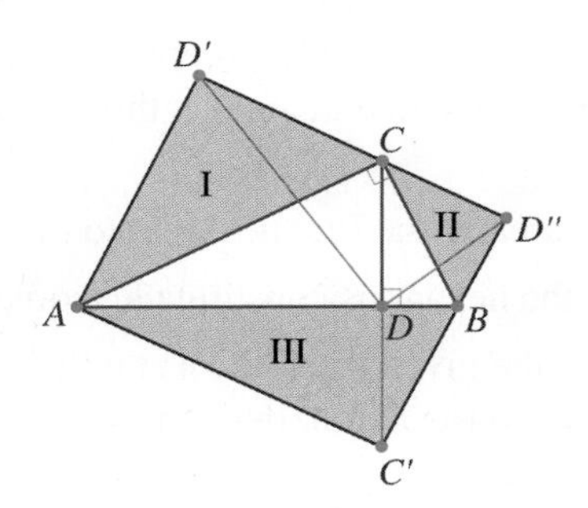

Figure 4.30

are constructed on the sides of $\triangle ABC$, we reflect each in its hypotenuse and obtain the shaded similar triangles in Figure 4.30.

Because the shaded triangles are congruent to $\triangle ACD$, $\triangle CDB$, and $\triangle ABC$, respectively, their areas satisfy the additive property—that is, $S_1 + S_2 = S_3$. Substituting this value in Equation 4.35, we get

$$1 = \frac{a^2 + b^2}{c^2}$$

$$c^2 = a^2 + b^2$$

Thus we have yet another proof of the Pythagorean Theorem.

Problem Set 4.3

1. In the interior of $\triangle ABC$, lines parallel to the sides of the triangle are drawn through a point P. They divide the triangle into six parts, three of which are triangles. If these triangles have areas S_1, S_2, and S_3, respectively, find the area of $\triangle ABC$.

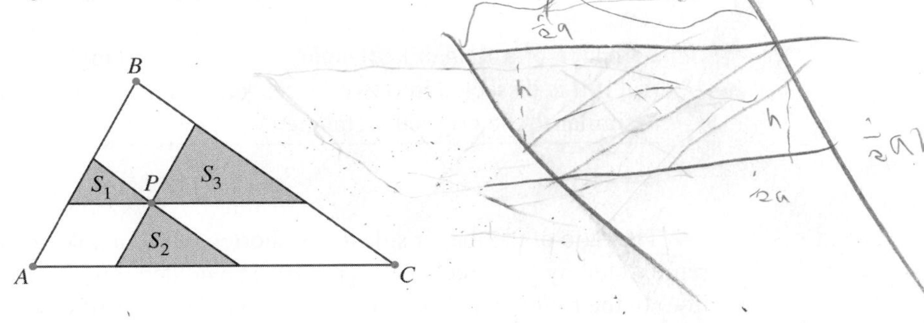

2. In $\triangle ABC$, $AB = BC$ and the angle bisector of $\angle A$ intersects $\overline{BC}$ at D. If the areas of $\triangle ABD$ and $\triangle ADC$ are S_1 and S_2, respectively, find AC in terms of S_1 and S_2. (*Hint:* Write three equations with three unknowns and solve for AC.)

3. Given a parallelogram, construct two lines parallel to one of the diagonals and divide the parallelogram into three parts of equal area.

4. Given an arbitrary triangle (construct your own scalene triangle), divide it into n parts of equal area using lines parallel to one of the sides. Solve the problem for $n = 2, 3, 4$, and 5.

•**5.** In a regular hexagon of side a, a second hexagon is inscribed by joining the midpoints of the sides of the first hexagon. In a like manner, a third hexagon is inscribed in the second, and so on indefinitely.

•**a.** What is the ratio of the area of the first hexagon to the area of the sixth hexagon?

b. Prove that the areas of the hexagons constitute a geometric sequence.

★**6.** Construct a line parallel to the given line (line n) that will divide $\triangle ABC$ into two parts of equal area. (Investigate the construction, describe your construction, and prove that it is valid.)

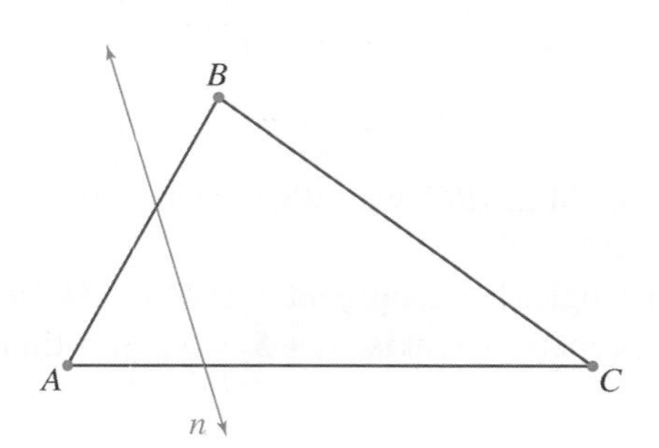

•**7.** •**a.** Prove that if a and b are sides of a parallelogram whose diagonals are d_1 and d_2, then $2(a^2 + b^2) = d_1^2 + d_2^2$.

b. Explain why the result in part (a) can be regarded as a generalization of the Pythagorean Theorem.

4.4 The Golden Ratio and the Construction of a Regular Pentagon

The front of the Parthenon of ancient Greece (built in 438 B.C.E.) has a shape that has been considered most pleasing to the eye. Similar rectangles have been used in art and architecture. A simple way to define such a triangle is as follows.

Definition of a Golden Rectangle A golden rectangle is a rectangle with the property that if it is dissected into two pieces—a square and a rectangle—then the new rectangle is similar to the original rectangle.

The ratio of the longer side to the shorter side of a golden rectangle is the **golden ratio**, often represented by the Greek letter ϕ (phi). The golden ratio is also referred to as the divine proportion. In the following problem, we calculate the value of ϕ.

Problem 4.2 Find the numerical value of ϕ.

Solution

Figure 4.31 shows a golden rectangle $ABCD$ dissected into the square $ABEF$ and the rectangle $ECDF$. The rectangles are similar if and only if the ratio of two adjacent sides in one rectangle equals the ratio of two corresponding adjacent sides in the other rectangle. Consider the ratio $\frac{a}{b}$ (see Figure 4.31), which we want to compute.

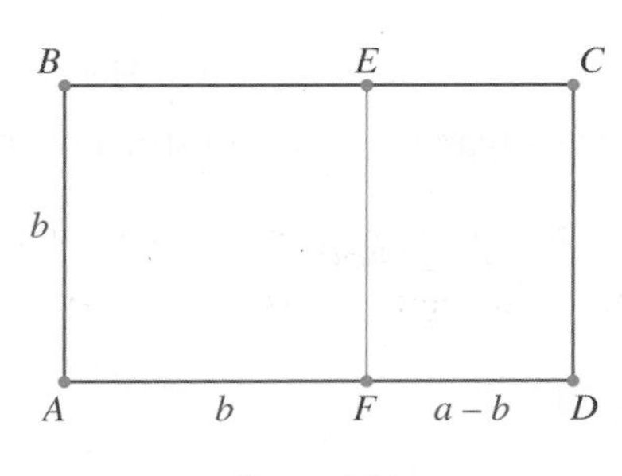

Figure 4.31

The possible ratios of adjacent sides in rectangle $ECDF$ are $\frac{a-b}{b}$ and $\frac{b}{a-b}$. Which one is equal to $\frac{a}{b}$? Notice that if $\frac{a}{b} = \frac{a-b}{b}$, then $a = a - b$, which is impossible. Thus

$$\frac{a}{b} = \frac{b}{a-b} \tag{4.36}$$

One way to find $\frac{a}{b}$ (which we decided earlier to call ϕ) is to divide the numerator and the denominator of the right side of Equation 4.36 by b. We get

$$\frac{b}{a-b} = \frac{\frac{b}{b}}{\frac{a-b}{b}} = \frac{1}{\frac{a}{b} - 1}$$

Substituting in Equation 4.36, we have

$$\frac{a}{b} = \frac{1}{\frac{a}{b} - 1}$$

$$\phi = \frac{1}{\phi - 1} \tag{4.37}$$

$$\phi^2 - \phi - 1 = 0$$

Using the quadratic formula, we obtain $\phi = \frac{1 \pm \sqrt{5}}{2}$. Since ϕ is positive, we conclude that $\phi = \frac{1+\sqrt{5}}{2}$.

Remark Notice that in Problem 4.2 for any given b (see Figure 4.31), we can find a segment of length a such that $\frac{a}{b}$ is the golden ratio.

Construction 4.1: Construction of a Golden Rectangle

Given the shorter side of a golden rectangle, construct the rectangle.

Solution Given a segment of length b, we need to construct a segment of length a such that $\frac{a}{b} = \frac{1+\sqrt{5}}{2}$ or $a = \frac{b}{2} + \frac{b\sqrt{5}}{2}$. Clearly, constructing $\frac{b}{2}$ poses no difficulty. To construct a segment of length $\frac{b\sqrt{5}}{2}$, let $x = \frac{b\sqrt{5}}{2}$. Then $x^2 = \frac{5}{4}b^2 = b^2 + \left(\frac{b}{2}\right)^2$. Thus x is the hypotenuse of a right triangle with sides b and $\frac{b}{2}$.

In Figure 4.32, $\overline{AB}$ has length b, $AE = \frac{b}{2}$, and $\overline{AE} \perp \overline{AB}$. Hence $BE = \frac{b\sqrt{5}}{2}$. We mark $\overline{ED} \cong \overline{BE}$ and, therefore, $AD = \frac{b}{2} + \frac{b\sqrt{5}}{2}$. It is sufficient now to construct the point C such that $BC = AD$ and $CD = AB$. □

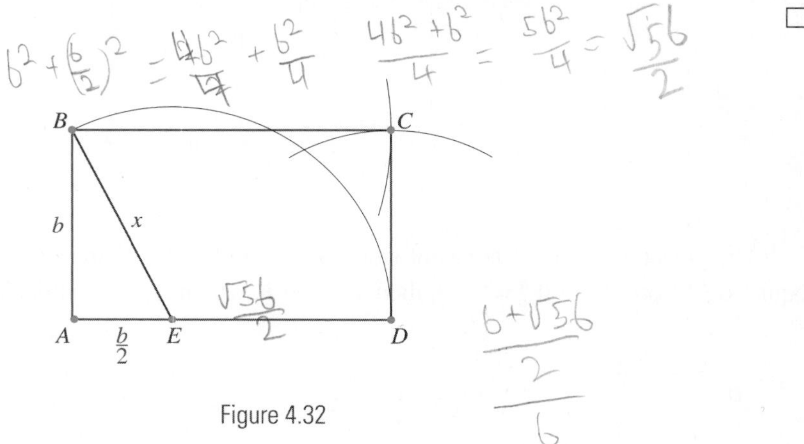

Figure 4.32

Notice that in the above construction we did not distinguish among investigation, construction, and proof of the construction. These parts are embedded in the solution.

Now Solve This 4.12

1. Construct a segment AB of any length. Now construct the point C on the segment so that $\frac{AC}{BC} = \phi$. Write your solution in three stages: investigation, construction, proof of the construction.
2. Prove that $\frac{AB}{CB} = \phi$.
3. Using your results from parts 1 and 2, state an alternative definition of ϕ.

Properties and Construction of a Regular Pentagon

Figure 4.33 shows a regular pentagon (all sides congruent and the five interior angles congruent) and two diagonals. Each of the exterior angles of the pentagon is $\frac{360°}{5}$ or 72°. Hence each interior angle is 180° − 72° or 108°. Let each side be 1 unit long. If we could express the length of a diagonal using rational numbers and radicals of rational numbers, we should be able to construct $\triangle ABE$—and hence the entire pentagon.

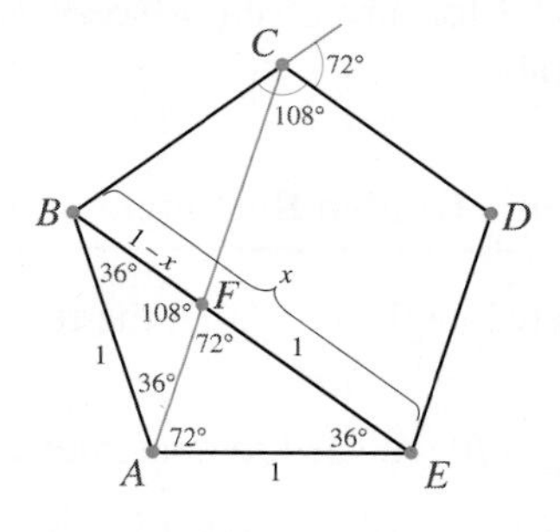

Figure 4.33

The diagonals BE and AC intersect at F. Notice that triangles ABC and BAE are congruent by SAS. To compute the length of a diagonal, we must search for similar triangles and, therefore, for congruent angles. Because each of the interior angles of the pentagon is 108°, the base angles of the isosceles triangles ABC and BAE are 36° each, as shown in Figure 4.33. Also, $\angle AFE$ is an exterior angle of $\triangle BFA$; hence it measures 36° + 36° or 72°. $\angle FAE$ is 108° − 36° or 72°; consequently, $\triangle FAE$ is isosceles, which implies $FE = 1$. If $BE = x$, then $BF = x - 1$. We get

$$\triangle ABF \sim \triangle EBA$$

$$\frac{AB}{EB} = \frac{BF}{BA}$$

$$\frac{1}{x} = \frac{x-1}{1}$$

or equivalently

$$x^2 - x = 1$$

$$x^2 - x - 1 = 0$$

whose positive solution is $x = \frac{1+\sqrt{5}}{2} = \phi$. Thus we have proved the following property:

Property of a Regular Pentagon The ratio of the length of a diagonal of a regular pentagon to the length of its side is the golden ratio.

Remark If the pentagon's side is a units long and a diagonal is b units long, we still have $\frac{b}{a} = \phi = \frac{1+\sqrt{5}}{2}$. Why?

Construction 4.2: Construction of a Pentagon with Any Given Side

The preceding property and remark can be used to construct a regular pentagon with any given side. To accomplish the construction, we need simply construct $\triangle ABE$ as shown in Figure 4.33. Given any side of length a, the diagonal is $(\frac{1+\sqrt{5}}{2})a$. Using the construction outlined in Problem 4.2, we know how to construct the diagonal and, consequently, $\triangle ABE$ by the SSS condition. To complete the construction of the pentagon, we construct vertex C (see Figure 4.33) by constructing $\triangle ABC$ congruent to $\triangle ABE$ (or by constructing $\angle B$ congruent to $\angle A$ and $\overline{BC}$ congruent to $\overline{AB}$), and similarly the rest of the vertices.

An alternative construction in which we find the center O of the circle that circumscribes the pentagon is shown in Figure 4.34. The main steps of the construction are as follows:

1. Given a side a of the pentagon, construct an isosceles triangle ABE with $AB = AE = a$ and $BE = a(\frac{1+\sqrt{5}}{2})$. (See Construction 4.1.)
2. Find the center O of the circle that circumscribes $\triangle ABE$.
3. Mark points C and D on the circle such that $BC = CD = a$.
4. Connect the points A, B, C, D, and E to obtain the required regular pentagon.

Notice that the circle with center O and radius a is not necessary for the construction. It is constructed in Figure 4.34 as a check that the construction is accurate.

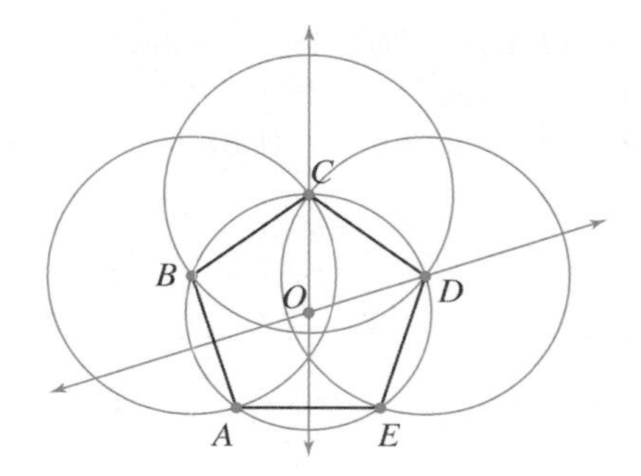

Figure 4.34

Now Solve This 4.13

Given a circle, construct a regular pentagon inscribed in the circle using the following approaches. (Describe the steps in each construction and prove that the resulting construction is valid.)

1. Use the construction of a regular pentagon with a given side and the concept of similarity.
2. Without using the construction of a regular pentagon, investigate the construction of a regular decagon (10-sided polygon) inscribed in a given circle. Figure 4.35 shows one of the 10 congruent triangles that are created by connecting the vertices of the decagon with the center of the circle. $\angle BAO$ has been bisected, and the intersection of the bisector with $\overline{BO}$ is C. Express x in terms of R.

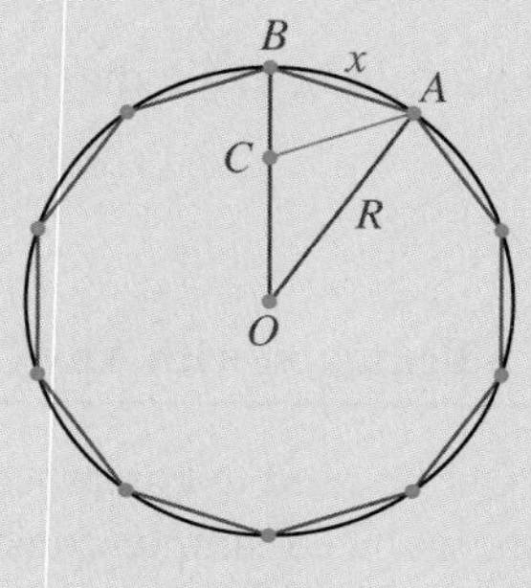

Figure 4.35

Construction of Other Regular Polygons

The construction of a regular n-gon is equivalent to subdividing a circle into n equal parts, which in turn is equivalent to constructing an angle of $\frac{360°}{n}$. (Why?) The constructions of an equilateral triangle, a square, and a regular hexagon are quite simple; these constructions were investigated in Chapter 1 and will be reviewed in the problem set at the end of this section. So far in this section, we have seen how to construct regular pentagons and decagons. These constructions were known to Euclid. In Book IV of his *Elements*, Euclid discusses the construction of regular n-gons for $n = 3, 4, 5, 6$, and 15. By successive bisections of arcs or angles, the Greeks knew how to construct regular polygons of 2^n, $3 \cdot 2^n$, $5 \cdot 2^n$, and $15 \cdot 2^n$ sides. But despite their many attempts, no one knew how to construct other regular polygons.

In 1796, Gauss stated that *a regular n-gon* ($n \geq 3$) *can be constructed by straight edge and compass if and only if n is a prime number of the form* $F(n) = 2^{2^n} + 1$ *(Fermat prime) or the product of such primes and powers of* 2. For $n = 0, 1, 2, 3, 4$, $F(n) = 3, 5, 17, 257, 65537$ are all prime. No other Fermat primes have been found.

At age 19, Gauss proved that this condition is sufficient but gave no explicit proof that it is necessary. In 1836, the French mathematician Pierre Wantzel proved that the condition is necessary. These proofs are beyond the scope of this text. Gauss's proof can be found in some number theory texts (e.g., Ore, Dummit, and Foote, pp. 581–583).

Historical Note

Carl Friedrich Gauss (1777–1855), a German mathematician, astronomer, and physicist, made major contributions to geometry and other branches of mathematics as well as to statistics, electricity, magnetism, and gravitation. He completed his doctoral thesis when he was only 22. At age 19, he proved the result mentioned previously about constructability of regular polygons. While a student, Gauss also showed how to construct a regular 17-gon with only a compass and straight edge; he was so proud of this discovery that he asked that a regular 17-gon be engraved on his tombstone.

Many mathematical and physical terms are named after Gauss, including Gaussian curvature, Gaussian distribution (also commonly called normal distribution), Gaussian domain, Gaussian elimination, Gaussian field, Gaussian function, Gaussian integer, Gaussian plane, Gauss–Jordan elimination, Gauss's lemma, Gauss–Seidel iteration, Gauss's test, and Gauss's fundamental theorem of electrostatics. When Napoleon invaded Germany, he decided to spare Göttingen—the city where Gauss lived (and was a professor at the university there)—commenting that "the foremost mathematician of all times lives there."

In 1837, **Pierre Wantzel** (1814–1848), a French mathematician, became the first person to prove the impossibility of trisecting any given angle by means of only a straight edge and compass. He also proved the impossibility of several other constructions. At age 15, Wantzel gave a proof of a method for finding square roots that was widely used but previously unproved. In 1832, he placed first in the entrance examinations to both École Polytechnique and École Normale (two of the most prestigious institutions of higher learning in France)—something that no one else had ever achieved. His friend Saint-Venant wrote the following about Wantzel:

> One could reproach him for having been too rebellious against those counseling prudence. He usually worked during the evening, not going to bed until late in the night, then reading, and got few hours of agitated sleep, alternatively abusing coffee and opium, taking his meals, until his marriage, at odd and irregular hours.

Problem Set 4.4

All constructions should include an investigation, a description of the construction, and a proof that the construction is the required one.

1. Prove that the ratio of the length of the longer side to the length of the shorter side in a triangle with angles 36°, 72°, and 72° is the golden ratio. (*Hint:* Bisect one of the 72° angles.)

2. a. Given a segment AB, construct the point P on the segment that divides it into two segments such that the shorter is to the longer as the longer is to the whole. (You need P so that $\frac{PB}{AP} = \frac{AP}{AB}$.)

   b. Use your answer to part (a) to give an alternative definition of the golden ratio.

3. If ϕ is the golden ratio, show that $\frac{1}{\phi} = \phi - 1$.

4. An equilateral triangle ABC is inscribed in a circle. Points M and N are midpoints of sides AC and AB, respectively, and line MN meets the circle at P and Q. Prove that $\frac{MN}{NQ}$ is the golden ratio.

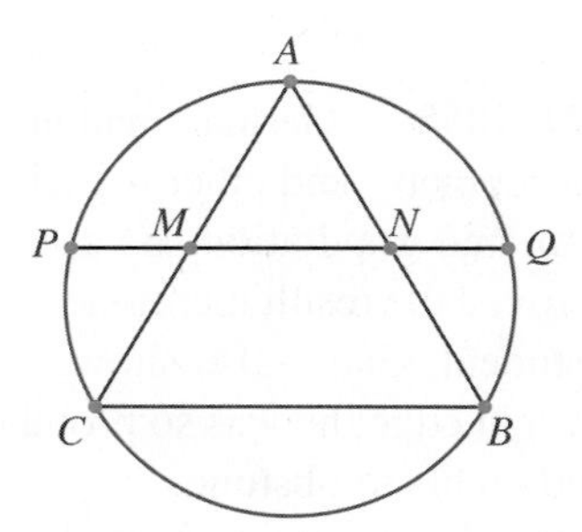

•**5.** $ABCDE$ is a regular pentagon. Sides CB and EA have been extended to meet at P. If each side of the pentagon is a units long, find BP in terms of a.

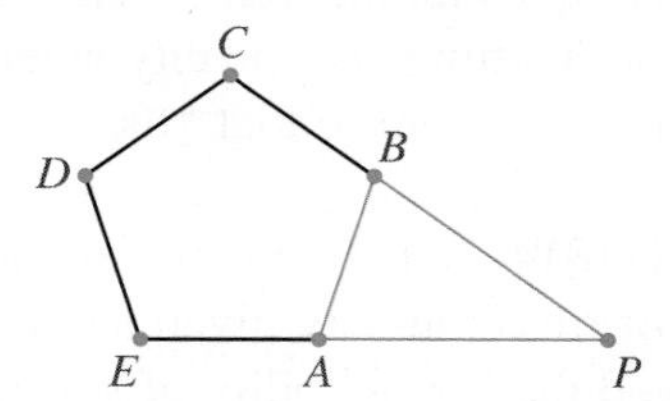

6. Referring to the figure below, the following is a step-by-step construction of a regular pentagon inscribed in a circle. Prove that the construction is valid.

a. Choose any point A on the circle and connect it to the center O.

b. Through O, construct the perpendicular line to $\overline{OA}$. Let P be one of the points where the perpendicular line intersects the circle.

c. Find the midpoint M of $\overline{OA}$. Connect M with P.

d. Construct the angle bisector of $\angle OMP$, and let N be the point where the angle bisector meets $\overline{OP}$.

e. At N, construct the perpendicular to $\overline{OP}$ intersecting the circle at B as shown. The segment BP is a side of the pentagon.

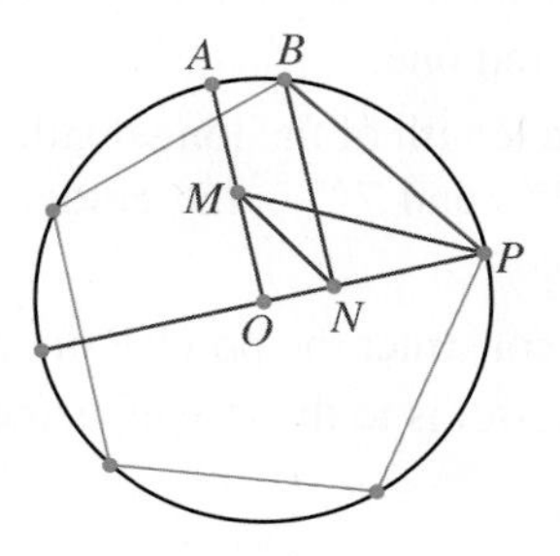

•**7.** Because we know how to construct an equilateral triangle and a regular pentagon, it is possible to construct a regular 15-gon. Explain how to accomplish the construction.

8. Take a strip of paper, make a knot, and tighten and flatten the knot as shown in the figure. Prove that the resulting shape is a regular pentagon.

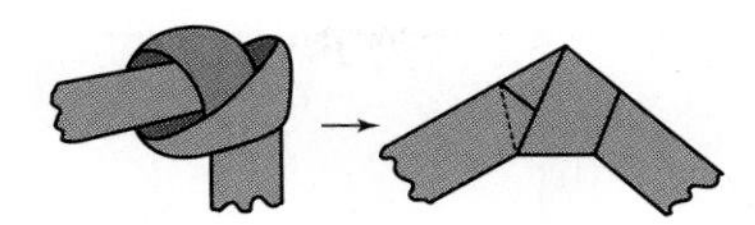

•**9.** Connect the vertices of a regular pentagon to each other to create the figure shown below.

a. Prove that the shaded figure is also a regular pentagon.

•**b.** Find the ratio of the area of the larger pentagon to the area of the smaller pentagon. Justify your answer.

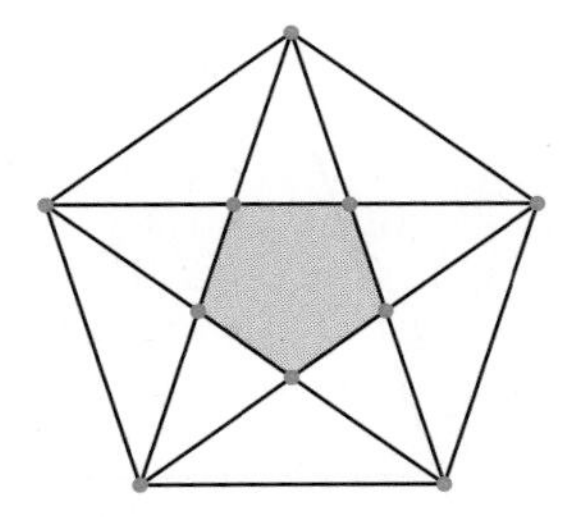

10. If you had a course in number theory or abstract algebra, you know that if m and n are positive integers whose greatest common divisor is 1, then it is possible to find integers x and y such that $mx + ny = 1$. Suppose that m and n are such integers and that it is possible to construct a regular m-gon and a regular n-gon. Prove that it is possible to construct a regular mn-gon. (Use the fact that a regular k-gon can be constructed if and only if an angle of $\frac{360}{n}$ degrees can be constructed.)

•**11.** This problem investigates the connection between the golden ratio and the **Fibonacci sequence**. The Fibonacci sequence 1, 1, 2, 3, 5, 8, 13, 21, . . . whose nth term is denoted by F_n is defined by $F_1 = F_2 = 1$ and for $n \geq 3$, $F_n = F_{n-1} + F_{n-2}$.

a. If ϕ is the golden ratio, show that

$$\phi^2 = 1 + 1 \cdot \phi$$
$$\phi^3 = 1 \cdot \phi + 1 \cdot \phi^2$$
$$\phi^4 = 1 \cdot \phi + 2 \cdot \phi^2$$
$$\phi^5 = 2 \cdot \phi + 3 \cdot \phi^3$$
$$\ldots$$

Conjecture an expression for ϕ^n in terms of ϕ, ϕ^2, and Fibonacci numbers.

b. Prove your conjecture in part (a).

•**c.** Compute the sequence of ratios $\frac{F_n}{F_{n-1}}$ for $n = 2, 3, 4, \ldots, 10$. What number does the sequence seem to approach?

•**d.** Assume that the sequence in part (c) has a limit and calculate that limit. (*Hint:* Use the relationship $F_n = F_{n-1} + F_{n-2}$.)

12. Given a rectangle, cut off three triangles from three corners of the rectangle such that the three cut-off triangles have the same area,

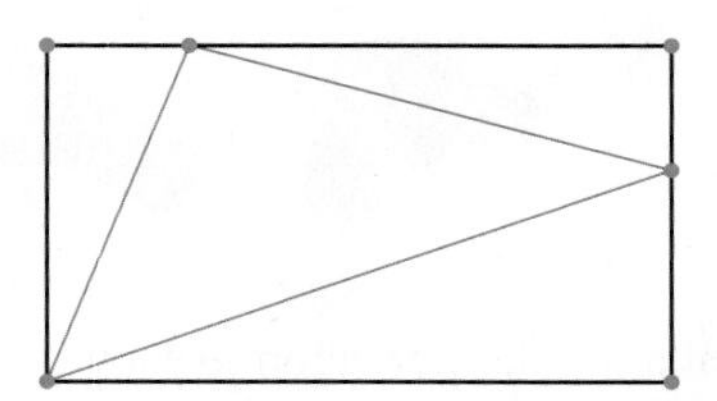

•**13.** In $\triangle ABC$, $BC = AC$. Also, D is a point on side AC such that $BD = AB$. Find the ratio $\frac{AB}{AD}$. Justify your answer.

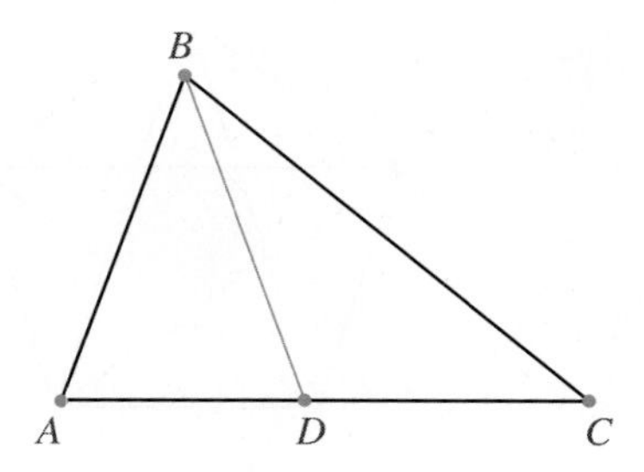

4.5 Circumference and Area of a Circle

In one of the propositions in the *Elements*, Euclid states that "the areas of any two circles are to each other as the squares of their diameters." In other words, if one circle has area A_1 and diameter D_1, and another circle has area A_2 and diameter D_2, then

$$\frac{A_1}{A_2} = \frac{D_1^2}{D_2^2} \tag{4.38}$$

This implies that $\frac{A_1}{D_1^2} = \frac{A_2}{D_2^2}$; that is, the ratio of the area of any circle to the square of its diameter is a constant. (The proof of Equation 4.38 will be explored in Problem Set 4.6.) If we designate this constant by k, then for any circle with diameter D and area A, we have

$$\frac{A}{D^2} = k \tag{4.39}$$

It is also known that the circumferences of two circles are related to each other as their corresponding diameters. That is, if C_1 and C_2 are the circumferences of any two circles, then

$$\frac{C_1}{C_2} = \frac{D_1}{D_2} \tag{4.40}$$

Consequently, $\frac{C_1}{D_1} = \frac{C_2}{D_2}$ and the ratio of the circumference of a circle to its diameter is constant. If we designate this constant by k', we find that for any circle with circumference C and diameter D.

$$\frac{C}{D} = k' \tag{4.41}$$

Now Solve This 4.14

Use the fact that a circle can be approximated by an inscribed (or circumscribed) regular polygon of many sides, your knowledge about the ratios of areas of similar polygons, and the ratio of their perimeters to give a plausible justification for Equations 4.38 and 4.40.

Around 225 B.C.E., Archimedes in his treatise *Measurement of a Circle* proved that the two constants k and k' in Equations 4.39 and 4.41 are equal. Each is, of course, the famous constant π. Here we will give a plausible argument for obtaining the formula for the area of a circle from the formula $C = 2\pi r$ for the circumference of a circle given its radius. We will also explore Archimedes's ingenious method for approximating the value of π and discuss other methods for evaluating π to any desired degree of accuracy.

$C = 2\pi r$ Implies $A = \pi r^2$

To obtain the formula for the area of a circle from its circumference, we consider a regular n-gon inscribed in the circle. We first express the area of the polygon in terms of its perimeter. For large n, the area of the n-gon approaches the area of the circle and the perimeter of the n-gon approaches the circumference of the circle. Figure 4.36 shows a regular octagon inscribed in a circle with center O and radius r. The area of the octagon is 8 times the area of $\triangle OAB$: $8 \cdot \frac{ah}{2}$. Similarly, the area of a regular n-gon inscribed in the circle is $n \cdot \frac{a_n \cdot h_n}{2}$, where a_n is the length of a side of the n-gon and h_n is the distance from O to a side. (Notice that the distance from O to each side of a regular n-gon is the same.) Let A_n be the area of the n-gon and p_n be its perimeter. Then

$$A_n = n\frac{a_n h_n}{2} = \frac{(na_n)h_n}{2} = \frac{p_n h_n}{2}$$

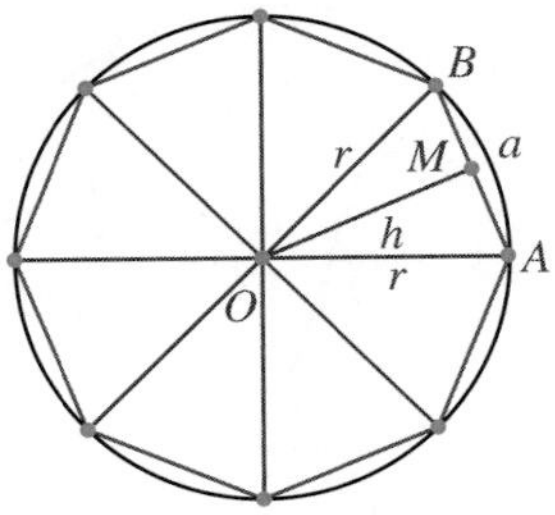

Figure 4.36

For large n, A_n approaches the area A of the circle, p_n approaches the circumference C of the circle, and h_n approaches the radius r of the circle. Hence, for the area of circle, we get $A = \frac{C \cdot r}{2}$. Substituting $2\pi r$ for C, we get $A = \frac{2\pi r \cdot r}{2} = \pi r^2$.

Now Solve This 4.15

Notice that we have not defined either the circumference or the area of a circle. Try to give the definition of each using the following concepts:

1. Limit
2. Least upper bound
3. Greatest lower bound

Archimedes's Evaluation of π

Archimedes showed that $\frac{223}{71} < \pi < \frac{22}{7}$. Using decimal notation (which was not known in Archimedes's time), this becomes

$$3.140845 < \pi < 3.142857\ldots$$

Thus π was computed to two decimal places' accuracy as 3.14.

Archimedes knew that $C = 2\pi r$ and hence $\pi = \frac{C}{2r}$, where C is the circumference of a circle and r is its radius. He noticed that the perimeters of regular polygons inscribed inside a circle as well as the perimeters of those circumscribed about a circle approximate the circumference of a circle; he also observed that this approximation gets better as the number of sides increases. Assuming that the circumference of a circle is greater than the perimeter of any inscribed polygon and smaller than the perimeter of any circumscribed polygon, we get

$$\frac{\text{Perimeter of inscribed polygon}}{2r} < \pi < \frac{\text{Perimeter of circumscribing polygon}}{2r} \tag{4.42}$$

Starting with an inscribed regular hexagon whose side must equal the radius of the circle (Why?), Archimedes calculated the perimeters of 12-, 24-, 48-, and 96-gons. He then developed formulas for the perimeters of circumscribing polygons in terms of the perimeters of the inscribed ones.

Before Archimedes, geometers gave a fairly crude approximation of π by calculating the area or perimeter of particular inscribed polygons. Archimedes's approach represented a completely new iterative approach in which we can achieve the desired level of accuracy by repeating the process and getting a new approximation using the previous one.

We will now investigate Archimedes's approach by starting in Figure 4.37 with a circle of radius 1. This choice is convenient but also legitimate because the ratio of the perimeter to the diameter is constant and, therefore, does not vary with the varying size of the diameter.

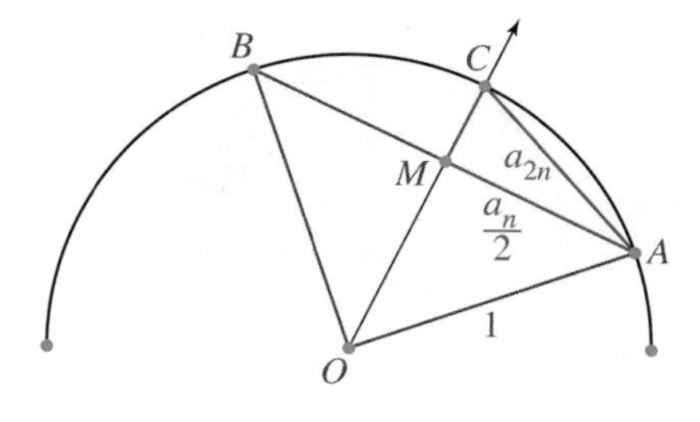

Figure 4.37

Let $AB = a_n$, the length of a side of an n-gon inscribed in the circle. The perpendicular bisector of $\overline{AB}$ contains the center O (Why?) and bisects $\overline{AB}$ at M and the arc AB at C. Because

$BC = CA$, $\overline{CA}$ is the side of a $2n$-gon inscribed in the circle. We express a_{2n} in terms of a_n by applying the Pythagorean Theorem first in $\triangle AMC$ and then in $\triangle OMA$:

$$CA^2 = CM^2 + MA^2 \tag{4.43}$$

$$a_{2n}^2 = CM^2 + \left(\frac{a_n}{2}\right)^2$$

$$CM = CO - OM = 1 - \sqrt{1 - MA^2} = 1 - \sqrt{1 - \left(\frac{a_n}{2}\right)^2}$$

Substituting the last expression for CM in Equation 4.43, we get

$$\begin{aligned} a_{2n}^2 &= \left(1 - \sqrt{1 - \left(\frac{a_n}{2}\right)^2}\right)^2 + \left(\frac{a_n}{2}\right)^2 \\ &= \left(1 - 2\sqrt{1 - \left(\frac{a_n}{2}\right)^2} + 1 - \left(\frac{a_n}{2}\right)^2\right) + \left(\frac{a_n}{2}\right)^2 \\ &= 2 - \sqrt{4 - a_n^2} \end{aligned}$$

Thus

$$a_{2n} = \sqrt{2 - \sqrt{4 - a_n^2}} \tag{4.44}$$

Using this recursive formula repeatedly and the fact that the length of the side of a regular hexagon inscribed in a circle equals the radius of the circle (we took $r = 1$), we get

$$\begin{aligned} a_6 &= 1 \\ a_{12} &= \sqrt{2 - \sqrt{4 - 1}} \\ a_{24} &= \sqrt{2 - \sqrt{4 - \left(\sqrt{\sqrt{2 - \sqrt{3}}}\right)^2}} \end{aligned}$$

After simplifying and using Equation 4.44 again and again:

$$\begin{aligned} a_6 &= 1 \\ a_{12} &= \sqrt{2 - \sqrt{3}} \\ a_{24} &= \sqrt{2 - \sqrt{2 + \sqrt{3}}} \\ a_{48} &= \sqrt{2 - \sqrt{2 + \sqrt{2 + \sqrt{3}}}} \\ a_{96} &= \sqrt{2 - \sqrt{2 + \sqrt{2 + \sqrt{2 + \sqrt{3}}}}} \\ &\ldots \end{aligned}$$

Based on the discussion preceding Equation 4.42, π can be approximated by dividing the perimeters of the inscribed polygons by the diameter of the circle, which is 2 in our case. Assuming that the perimeters of the polygons converge to the circumference of the circle, we get

$$\pi = \lim_{n\to\infty} \frac{6\cdot 2^n a_n}{2}, \text{ where } k = 6\cdot 2^n$$
$$\pi = \lim_{n\to\infty} 6\cdot 2^{n-1}\sqrt{2-\underbrace{\sqrt{2+\sqrt{2+\cdots+\sqrt{3}}}}_{n \text{ radicals}}} \tag{4.45}$$

Now Solve This 4.16

By starting with an inscribed square rather than a hexagon, derive a formula for π similar to Equation 4.45.

Not having computers, calculators, nor even decimal notation, Archimedes stopped after he found the perimeter of a 96-gon. However, he also used circumscribing polygons to obtain the right part of the inequality $\frac{223}{71} < \pi < \frac{22}{7}$ (He also estimated the involved square roots.) Notice in Table 4.1 that using only inscribed polygons we would not be able to tell with certainty which digits in the approximation are the correct digits of π. To overcome this problem, we use circumscribing polygons.

Figure 4.38 shows a side of an inscribed regular n-gon and half a side of a circumscribing regular n-gon. (The circumscribing polygon is constructed by drawing a tangent to the circle at each vertex of the inscribed polygon. The intersections of the tangents at the vertices of the regular inscribed n-gon determine the vertices of the circumscribing polygon.)

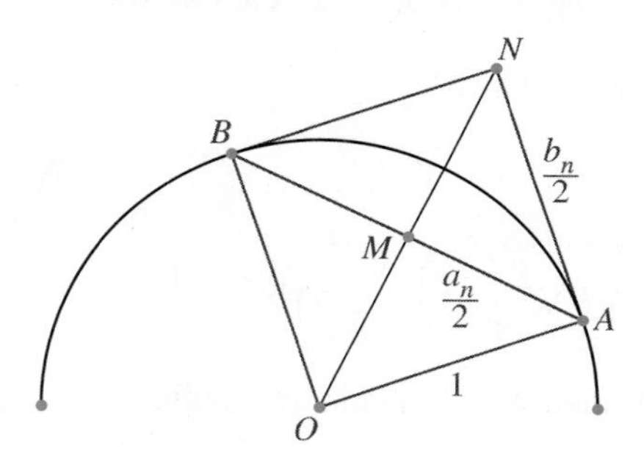

Figure 4.38

From $\triangle AMN \sim \triangle OMA$, we get

$$\frac{AM}{OM} = \frac{AN}{OA}$$

Therefore

$$\frac{\frac{a_n}{2}}{\sqrt{1-\left(\frac{a_n}{2}\right)^2}} = \frac{\frac{b_n}{2}}{1} \tag{4.46}$$

$$b_n = \frac{2a_n}{\sqrt{4-a_n^2}}$$

Table 4.1 Approximation of π

Sides	Inscribed Perimeter	Circumscribed Perimeter	Approximation of Pi
6	3.0000000000000000000	3.4641016151377545871	3.
12	3.1058285412302491482	3.2153903091734724777	3.
24	3.1326286132812381972	3.1596599420975004833	3.1
48	3.1393502030468672071	3.1460862151314349711	3.1
96	3.1410319508905096381	3.1427145996453682982	3.14
192	3.1414524722854620755	3.1418730499798238717	3.141
384	3.1415576079118576455	3.1416627470568485262	3.141
768	3.1415838921483184087	3.1416101766046895388	3.1415
1536	3.1415904632280500957	3.1415970343215261520	3.14159
3072	3.1415921059992715505	3.1415937487713520280	3.14159
6144	3.1415925166921574476	3.1415929273850970335	3.141592
12288	3.1415926193653839552	3.1415927220386138183	3.141592
24576	3.1415926450336908967	3.1415926707019980479	3.1415926
49152	3.1415926514507676517	3.1415926578678444198	3.14159265
98304	3.1415926530550368417	3.1415926546593060325	3.14159265
196608	3.1415926534561041393	3.1415926538571714369	3.141592653
393216	3.1415926535563709637	3.1415926536566377881	3.141592653
786432	3.1415926535814376698	3.1415926536065043759	3.141592653
1572864	3.1415926535877043463	3.1415926535939710228	3.1415926535
3145728	3.1415926535892710154	3.1415926535908376846	3.1415926535
6291456	3.1415926535896626827	3.1415926535900543500	3.1415926535
12582912	3.1415926535897605995	3.1415926535898585163	3.141592653589
25165824	3.1415926535897850787	3.1415926535898095579	3.141592653589
50331648	3.1415926535897911985	3.1415926535897973183	3.14159265358979
100663296	3.1415926535897927285	3.1415926535897942584	3.14159265358979
201326592	3.1415926535897931110	3.1415926535897934935	3.141592653589793
402653184	3.1415926535897932066	3.1415926535897933022	3.141592653589793
805306368	3.1415926535897932305	3.1415926535897932544	3.1415926535897932
1610612736	3.1415926535897932365	3.1415926535897932424	3.1415926535897932
3221225472	3.1415926535897932380	3.1415926535897932395	3.1415926535897932
6442450944	3.1415926535897932383	3.1415926535897932387	3.141592653589793238

Using Equations 4.44 and 4.46 and starting with $a_6 = 1$, we can compute the perimeters of the inscribed regular polygons in the unit circle and the perimeters of the corresponding circumscribing polygons with $6, 6 \cdot 2, 6 \cdot 2^2, 6 \cdot 2^3, \ldots$ number of sides. When we divide the perimeter by the diameter of the circle (i.e., by 2), we get an approximation of π from below and from above; see Equation 4.42 and Table 4.1.

Now Solve This 4.17

Use Equations 4.44 and 4.46 to show that

$$3 < \pi < 2\sqrt{3}$$

$$3.11 = 6\sqrt{2 - \sqrt{3}} < \pi < 12(2 - \sqrt{3}) = 3.22$$

Conclude that 3.1 is an approximation of π that is correct to one decimal point.

Historical Note: Archimedes

Archimedes, who was born in the Greek city of Syracuse, in what is now Sicily (circa 287–212 B.C.E.), was a Greek mathematician, physicist, and inventor. He is regarded by many as the greatest mathematician of antiquity—indeed, as one of the greatest mathematicians of all time. Archimedes invented mechanical devices that enabled the besieged city of Syracuse to delay the Roman conquest of the city for several years. Archimedes also discovered the law of the lever. Based on his "law," Archimedes presumably said, "Give me a place to stand and I will move the Earth." Another famous story tells of an occasion when he discovered the principle of buoyancy while in his bathtub: In his excitement, Archimedes got out of the tub and ran through the streets naked shouting, "Eureka! Eureka!" ("I have found it!"). In his work on areas and volumes, Archimedes also used methods similar to the ones used in integral calculus.

Many historical organizations and publications are named after Archimedes. For example, the Cambridge Society (in the United Kingdom) named the Archimedeans publishes the journal *Eureka*. For a large amount of information and illustrations on Archimedes, see the Archimedes home page at http://www.mcs.drexel.edu/~crorres/ and the program description on PBS's *Nova* (September 30, 2003) at www.pbs.org/wgbh/nova/archimedes/about.html.

Problem Set 4.5

•**1.** In the following figure, find the length of the indicated path from A to B in terms of r, the radius of the largest semicircle.

a. The first path along the large semicircle with radius r

•**b.** The second path along the two semicircles, each of which has radius $\frac{r}{2}$

•**c.** The third path along the four congruent semicircles

d. The nth path along 2^{n-1} congruent semicircles

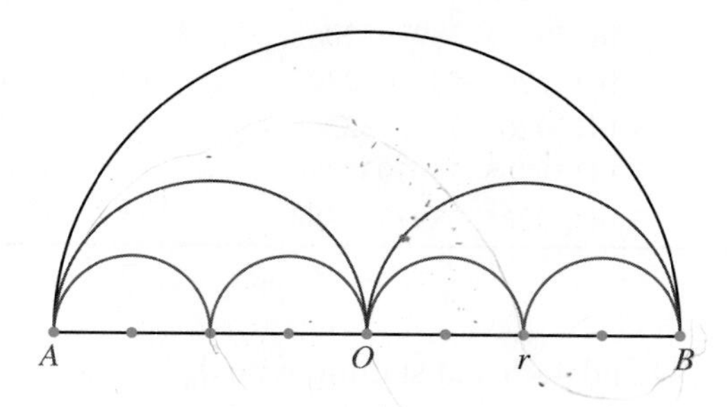

2. Imagine that a rope is tightly stretched around the Earth's equator. Assuming that the Earth is a perfect sphere, imagine that the rope is taken off and 100 feet are added to it. The extended rope is put back around the equator so that the "gap" created between the new circle and the equator is the same all around.

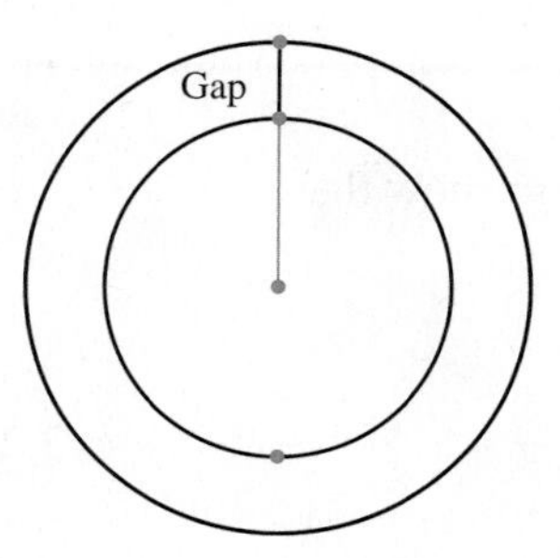

a. Without making any calculations, guess whether a cat could walk through the "gap" between the two circles.

b. Find the size of the "gap"—that is, the distance between the two circles. Will the answer be different if the rope is stretched around the equator of a different planet?

•**3.** **a.** Four congruent circles are cut out from a square sheet of tin as shown. What percentage of the tin is wasted?

•**b.** What percentage of the tin is wasted if nine congruent circles are cut out of the sheet?

c. What percentage of the tin is wasted if n^2 congruent circles are cut out of the sheet? Prove your answer.

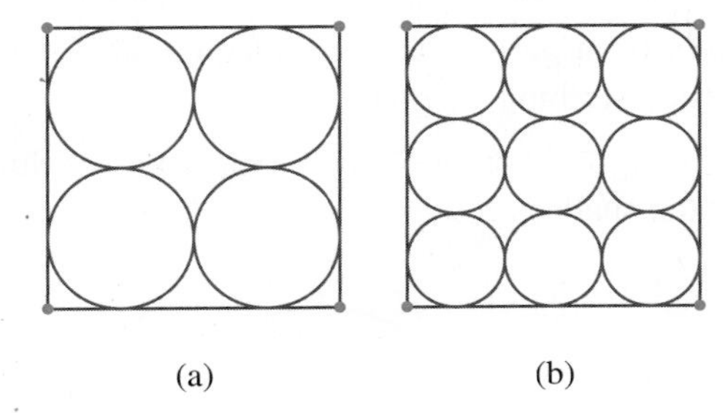

(a) (b)

4. For each of the following, find the ratio of the area of the circumscribing regular polygon to the area of the inner circle inscribed in the polygon.

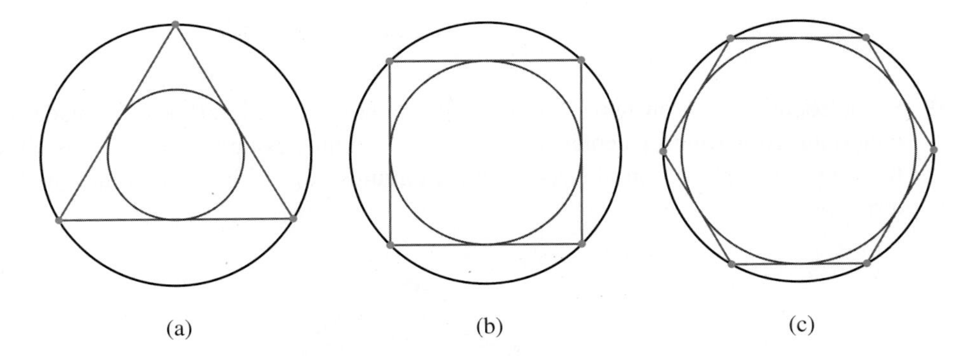

(a) (b) (c)

5. Prove that the sum of the areas of the semicircles constructed on the legs of a right triangle equals the area of the semicircle constructed on the hypotenuse of the triangle.

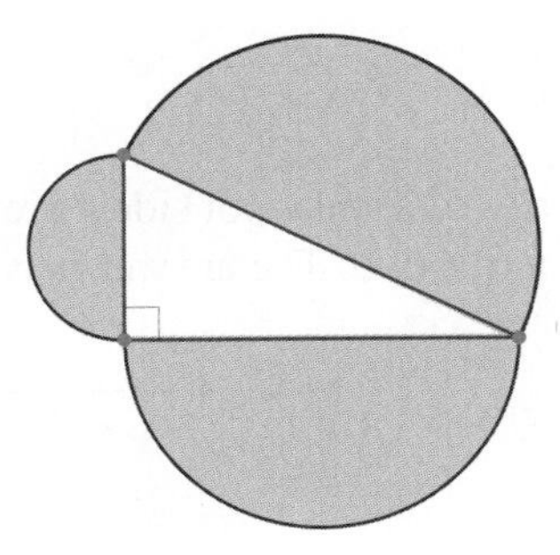

6. $\triangle ABC$ is inscribed in a semicircle with diameter AB. With the legs of the triangle as diameters, two semicircles are constructed. Prove that the shaded area between the semicircles equals the area of the triangle.

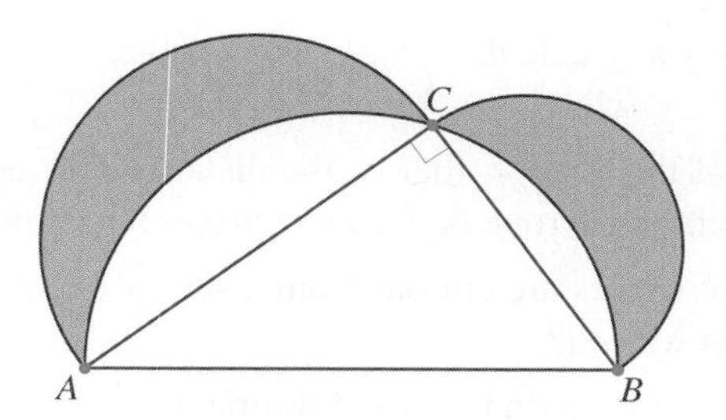

7. **a.** Find formulas involving radicals for the side of a regular 48-gon inscribed in a unit circle and the side of a regular 48-gon circumscribing the unit circle.

   b. Use your results from part (a) to write inequalities for π involving perimeters of inscribed and circumscribing 48-gons.

8. Each of the three congruent circles shown below passes through the centers of the other two circles. Find the shaded area in terms of the radius r of the circles.

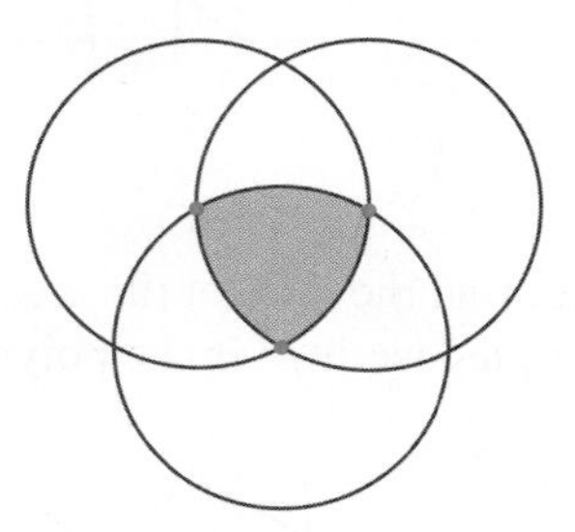

•**9.** In the regular hexagon with side a, six petals are drawn. Each petal is made of arcs of congruent circles whose centers are the vertices of the hexagon. All of the arcs intersect at the center of the hexagon (the center of the circumscribing circle). Find the shaded area in terms of a.

★**10.** The four arcs in the interior of a square of side a are centered at the vertices of the square. Find the shaded area in terms of a. The answer may surprise you. Why?

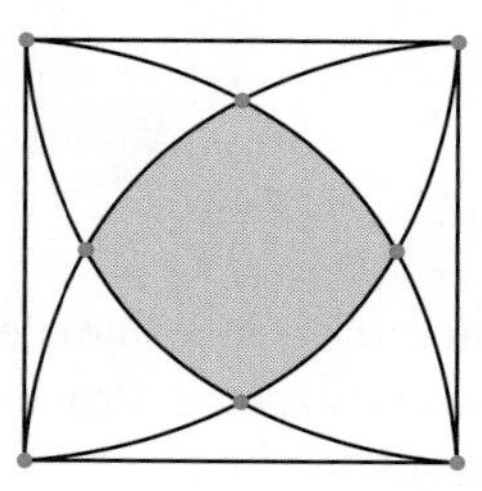

- **11.** Draw a circle and construct two new concentric circles that will divide the original circle into three non-overlapping regions of equal area.

12. In a circle, an arc that subtends a chord of length a is twice the length of an arc that subtends a chord of length b. Find the radius of the circle in terms of a and b.

13. A remarkable way to estimate the value of π using probability is the Buffon's Needle problem. Investigate this problem using references from the Internet and write a paper describing the problem, its solution, and its history.

*4.6 Other Recursive Formulas for Evaluating π

Because Archimedes's approach for evaluating π was extremely laborious (the 96-gon provides only the first two decimal places), mathematicians began looking for more efficient, recursive formulas for the perimeters of inscribed and circumscribing polygons. The following approach is attributed to two famous Dutch physicists and mathematicians: Huygens (1629–1695) and Snell (1581–1626). It is interesting in its own right, but even more so because the recursive formulas are similar to ones used in more modern approaches.

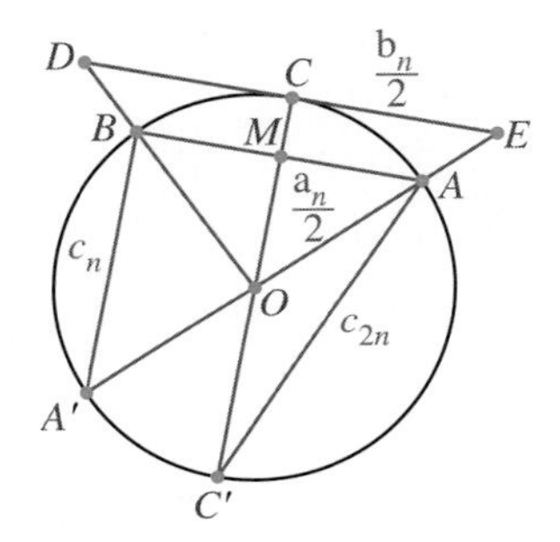

Figure 4.39

In Figure 4.39 (as in Figure 4.38), $MA = \dfrac{a_n}{2}$ and $CE = \dfrac{b_n}{2}$. We introduce a new point A' that is diametrically opposite to A, and label $BA' = c_n$. Because $\triangle ABA'$ is a right triangle,

$$c_n = \sqrt{4 - a_n^2} \tag{4.47}$$

We can develop a recursive formula for c_{2n} in terms of c_n by noticing that $CA = a_{2n}$ and $AC' = c_{2n}$.

In the right triangle CAC', $(AC')^2 = C'M \cdot CC'$. Because OM is the midsegment in $\triangle ABA'$,

$$c_{2n}^2 = 2(1 + OM) = 2\left(1 + \frac{1}{2}BA'\right) = 2\left(1 + \frac{1}{2}c_n\right)$$

Consequently,

$$c_{2n} = \sqrt{2 + c_n} \tag{4.48}$$

*Optional topic.

From this and Equation 4.47, we can find the same recursive formula for a_{2n} in terms of a_n as we found earlier in Equation 4.44:

$$\begin{aligned} a_{2n} &= \sqrt{4 - c_{2n}^2} \\ &= \sqrt{4 - (2 + c_n)} \\ &= \sqrt{2 - c_n} \\ &= \sqrt{2 - \sqrt{4 - a_n^2}} \end{aligned}$$

Hence

$$a_{2n} = \sqrt{2 - \sqrt{4 - a_n^2}} \tag{4.49}$$

A simple formula expressing c_{2n} in terms of a_n and c_n can be developed now from Equations 4.48 and 4.49 or from $\triangle CAC' \sim \triangle CMA$ (show that these triangles are indeed similar):

$$\frac{AM}{AC} = \frac{AC'}{CC'} \tag{4.50}$$

$$\begin{aligned} AM \cdot CC' &= AC \cdot AC' \\ \frac{a_n}{2} \cdot 2 &= a_{2n} \cdot c_{2n} \\ c_{2n} &= \frac{a_n}{a_{2n}} \end{aligned}$$

Next we express b_n in terms of a_n and c_n:

$$\frac{DE}{AB} = \frac{OC}{OM} = \frac{OC}{\frac{1}{2}A'B} = \frac{1}{\frac{1}{2}c_n}$$

$$\frac{b_n}{a_n} = \frac{2}{c_n} \tag{4.51}$$

$$b_n = \frac{2a_n}{c_n}$$

or equivalently

$$c_n = \frac{2a_n}{b_n}$$

A simple relationship between a_n and b_n, not involving c_n, follows from Equations 4.50 and 4.51:

$$\begin{aligned} c_{2n} &= \frac{a_n}{a_{2n}} \\ c_{2n} &= \frac{2a_{2n}}{b_{2n}} \\ \frac{a_n}{a_{2n}} &= \frac{2a_{2n}}{b_{2n}} \\ 2a_{2n}^2 &= a_n b_{2n} \end{aligned} \tag{4.52}$$

We know that $a_6 = 1$. We can find b_6 from Equations 4.51 and 4.47:

$$b_6 = \frac{2a_6}{c_6} = \frac{2 \cdot 1}{\sqrt{4 - a_6^2}} = \frac{2}{\sqrt{3}}$$

Notice that Equation 4.52 with the above values for a_6 and b_6 is not sufficient for determining the sequences $a_6, a_{12}, a_{24}, \ldots$ and $b_6, b_{12}, b_{24}, \ldots$. Using Equation 4.49 would suffice, but instead we will derive a simpler relationship not involving square roots. For that purpose, we use Equation 4.48 in conjunction with Equations 4.50 and 4.51 as follows:

$$\begin{aligned} c_{2n} \cdot c_{2n} &= 2 + c_n \\ \frac{2a_{2n}}{b_{2n}} \cdot \frac{a_n}{a_{2n}} &= 2 + \frac{2a_n}{b_n} \\ \frac{a_n}{b_{2n}} &= 1 + \frac{a_n}{b_n} \\ a_n b_n &= b_n b_{2n} + a_n b_{2n} \\ b_{2n} &= \frac{a_n b_n}{a_n + b_n} \end{aligned} \tag{4.53}$$

Notice that Equation 4.52 and 4.53 involve geometric and harmonic means. For convenience in computing π, and to obtain elegant formulas involving exactly geometric and harmonic means, we develop corresponding formulas for p_n and t_n, the perimeters of inscribed and circumscribed n-gons, respectively, where $n = 6 \cdot 2^k$, for $k = 0, 1, 2, 3, \ldots$. Notice that $p_n = 6 \cdot 2^k a_n$ and $t_n = 6 \cdot 2^k b_n$. With this in mind, we multiply each side of Equation 4.52 by $(6 \cdot 2^{k+1})^2$:

$$\begin{aligned} 2a_{2n}^2 &= a_n b_{2n} \\ 2 \cdot (6 \cdot 2^{k+1})^2 a_{2n}^2 &= 6 \cdot 2^{k+1} a_n \cdot 6 \cdot 2^{k+1} b_{2n} \\ 2p_n^2 &= 2(6 \cdot 2^k a_n)(6 \cdot 2^{k+1} b_{2n}) \\ p_{2n}^2 &= p_n t_{2n} \end{aligned} \tag{4.54}$$

Similarly, we multiply both sides of Equation 4.53 by $6 \cdot 2^{k+1}$ and the right side by $\frac{6 \cdot 2^n}{6 \cdot 2^n}$ as well:

$$6 \cdot 2^{k+1} b_{2n} = 6 \cdot 2^{k+1} \frac{a_n b_n}{a_n + b_n} \cdot \frac{6 \cdot 2^k}{6 \cdot 2^k} \tag{4.55}$$

$$t_{2n} = \frac{2 p_n t_n}{p_n + t_n}$$

Notice the elegant geometric and harmonic means in Equations 4.54 and 4.55. These two equations, with the initial conditions $p_1 = 6$ and $t_1 = 6 \cdot \frac{2}{\sqrt{3}} = 4\sqrt{3}$, determine the sequence of perimeters for the inscribing and circumscribing polygons. Dividing the perimeters by 2, we obtain the sequences for approximating π.

Now Solve This 4.18

Use Mathematica, a spreadsheet, or a calculator and Equations 4.54 and 4.55 to find an approximation of π accurate to four places after the decimal point. Compare your results to Table 4.1.

More Modern Ways of Evaluating π

The Chinese mathematician Tsu Chung Chi (430–501 C.E.) gave the rational approximation for π of $\frac{355}{113}$, which is correct to six decimal places. Thus he held the world record for the longest approximation of π for 800 years. He also proved that $3.1415926 < \pi < 3.1415927$. Details on how Tsu Chung Chi proved his results are unknown because his manuscript, which he co-wrote with his son, has been lost.

The next improvement was accomplished in 1424 by the Moslem mathematician Jamshid al-Kashi (born about 1380 in Kashan, Iran; died in 1429 in Sumarkand, now Uzbekistan), who used a method similar to Archimedes's strategy to compute the value of π to 14 correct digits. Ludolf van Ceulen (1540–1610 C.E.) used Archimedes's polygon method to calculate π to 35 decimal places. Equation 4.49 is often referred to as the *van Ceulen doubling formula*. In Europe, for a long time π was called the *Ludolphine number*.

In the 1600s, with the discovery of calculus by Newton and Leibniz, new formulas for π were discovered. The Taylor formula applied to arctan x gives

$$\arctan x = x - \frac{x^3}{3} + \frac{x^5}{5} - \frac{x^7}{7} + \cdots = \sum_{n=0}^{\infty} \frac{(-x)^{2n+1}}{2n+1} \tag{4.56}$$

Substituting $x = 1$, we get

$$\frac{\pi}{4} = 1 - \frac{1}{3} + \frac{1}{5} + \frac{1}{7} + \frac{1}{9} - \frac{1}{11} + \cdots$$

In spite of its elegant form, this series for $\frac{\pi}{4}$ is not useful because it converges very slowly. To obtain π accurate to only two digits, about 1000 terms are required. Nevertheless, the series in Equation 4.56 converges more rapidly for values smaller than 1. By using the identity

$$\frac{\pi}{4} = \arctan\frac{1}{2} + \arctan\frac{1}{3} \tag{4.57}$$

and the series in Equation 4.56 for $x = \frac{1}{2}$ and $x = \frac{1}{3}$, we can obtain a faster and better approximation of π. In 1706, John Machin discovered the identity

$$\frac{\pi}{4} = 4\arctan\frac{1}{5} - \arctan\frac{1}{239}$$

and with Equation 4.56 used it to calculate the first 100 digits of π.

Euler (1707–1783), one of the most prolific mathematicians of all time, derived many expressions for π, including the following:

$$\frac{\pi^2}{6} = \sum_{k=1}^{\infty} \frac{1}{k^2} \quad \text{and} \quad \frac{2^2\pi^4}{5!3} = \sum_{k=1}^{\infty} \frac{1}{k^4}$$

In 1910, the prodigious Indian mathematician Ramanujan (1887–1920) discovered the following remarkable formula for π:

$$\frac{1}{\pi} = \frac{2\sqrt{2}}{9801} \sum_{k=0}^{\infty} \frac{(4k)!(1103 + 26390k)}{(k!)^4 396^{4k}}$$

Each term in this series produces eight additional correct digits, In 1985, the formula was used to compute π to 17 million correct digits.

In 1976, Salamin and Brent independently discovered a new algorithm that produces approximations of π that converge more rapidly than in any previous approach. The algorithm may be stated as follows (see Bailey et al., 1997). We let $a_0 = 1$, $b_0 = \frac{\sqrt{2}}{2}$, and $s_0 = \frac{1}{2}$:

$$\begin{aligned} a_k &= \frac{a_{k-1} + b_{k-1}}{2} \\ b_k &= \sqrt{a_{k-1}b_{k-1}} \\ c_k &= a_k^2 - b_k^2 \\ s_k &= s_{k-1} - 2^k c_k \\ p_k &= \frac{2a_k^2}{s_k} \end{aligned}$$

Each iteration of this algorithm approximately doubles the number of correct digits. Only 25 iterations are necessary to compute π to more than 45 million correct digits. Notice the similarity of these formulas to Equations 4.54 and 4.55! In 1994, the Chadnovsky brothers computed π to more than 4 billion digits using a formula for $\frac{1}{\pi}$ similar to Ramanujan's formula given earlier.

*4.7 Trigonometric Functions

Trigonometry (often abbreviated as "trig") is not only indispensable in applications of mathematics, but also useful in solving geometrical problems, as we shall soon see. In this section we give a brief review of trigonometry by emphasizing the main ideas of the subject and asking you to complete the suggested investigations in several Now Solve This inquiries.

From our study of similar triangles, we know that given two similar triangles, the ratio of a pair of sides in one triangle equals the ratio of the pair of corresponding sides in the other triangle. Thus, in a right triangle with an angle α, the ratio of the side opposite α to the hypotenuse is the same for all right triangles with the same angle α. That ratio depends only on the size of α; hence it is a function of α. This function is **sin** ($\boldsymbol{\alpha}$), commonly abbreviated as sin α. Similarly, we define other trigonometric functions as shown in Figure 4.40.

$$\sin\alpha = \frac{a}{c}$$

$$\cos\alpha = \frac{b}{c}$$

$$\tan\alpha = \frac{a}{b}$$

$$\cot\alpha = \frac{b}{a}$$

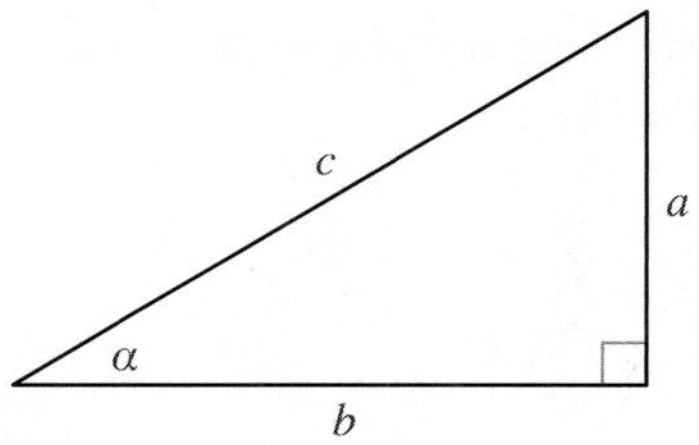

Figure 4.40

Now Solve This 4.19

1. Prove that $(\sin\alpha)^2 + (\cos\alpha)^2 = 1$. *Note:* $(\sin\alpha)^2$ is commonly written as $\sin^2\alpha$.
2. Find, if possible, the four trig functions defined in Figure 4.40 for $\alpha = 0, 30°, 45°, 60°, 90°$. Express your answers using rational numbers or radicals.
3. Angles are measured in degrees and in **radians**. Define the radian measure, and explain why it is preferred in calculus and therefore in science.

Notice that our first definitions of trig functions were in the context of a right triangle, and hence applied to angles greater than 0 and less than $\frac{\pi}{2}$ (90°). We now extend the definitions of these functions for any angle. One way to motivate the new definitions is through periodic motion on a circle. Let $P(x, y)$ be a point that moves on a circle of radius r centered at the origin, as in Figure 4.41. Suppose P starts at $A(r, 0)$ and moves around the circle, sweeping through an angle θ. We say that θ is in **standard position** and that $\overrightarrow{OA}$ is its **initial side** and $\overrightarrow{OP}$ is its **terminal side**. If the motion is counterclockwise, we define θ to be positive ($\theta > 0$); if the motion is clockwise, then $\theta < 0$.

*Optional topic.

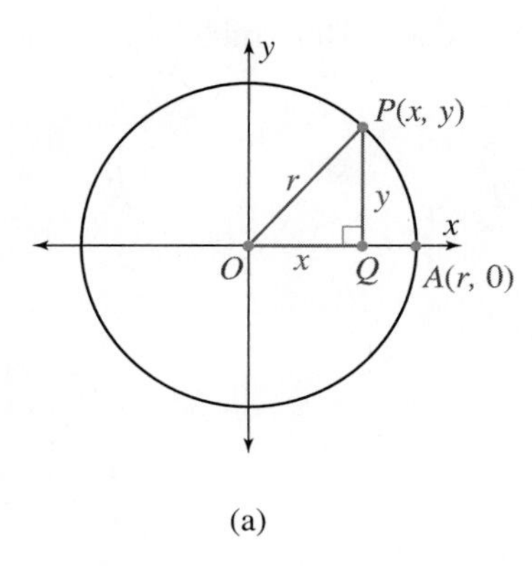

(a)

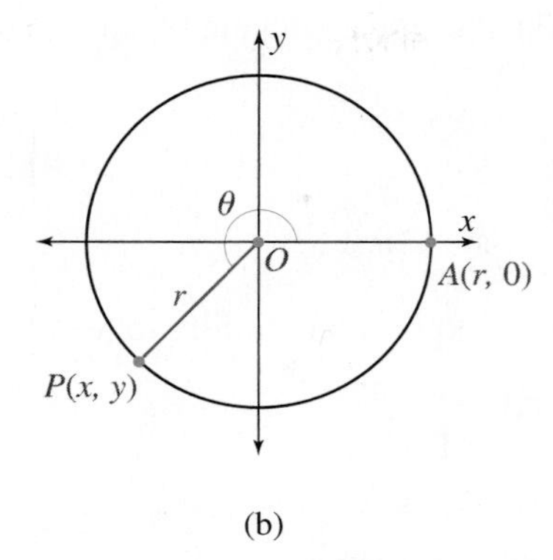

(b)

Figure 4.41

From Figure 4.41a and our original definition of trig functions in a right triangle, we have

$$\begin{aligned} \sin\theta &= \frac{y}{r} \\ \cos\theta &= \frac{x}{r} \\ \tan\theta &= \frac{y}{x} \\ \cot\theta &= \frac{x}{y} \end{aligned} \tag{4.58}$$

From properties of similar triangles, it follows that these ratios are independent of the radius of the circle and hence are well defined. When the point P moves counterclockwise around the circle starting at A and returning to A, its projection Q on the x-axis moves on the x-axis from $A(r, 0)$ to $(-r, 0)$ and back to $A(r, 0)$. This motion of Q, which is called **harmonic motion**, is of major importance in physics. The position of the point Q is a function of θ. It is natural to define the trig functions as in Equation 4.58 for all angles θ ($|\theta|$ can be greater than 2π when P makes more than one full rotation around the circle). We then have

$$\begin{aligned} x &= r\cos\theta \\ y &= r\sin\theta \end{aligned} \tag{4.59}$$

Now Solve This 4.20

1. Why it is legitimate to define the trig functions in a circle of radius 1 (unit circle)? If the radius is 1, explain why $\cos\theta = x$ and $\sin\theta = y$, where $P(x, y)$ is the point of the terminal side of θ on the circle.
2. Justify each of the following by referring to an appropriate drawing.
   (a) $\cos(-\theta) = \cos\theta$
   (b) $\sin(-\theta) = -\sin\theta$

(c) $\sin(\pi - \theta) = \sin\theta$

(d) $\cos(\pi - \theta) = -\cos\theta$

3. Each of the following equals $\sin\theta$, $-\sin\theta$, $\cos\theta$, or $-\cos\theta$. Determine which one and why.

(a) $\sin\left(\frac{\pi}{2} + \theta\right)$, $\sin\left(\frac{\pi}{2} - \theta\right)$, $\sin(\pi + \theta)$, $\sin\left(\frac{3\pi}{2} - \theta\right)$, $\sin\left(\frac{3\pi}{2} + \theta\right)$, $\sin(2\pi - \theta)$

(b) Expressions similar to the ones in part (a) for $\cos\theta$

4. Explain why

(a) $-1 \leq \sin\theta \leq 1$

(b) $-1 \leq \cos\theta \leq 1$

(c) $-\infty < \tan\theta < \infty$

Solving Triangles

All triangles have six parts: three sides and three angles. Finding these six parts of a triangle is often called "solving the triangle." Frequently, a triangle is determined by three of its parts: By ASA, a side and two adjacent angles determine a unique triangle. Using trig functions, we can then find the lengths of the other two sides of the triangle. This concept is explored in the following problem.

Problem 4.3 In $\triangle ABC$, side a and the three angles α, β, and γ are given. Find the other sides of the triangle in terms of a and trig functions of α, β, or γ.

Solution

To find side b in $\triangle ABC$ in Figure 4.42, we try to create a right triangle that has b as one of its sides. This can be readily accomplished by constructing the altitude from C to side AB. Let $CD = h$. Then

$$\sin\alpha = \frac{h}{b} \quad \text{or} \quad b = \frac{h}{\sin\alpha} \tag{4.60}$$

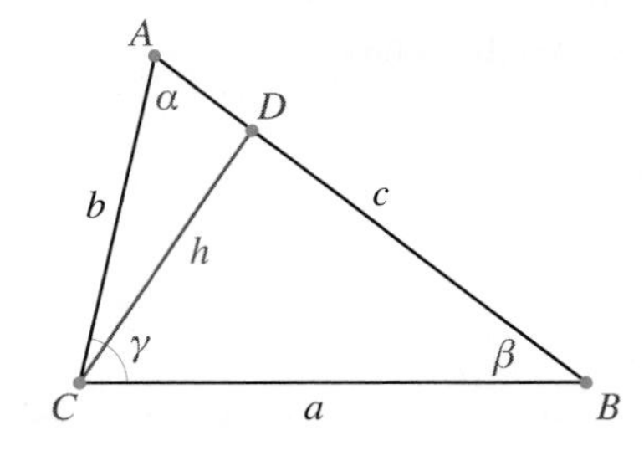

Figure 4.42

Thus we would know b if h were known. However, h can be found in $\triangle CDB$, the other right triangle containing h as a side. We get

$$\sin\beta = \frac{h}{a} \quad \text{or} \quad h = \alpha\sin\beta \tag{4.61}$$

Substituting for h in Equation 4.60, we get

$$b = \frac{a \sin \beta}{\sin \alpha} \tag{4.62}$$

Similarly, we get

$$c = \frac{a \sin \gamma}{\sin \alpha} \tag{4.63}$$

These solutions can be written in a more appealing symmetric form that is easy to remember. From Equations 4.62 and 4.63, dividing each side of the first equation by sin β and the second equation by sin γ, we get

$$\frac{b}{\sin \beta} = \frac{a}{\sin \alpha} \tag{4.64}$$

$$\frac{c}{\sin \gamma} = \frac{a}{\sin \alpha}$$

These equations can also be written as

$$\frac{a}{\sin \alpha} = \frac{b}{\sin \beta} = \frac{c}{\sin \gamma} \tag{4.65}$$

The last equation is the famous **law of sines**.

Another Proof of the Law of Sines

Important discoveries often occur as a by-product of seemingly unrelated activities. Consider the following problem.

Problem 4.4 Given the three sides a, b, c of a triangle and its angles α, β, γ, find the radius R of the circumscribing circle.

Solution

Consider Figure 4.43. One way to find the radius of the circle is to connect any vertex to the center and then extend the radius to obtain a diameter. A diameter is especially useful because an inscribed angle that subtends a diameter is a right angle. In $\triangle A_1BC$, $\angle A_1$ intercepts the same arc as does $\angle A$, which implies that the angles are congruent and therefore $m(\angle A_1) = \alpha$. In the right triangle A_1BC, the hypotenuse $\overline{BA_1}$ is a diameter. Thus

$$\sin \alpha = \frac{BC}{BA_1} = \frac{a}{2R} \tag{4.66}$$

$$R = \frac{a}{2 \sin \alpha}$$

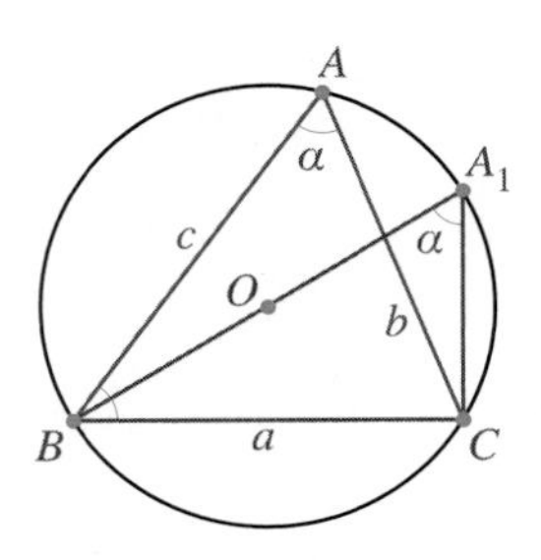

Figure 4.43

Notice that Equation 4.66 is the answer to Problem 4.4, but side a and its opposite angle α could have been side b and angle β, or side c and angle γ. Hence we have

$$\frac{a}{2\sin\alpha} = \frac{b}{2\sin\beta} = \frac{c}{2\sin\gamma} = R$$

The Extended Law of Sines $$\frac{a}{\sin\alpha} = \frac{b}{\sin\beta} = \frac{c}{\sin\gamma} = 2R$$

Now Solve This 4.21

1. Prove the law of sines in each of the same ways as in Problem 4.4 for an obtuse triangle. [*Hint:* $\sin(180° - \alpha) = \sin\alpha$.]
2. Prove that the area of a triangle is equal to one-half the product of any two sides of the triangle times the sine of the included angle. That is, if A is the area, then

$$A = \frac{1}{2}bc\sin\alpha = \frac{1}{2}ac\sin\beta = \frac{1}{2}ab\sin\gamma$$

3. Divide the equations in part 2 by abc to obtain yet another proof of the law of sines.
4. Prove that the area A of a triangle is also given by

$$A = \frac{abc}{4R}$$

Generalization of the Pythagorean Theorem: The Law of Cosines

We have seen that a triangle is uniquely determined by the ASA condition and developed the law of sines to find the other two sides of the triangle. Similarly, a unique triangle is determined by the SAS condition. If the sides are a and b and the included angle γ is a right angle, then side c is the hypotenuse in a right triangle; hence, by the Pythagorean Theorem, $c^2 = a^2 + b^2$. In the following problem we generalize the Pythagorean Theorem for any included angle γ.

Problem 4.5 In Figure 4.44, the sides a and b and the included angle γ of $\triangle ABC$ are given. Find side c in terms of a, b, and γ.

Solution

One way to approach this problem is to think about c as being the distance between A and B. For that purpose, we need to set up a coordinate system and find the coordinates of A and B in terms of a, b, and γ. We choose the coordinate system in Figure 4.44b so that γ will be in a standard position. This can be accomplished if the vertex C is at the origin and the side CA is on the x-axis. Then the coordinates of A are $A(b, 0)$. To find the coordinates of B, we notice that B is on a circle centered at the origin (vertex C), whose radius is a as shown in Figure 4.44. By the definitions of the cosine and sine functions, if B has coordinates x and y, we get

$$\cos\gamma = \frac{x}{a} \quad \text{and} \quad \sin\gamma = \frac{y}{a}$$

$$x = a\cos\gamma \quad \text{and} \quad y = a\sin\gamma$$

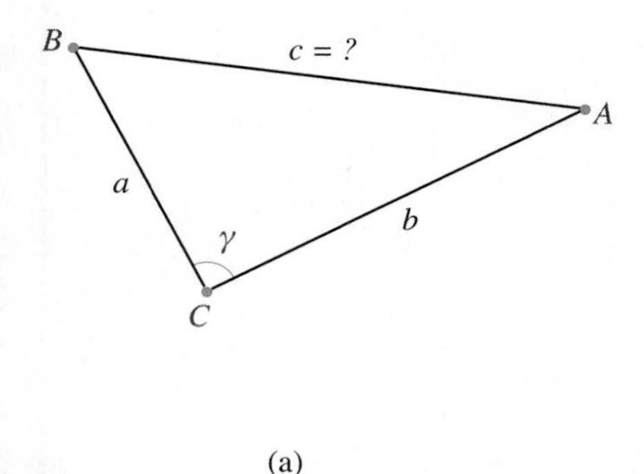

(a)

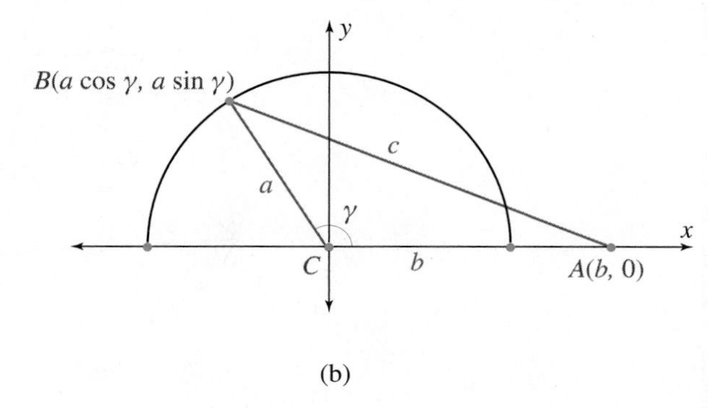

(b)

Figure 4.44

Applying the distance formula to the distance AB where $B(a\cos\gamma, a\sin\gamma)$ and $A(b, 0)$, we get

$$c^2 = (AB)^2 = (a\cos\gamma - b)^2 + (a\sin\gamma - 0)^2 \tag{4.67}$$

Equation 4.67 answers the question in Problem 4.5, but we can simplify this answer by using the fact that $\cos^2\gamma + \sin^2\gamma = 1$:

$$\begin{aligned} c^2 &= a^2\cos^2\gamma - 2ab\cos\gamma + b^2 + a^2\sin^2\gamma \\ &= a^2(\cos^2\gamma + \sin^2\gamma) + b^2 - 2ab\cos\gamma \\ &= a^2 + b^2 - 2ab\cos\gamma \end{aligned}$$

Solving Problem 4.5 we have proved the law of cosines:

The Law of Cosines In a triangle with sides a and b and the included angle γ, the third side c is

$$c^2 = a^2 + b^2 - 2ab\cos\gamma \tag{4.68}$$

Now Solve This 4.22

1. Think about the law of cosines in words: "The square of a side of a triangle is equal to the sum of the squares of the other two sides minus twice the product of these sides times the cosine of the included angle."
   (a) State the law of cosines in the format of Equation 4.68 for a^2 and for b^2 (name the angle opposite a as α and the angle opposite b as β).

(b) Use your answer to part (a) to find cos α and cos β in terms of the sides a, b, and c.

2. Evaluate Equation 4.68 when $\gamma = 90°$.
3. Show in Equation 4.68 that when $\gamma = 180°$, then $c = a + b$. Explain this result geometrically.
4. Find another proof for the law of cosines (you may want to consult the Internet).
5. Prove Brahmagupta's Theorem (named after the Hindu mathematician Brahmagupta, 598–668 C.E.): If a quadrilateral with sides a, b, c, and d is inscribed in a circle, then the diagonal e (shown in the Figure 4.45) is given by

$$e^2 = \frac{(ab + cd)(ac + bd)}{bc + ad}$$

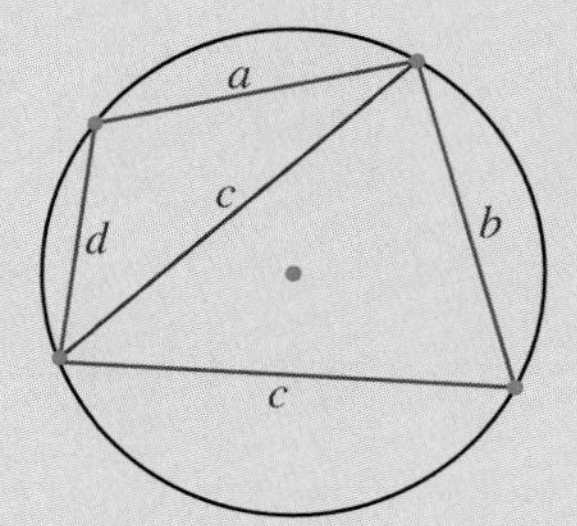

Figure 4.45

Addition Formulas

In Figure 4.46, $\triangle ABC$ is inscribed in a circle with diameter 1. Using the extended law of sines, we get

$$\frac{a}{\sin \alpha} = \frac{b}{\sin \beta} = \frac{c}{\sin \gamma} = 2R = 1 \quad (4.69)$$

$$a = \sin \alpha, \quad b = \sin \beta, \quad \text{and} \quad c = \sin \gamma$$

Since $\gamma = 180° - (\alpha + \beta)$, $\sin \gamma = \sin[(180° - (\alpha + \beta)] = \sin(\alpha + \beta)$. Consequently, $c = \sin(\alpha + \beta)$.

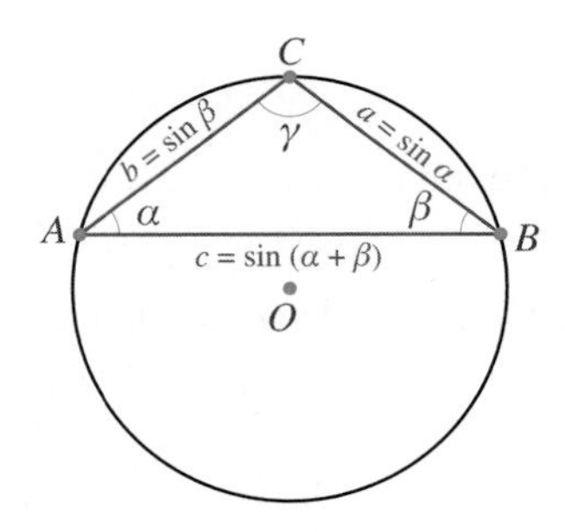

Figure 4.46

Applying the triangle inequality to $\triangle ABC$ tells us that $c < a + b$ and hence that

$$\sin(\alpha + \beta) < \sin \alpha + \sin \beta$$

Nevertheless, $\sin(\alpha + \beta)$ can be expressed in terms of the trig functions of α and β. This restatement can be accomplished in a variety of ways, one of which is described in Figure 4.47. The circle

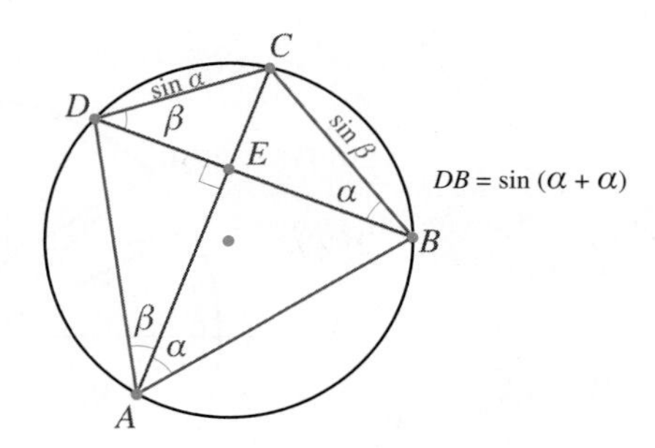

Figure 4.47

in the figure has diameter equal to 1, and *ABCD* is a quadrilateral with perpendicular diagonals *AC* and *BD*. As in Equation 4.69, $DC = \sin\alpha$, $BC = \sin\beta$, and $DB = \sin(\alpha + \beta)$.

In $\triangle DCE$, $\cos\beta = \dfrac{DE}{DC} = \dfrac{DE}{\sin\alpha}$ and hence

$$DE = \sin\alpha\cos\beta \tag{4.70}$$

Similarly, in $\triangle CBE$, $\cos\alpha\, \dfrac{EB}{CB} = \dfrac{EB}{\sin\beta}$ and therefore

$$EB = \sin\alpha\cos\beta \tag{4.71}$$

Since $\sin(\alpha + \beta) = DB = DE + EB$, we use Equations 4.70 and 4.71 and obtain the following result:

$$\sin(\alpha + \beta) = \sin\alpha\cos\beta + \sin\beta\cos\alpha \tag{4.72}$$

Equation 4.72 was proved only for acute angles α and β. It is, however, true for all angles—a fact that can be proved using some of the ideas in Now Solve This 4.23. For now, let's prove similar formulas for $\sin(\alpha - \beta)$, $\cos(\alpha - \beta)$, and $\cos(\alpha + \beta)$.

Formula for $\cos(\alpha - \beta)$

Figure 4.48a shows the angles α, β, and $\alpha - \beta$ for $\alpha > \beta$ and $\beta > 0$; Figure 4.48b shows a similar configuration when $\beta < 0$. In each figure, the circle is a unit circle (of radius 1).

In both figures, α and β are in the standard positions. Also, $\angle AOB$ measures $\alpha - \beta$ (in Figure 4.48b, $\beta < 0$). To use the definition of $\cos(\alpha - \beta)$ in terms of coordinates, we need to place $\alpha - \beta$ in the standard position. In each part of Figure 4.48, we find point *P* such that $\angle COP$ (which is in the standard position) is $\alpha - \beta$. Because each of the central angles $\angle AOB$ and $\angle POC$ is $\alpha - \beta$, the corresponding chords *AB* and *PC* have the same lengths. We use the distance formula to equate the squares of their lengths:

$$(AB)^2 = (PC)^2$$

$$(\cos\alpha - \cos\beta)^2 + (\sin\alpha - \sin\beta)^2 = [\cos(\alpha - \beta) - 1]^2 + [\sin(\alpha - \beta) - 1]^2$$

$$(\cos^2\alpha - 2\cos\alpha\cos\beta + \cos^2\beta) + (\sin^2\alpha - 2\sin\alpha\sin\beta + \sin^2\beta)$$

$$= \cos^2(\alpha - \beta) - 2\cos(\alpha - \beta) + 1 + \sin^2(\alpha - \beta)$$

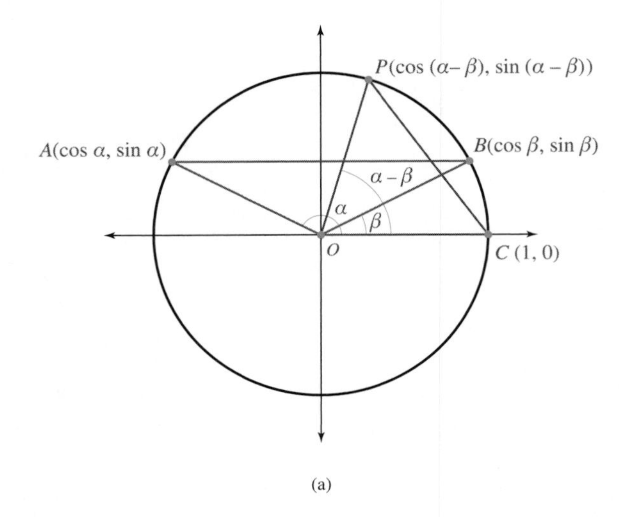

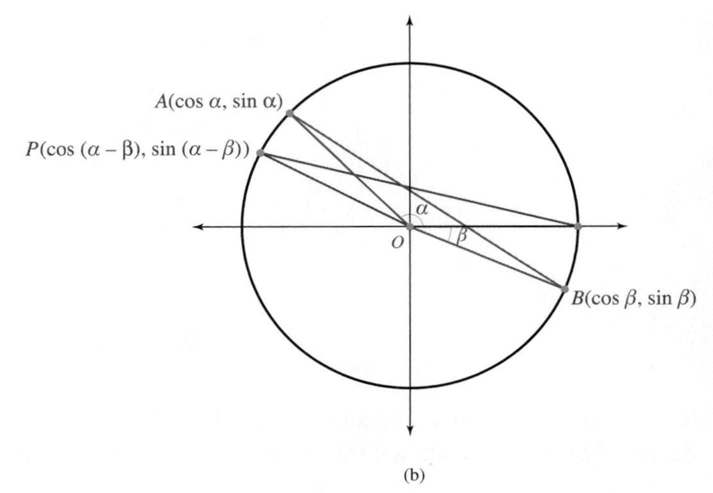

Figure 4.48

Because

$$\cos^2 \alpha + \sin^2 \alpha = \cos^2 \beta + \sin^2 \beta$$
$$= \cos^2(\alpha - \beta) + \sin^2(\alpha - \beta)$$
$$= 1$$

we get

$$2 - 2\cos\alpha\cos\beta - 2\sin\alpha\sin\beta = 2 - 2\cos(\alpha - \beta)$$

or

$$\cos(\alpha - \beta) = \cos\alpha\cos\beta + \sin\alpha\sin\beta \qquad (4.73)$$

This formula was proved for both positive and negative β. In fact, using figures like Figure 4.48, we can show that Equation 4.73 is true for any α and β.

Now Solve This 4.23

1. (a) Use diagrams to show that the image of $P(x, y)$ under rotation about the origin by $\frac{\pi}{2}$ (counterclockwise) is the point $P(-y, x)$.

   (b) In a right triangle, we can see that $\cos\left(\frac{\pi}{2} - \alpha\right) = \sin\alpha$ and $\sin\left(\frac{\pi}{2} - \alpha\right) = \cos\alpha$, where α is an acute angle. To prove that these identities are true for any angle α, write $\frac{\pi}{2} - \alpha$ as $-\alpha + \frac{\pi}{2}$ and use your result from part (a) plus the identities $\cos(-\alpha) = \cos\alpha$ and $\sin(-\alpha) = -\sin\alpha$.

2. Use the fact that Equation 4.73 holds true for any angles α and β to prove the following:

   (a) $\cos(\alpha + \beta) = \cos\alpha\cos\beta - \sin\alpha\sin\beta$

(b) $\sin(\alpha + \beta) = \sin \alpha \cos \beta + \sin \beta \cos \alpha$ [start with $\sin(\alpha + \beta) = \cos\left(\frac{\pi}{2} - (\alpha + \beta)\right)$ $= \cos\left(\left(\frac{\pi}{2} - \alpha\right) - \beta\right)$.]

(c) $\sin(\alpha - \beta) = \sin \alpha \cos \beta - \sin \beta \cos \alpha$

3. Prove the following theorem, which is attributed to Ptolemy of Alexandria (circa 85–165 C.E.):

> If a quadrilateral is inscribed in a circle, then the product of the lengths of the diagonals is equal to the sum of the products of the lengths of pairs of opposite sides.

In Figure 4.49, $ABCD$ is inscribed in a circle. Congruent inscribed angles have been marked by the same Greek letter.

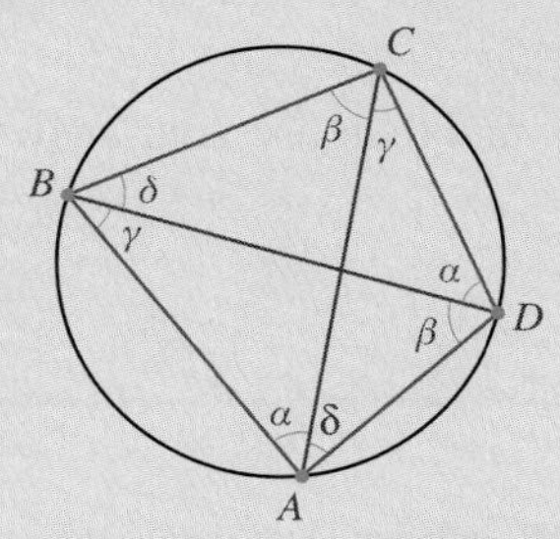

Figure 4.49

Ptolemy's Theorem asserts that

$$AC \cdot BD = AB \cdot CD + BC \cdot AD$$

(a) Choose the diameter of the circle in Figure 4.49 to be 1, and show that Ptolemy's Theorem is equivalent to

$$\sin(\gamma + \delta) \sin(\gamma + \beta) = \sin \gamma \sin \alpha + \sin \delta \sin \beta$$

(b) Show that

$$\begin{aligned}\sin(\gamma + \delta) \sin(\gamma + \beta) &= \frac{1}{2}[\cos(\delta - \beta) - \cos(2\gamma + \delta + \beta)] \\ &= \frac{1}{2}[\cos(\delta - \beta) + \cos(\gamma - \alpha)]\end{aligned}$$

(Use the fact that $\alpha + \beta + \gamma + \delta = \pi$.)

(c) Show that

$$\begin{aligned}\cos(\delta - \beta) + \cos(\gamma - \alpha) &= \cos\delta\cos\beta + \sin\delta\sin\beta + \cos\gamma\cos\alpha + \sin\gamma\sin\alpha \\ &= 2(\sin\gamma\sin\alpha + \sin\delta\sin\beta)\end{aligned}$$

by showing that

$$\cos(\delta + \beta) = -\cos(\gamma + \alpha)$$

(d) Use your results from parts (b) and (c) to prove the equivalent form of Ptolemy's Theorem in part (a).

4. Use Figure 4.50 to prove that Ptolemy's Theorem implies the addition formula for $\sin(\alpha + \beta)$.

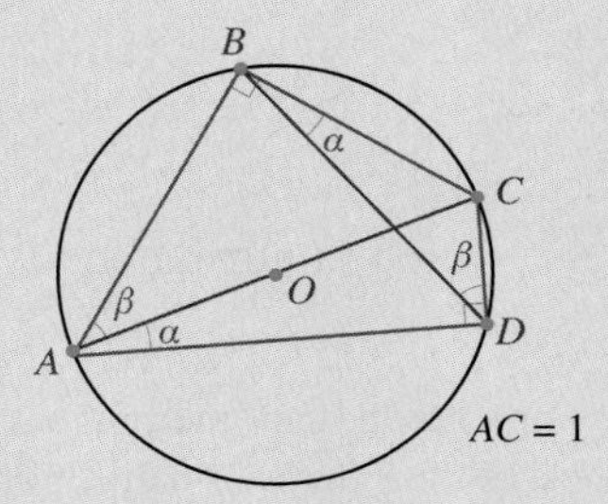

Figure 4.50

5. Use the addition formulas for $\sin(\alpha + \beta)$ and $\cos(\alpha + \beta)$ to express $\tan(\alpha + \beta)$ in terms of $\tan \alpha$ and $\tan \beta$.
6. Show that $\tan(-\theta) = -\tan\theta$, and then use your answer to part 5 to prove that

$$\tan(\alpha - \beta) = \frac{\tan\alpha - \tan\beta}{1 + \tan\alpha\tan\beta}$$

Morley's Theorem

In 1899, Frank Morley, an English-born American mathematician, discovered an amazing property of triangles (also introduced in Chapter 0):

> The points of intersection of the adjacent angle trisectors of the angles of any triangle ($\triangle ABC$ in Figure 4.51) are the vertices of an equilateral triangle ($\triangle PQS$).

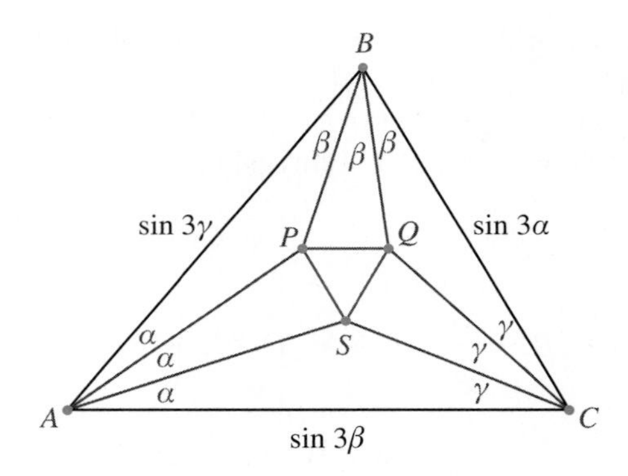

Figure 4.51

This surprising result is frequently referred to as "Morley's miracle." Many different proofs of the theorem have been published, and new proofs are posted on the web from time to time. Synthetic Euclidean-type proofs start with an equilateral triangle and then reconstruct the trisectors. Proofs that start with an arbitrary triangle and then verify that the triangle created by the trisectors is equilateral use trigonometry. Here is one such proof.

In Figure 4.51, $\triangle ABC$ is an arbitrary triangle with angles 3α, 3β, and 3γ. For convenience, we choose the diameter of the circumscribing circle to be 1. Then, using the law of sines, we find the sides of $\triangle ABC$ to be $\sin 3\alpha$, $\sin 3\beta$, and $\sin 3\gamma$. Our plan is to use the law of sines again to find AP in $\triangle APB$ in terms of α, β, and γ and to similarly find AS in $\triangle ASC$. Then we can use the law of cosines to find PS in $\triangle APS$. Because we anticipate getting the same expression for PQ and QS, we endeavor to obtain an expression that is symmetric in α, β, and γ for PS (i.e., an expression in which a permutation of the angles will not change its value).

In $\triangle ABP$,

$$\frac{AP}{\sin\beta} = \frac{\sin 3\gamma}{\sin(\angle APB)} \tag{4.74}$$

Because $sin(\angle APB) = \sin[\pi - (\alpha + \beta)] = \sin(\alpha + \beta)$, and $3\alpha + 3\beta + 3\gamma = \pi$, Equation 4.74 implies

$$AP = \frac{\sin\beta \sin 3(\alpha + \beta)}{\sin(\alpha + \beta)} \tag{4.75}$$

Similarly, in $\triangle ASC$,

$$\frac{AS}{\sin\gamma} = \frac{\sin 3\beta}{\sin(\alpha + \gamma)}$$

or

$$AS = \frac{\sin\gamma \sin 3(\alpha + \gamma)}{\sin(\alpha + \gamma)}. \tag{4.76}$$

Notice that Equations 4.74 and 4.76 contain ratios of the form $\frac{\sin 3x}{\sin x}$. In an attempt to simplify the expressions for AP and AS, we write $\sin 3x$ in terms of functions of x:

$$\begin{aligned}\sin 3x = \sin(2x + x) &= \sin 2x \cos x + \cos 2x \sin x \\ &= (2\sin x \cos x)\cos x + (\cos^2 x - \sin^2 x)\sin x \\ &= \sin x(3\cos^2 x - \sin^2 x)\end{aligned}$$

Thus

$$\frac{\sin 3x}{\sin x} = 3\cos^2 x - \sin^2 x \tag{4.77}$$

Because $\alpha + \beta + \gamma = \frac{\pi}{3}$ and $\sin\frac{\pi}{3} = \frac{\sqrt{3}}{2}$, we write Equation 4.77 as a product as follows:

$$\begin{aligned}3\cos^2 x - \sin^2 x &= 4\left[\left(\frac{\sqrt{3}}{2}\cos x\right)^2 - \left(\frac{1}{2}\sin x\right)^2\right] \\ &= 4\left(\sin\frac{\pi}{2}\cos x - \cos\frac{\pi}{3}\sin x\right)\left(\sin\frac{\pi}{3}\cos x + \cos\frac{\pi}{3}\sin x\right) \\ &= 4\sin\left(\frac{\pi}{3} - x\right)\sin\left(\frac{\pi}{3} + x\right)\end{aligned}$$

Hence

$$\frac{\sin 3x}{\sin x} = 4\sin\left(\frac{\pi}{3} - x\right)\sin\left(\frac{\pi}{3} + x\right) \tag{4.78}$$

To simplify Equation 4.75, we substitute $x = \alpha + \beta$:

$$\begin{aligned}\frac{\sin 3(\alpha + \beta)}{\sin(\alpha + \beta)} &= 4\sin\left(\frac{\pi}{3} + (\alpha + \beta)\right)\sin\left(\frac{\pi}{3} - (\alpha + \beta)\right)\\ &= 4\sin\left(\frac{2\pi}{3} - \gamma\right)\sin\gamma\\ &= 4\sin\left[\pi - \left(\frac{2\pi}{3} - \gamma\right)\right]\sin\gamma\\ &= 4\sin\left(\frac{\pi}{3} + \gamma\right)\sin\gamma\end{aligned}$$

Thus

$$AP = 4\sin\beta\sin\gamma\sin\left(\frac{\pi}{3} + \gamma\right) \tag{4.79}$$

Similarly,

$$AS = 4\sin\beta\sin\gamma\sin\left(\frac{\pi}{3} + \beta\right) \tag{4.80}$$

Next, we apply the law of cosines in $\triangle APS$ and substitute for AP and AS the expressions in Equations 4.79 and 4.80:

$$(PS)^2 = (AP)^2 + (AS)^2 - 2(AP)(AS)\cos\alpha \tag{4.81}$$

$$(PS)^2 = 4^2\sin^2\beta\sin^2\gamma\left[\sin^2\left(\beta + \frac{\pi}{3}\right) + \sin^2\left(\gamma + \frac{\pi}{3}\right) - 2\sin\left(\beta + \frac{\pi}{3}\right)\sin\left(\gamma + \frac{\pi}{3}\right)\right]$$

Because we expect to obtain an expression that is symmetric in α, β, and γ, we anticipate that the quantity in the brackets in Equation 4.81 can be simplified as $\sin^2\alpha$. This can be shown to be the case by manipulating (with some effort) the expression and using the fact that $\alpha + \beta + \gamma = \frac{\pi}{3}$. Alternatively,

$$\left(\beta + \frac{\pi}{3}\right) + \left(\gamma + \frac{\pi}{3}\right) + \alpha = (\alpha + \beta + \gamma) + \frac{2\pi}{3} = \frac{\pi}{3} + \frac{2\pi}{3} = \pi$$

Thus there exists a triangle with angles $\left(\beta + \frac{\pi}{3}\right)$, $\left(\gamma + \frac{\pi}{3}\right)$, and α. Because all such triangles are similar, we can choose one whose circumscribing circle has a diameter of length 1. Then the side opposite to any of the angles of the triangle equals the sine of that angle. Hence we have the sides whose lengths are shown in Figure 4.52.

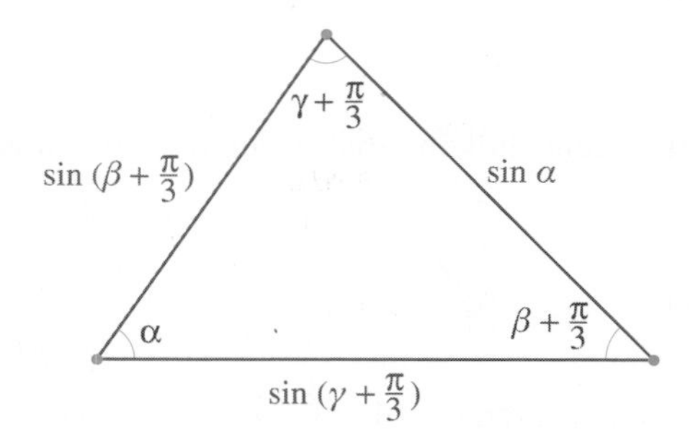

Figure 4.52

Applying the law of cosines to the triangle in Figure 4.52, we get

$$\sin^2 \alpha = \sin^2\left(\beta + \frac{\pi}{3}\right) + \sin^2\left(\gamma + \frac{\pi}{3}\right) - 2\sin\left(\beta + \frac{\pi}{3}\right)\sin\left(\gamma + \frac{\pi}{3}\right)\cos \alpha$$

Thus Equation 4.81 becomes

$$\begin{aligned} (PS)^2 &= 4^2 \sin^2 \beta \sin^2 \gamma \sin^2 \alpha \\ PS &= 4 \sin \beta \sin \gamma \sin \alpha \end{aligned} \tag{4.82}$$

Because of the symmetry of Equation 4.82, each side of $\triangle PQS$ (Figure 4.51) equals $4 \sin \alpha \sin \beta \sin \gamma$. □

Historical Note: Frank Morley

Frank Morley (1860–1937) was born into a Quaker family in England. In 1887, he took a position as an instructor at the Quaker College in Haverfold, Pennsylvania (now Haverfold College), but was promptly promoted to professor. On his informative website, Alexander Bogomonly writes:

> In 1899, Frank Morley, then professor of mathematics at Haverfold College, came across a result so surprising that it entered mathematical folklore under the name "Morley's Miracle." Morley's original proof stemmed from his results on algebraic curves tangent to a given number of lines.

In 1900, Morley became professor of mathematics at Johns Hopkins University. During his tenure there, Morley supervised 48 doctoral students, was the editor of the *American Journal of Mathematics* for 30 years, and served as president of the American Mathematical Society.

Christopher Morley (1890–1957), one of Frank Morley's three sons, was a well-known and highly regarded American novelist.

Now Solve This 4.24

1. Prove that the side of the equilateral triangle in Morley's Theorem is equal to $8R \sin \alpha \sin \beta \sin \gamma$, where R is the radius of the circumscribing circle of the original triangle.
2. If the original triangle is equilateral, find the ratio of the side of the smaller triangle to the side of the original triangle (use a calculator to find the value of $\sin \frac{\pi}{9}$).

Problem Set 4.7

1. a. Give a plausible argument showing that, when α is measured in radians,

$$\lim_{a\to 0}\frac{\sin\ \alpha}{\alpha}=1$$

b. Prove that when x is measured in radians, $\frac{d}{dx}(\sin x) = \cos x$.

c. Find the derivative of $\sin x$ when x is measured in degrees.

•**2.** In a circle of radius R, central angle α intercepts an arc of length ℓ.

•**a.** Find ℓ in terms of α and R when α is measured (i) in radians and (ii) in degrees.

b. Find the area of the shaded sector in terms of α and R when α is measured (i) in radians and (ii) in degrees.

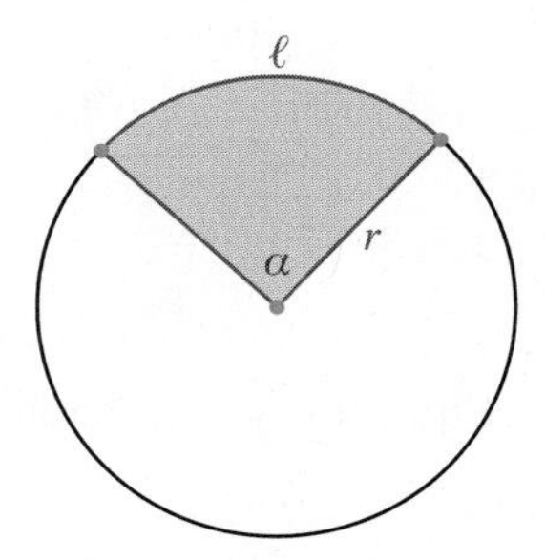

3. When a problem is solved in two different ways and the two answers are set equal to each other, a useful result is often obtained. For example, consider finding x in the figure below. Show that $x = AC - DC = \cot\beta - \cot\alpha$. Then find an alternative solution by applying the law of sines to $\triangle ABD$ to show that $\frac{x}{\sin(\alpha-\beta)} = \frac{1}{\sin\alpha\,\sin\beta}$. Then derive the formula for $\sin(\alpha - \beta)$ in terms of functions of α and β.

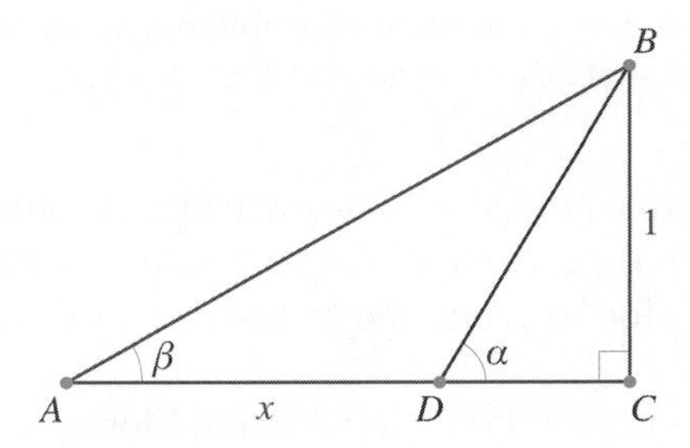

•**4.** Without using a calculator, prove that

a. If $AE = ED = DC = BC$, then $\alpha + \beta + \gamma = 90°$.

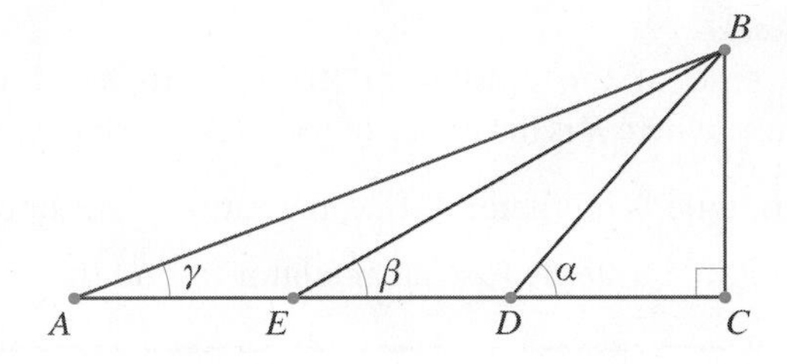

b. If α and β are angles in the following figure, then $\alpha + 2\beta = 45°$.

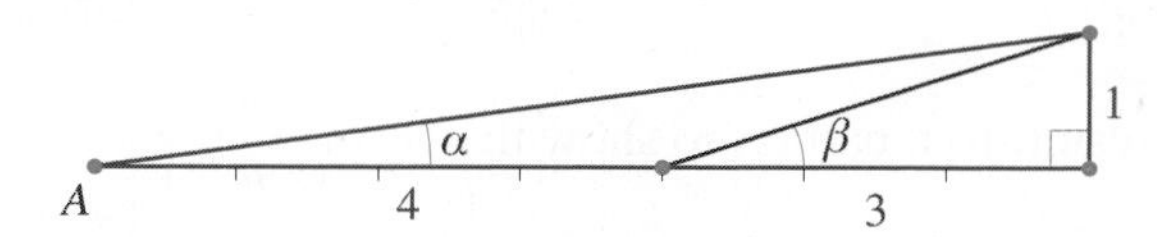

•**5.** Prove that

a. $\tan(\alpha + k\pi) = \tan \alpha$ for any integer k.

b. $\tan(\pi - \alpha) = \tan(-\alpha) = -\tan \alpha$.

•**c.** If $\alpha + \beta + \gamma = \pi$, then $\tan \alpha + \tan \beta + \tan \gamma = \tan \alpha \tan \beta \tan \gamma$.

6. Let $A(x_1, y_1)$ and $B(x_2, y_2)$ be any two points on a nonvertical line, and let α be the angle that the line makes with the x-axis as shown in the figure. Prove that $\tan \alpha = \dfrac{y_2 - y_1}{x_2 - x_1}$; in other words, prove that $\tan \alpha$ is the slope of the line. Distinguish between the case when α is acute and the case when α is obtuse.

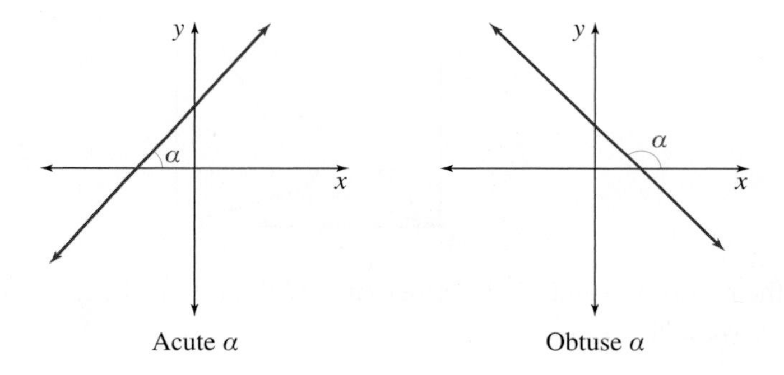

Acute α Obtuse α

7. Lines ℓ_1 and ℓ_2 with slopes m_1 and m_2, respectively, make angles α_1 and α_2, respectively, with the positive x-axis. If θ is an angle between the two lines as shown in the figure, prove the following:

a. If the lines are perpendicular, then $\tan \alpha_1 = \tan\left(\dfrac{\pi}{2} + \alpha_2\right)$ and therefore $m_1 m_2 = -1$.

b. $\tan \theta = \dfrac{m_1 - m_2}{1 + m_1 m_2}$

c. For $\theta = 90°$, part (b) implies $m_1 m_2 = -1$.

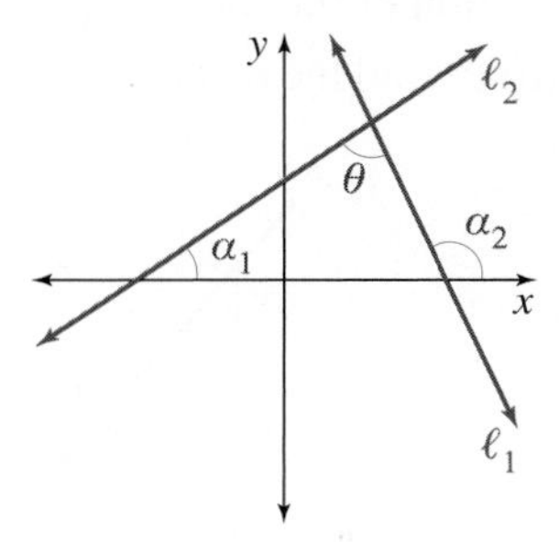

8. **a.** Prove that the perimeter of a regular n-gon inscribed in a circle of radius r is $2rn\left(\sin\frac{180°}{n}\right)$ and the perimeter of the circumscribed regular polygon is $2rn\left(\tan\frac{180°}{n}\right)$.

b. Use your result from part (a) to show that $n\sin\frac{180°}{n} < n\tan\frac{180°}{n}$.

c. Use the inequalities in part (b) and a calculator to approximate π to eight correct digits after the decimal point.

d. In view of this method for calculating π, why do you think that in Section 4.5 we used more complicated formulas involving radicals?

•**9.** Find each of the following in terms of radicals:

•**a.** $\sin 15°$ **b.** $\cos 75°$ **c.** $\sin 7.5°$

10. **a.** Prove that $\cos 3\alpha = 4\cos^3\alpha - 3\cos\alpha$.

b. Find the value of $\sin 18°$ in terms of radicals by noticing that $\sin 36° = \cos(90° - 36°) = \cos 54°$, $36 = 2 \cdot 18$, and $54 = 3 \cdot 18$.

•**11.** In the figure, $ABCD$ is a square. $\angle ADE$ and $\angle EBC$ are 75° each. Prove that $\triangle DEC$ is equilateral.

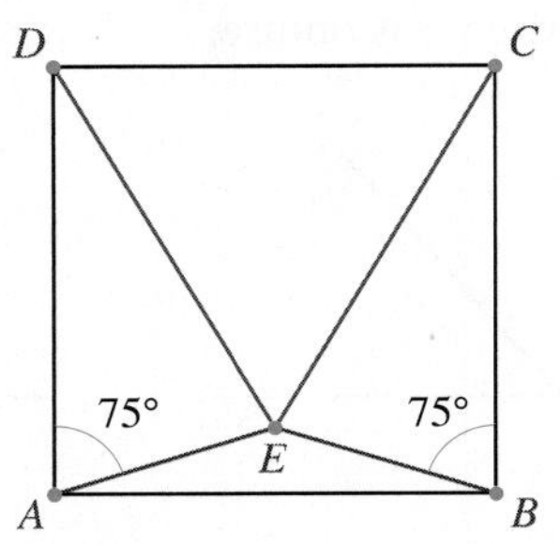

12. If r is the radius of the circle inscribed in $\triangle ABC$ and R is the radius of the circumscribed circle, prove that

$$\frac{r}{R} = 4\sin\frac{\alpha}{2}\sin\frac{\beta}{2}\sin\frac{\gamma}{2}$$

where α, β, and γ are the angles of the triangle.

•**13.** **a.** Prove that the equation of the straight line through $(a, 0)$ and $(0, b)$ can be written in the form $\frac{x}{a} + \frac{y}{b} = 1$.

b. A line not through the origin can be determined by the perpendicular $\overline{OP}$ and the angle θ it makes with the x-axis, as shown in the figure. If $OP = d$, use your result from part (a) to show that the equation of the line is $x\cos\theta + y\sin\theta - d = 0$.

•**c.** Use the result in part (b) to find the distance from the line $Ax + By + C = 0$ to the origin in terms of A, B, and C.

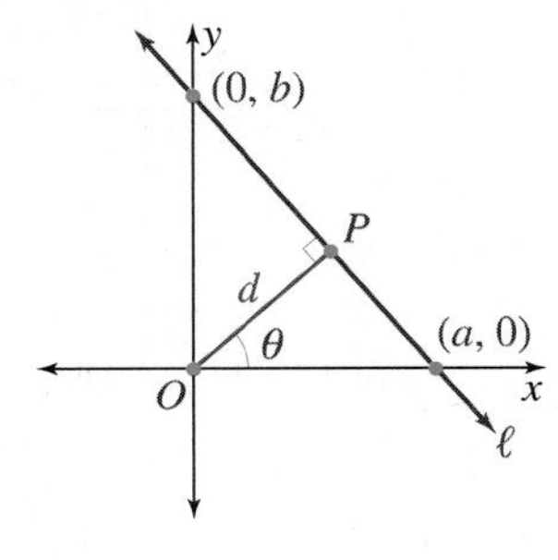

5 Isometries and Size Transformations

To gain access to a greater number of designs, I used transformational geometry techniques including reflections, glide reflections, translations, and rotations.

—M. C. Escher, 1898–1972,
artist and mathematician at heart

Introduction

In *The Elements* (about 300 B.C.E.), Euclid attempted to prove the congruence of two triangles by moving one triangle so that it coincides with the other. However, such "motions" were not defined and had only an intuitive meaning for an individual figure. In 1827, the German astronomer and mathematician A. F. Möbius (1790–1868) gave a meaning to the composition of motions by extending the idea of motion to the whole plane. In this chapter, we define a "motion" as an isometry in the plane and investigate properties of three types of isometries: translations, reflections, and rotations. We also discover a fourth isometry—glide reflection—and use isometries to solve problems that were investigated in the eighteenth and nineteenth centuries and that are much more difficult to solve using traditional Euclidean geometry.

5.1 Reflections, Translations, and Rotations

Symmetry is perhaps one of the most recognizable properties of figures. One type of symmetry is shown in Figure 5.1. We say that each figure is symmetrical about the respective line, or that it has a line symmetry. (A formal definition will be given later.)

If a mirror was placed perpendicular to the plane of a drawing, then the image of each drawing would be the drawing itself. A convenient device that behaves like a mirror is a Mira. A Mira is a transparent plastic device that enables one to see the reflection of a given figure behind the Mira. (For more information on the Mira, see http://homepage.mac.com/efithian/Geometry/Activity-05.html.)

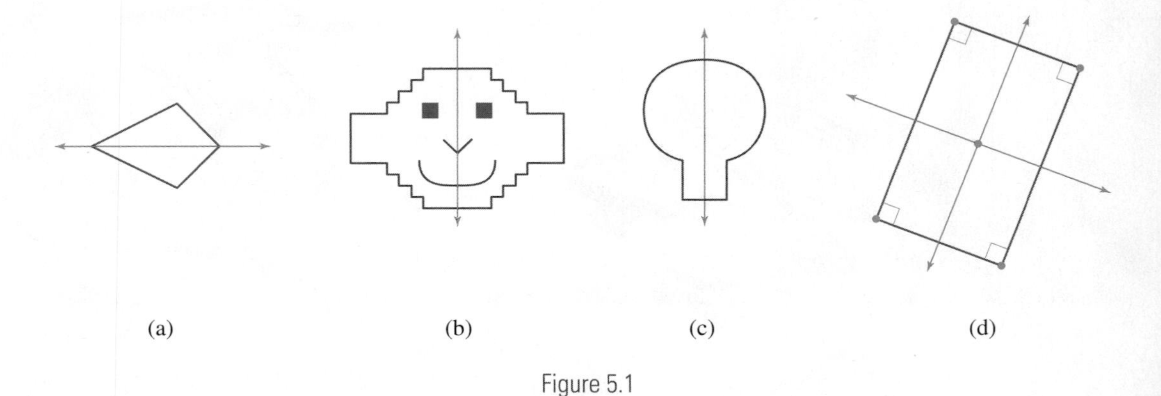

Figure 5.1

Reflections

To investigate symmetries, and in particular line symmetry, we need a definition of a reflection in a line. In Figure 5.2a, point P is reflected in line ℓ, and the image P' is such that $\overline{PP'}$ is perpendicular to ℓ and P' is the same distance from ℓ as is P; in other words, ℓ is the perpendicular bisector of $\overline{PP'}$. If P is on ℓ in Figure 5.2b, its image P' is P itself.

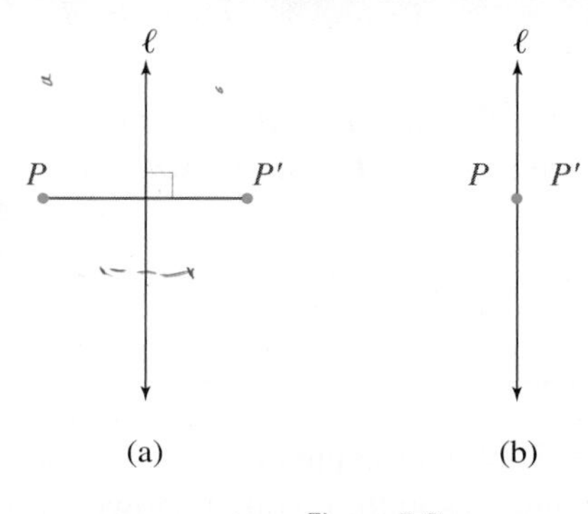

Figure 5.2

The preceding discussion leads to the following definition of a **reflection in a line**.

Definition of a Reflection in a Line A reflection in a line ℓ is a correspondence that pairs each point P in the plane and not on ℓ with point P' such that ℓ is the perpendicular bisector of $\overline{PP'}$. If P is on ℓ, then P is paired with itself; that is, $P' = P$.

Notice that reflection in a line is a function that assigns points in the plane to other points in the plane. The domain of this function is the plane (i.e., the set of points in the plane), and the values of the function are points in the plane. In geometry and other areas of mathematics, a function is frequently referred to as a **mapping**. Thus reflection is a mapping from the plane to the plane.

If we denote the reflection in line ℓ by M_ℓ (M for mirror and ℓ for the line of reflection), then if P is assigned to P' we may write $M_\ell(P) = P'$. The point P' is the value of the function at P and is commonly referred to as the **image** of P. The point P is called the **preimage** of P'. Given any point B in the plane, we can always find its preimage—that is, a point A whose image under M_ℓ is B or $M_\ell(A) = B$. Because every point in the plane has a preimage, the **range** of M_ℓ (the set of all images) is the whole plane and the mapping is **onto**. Consequently, M_ℓ maps the plane onto the entire plane. Also, notice that different points in the plane have different images under M_ℓ;

that is, whenever $A \neq B$, we have $M_\ell(A) \neq M_\ell(B)$ (or equivalently $A' \neq B'$). Thus M_ℓ is a one-to-one mapping. A reflection in a line is an example of a **transformation of the plane**.

Definition of a Transformation of the Plane A transformation of the plane is a one-to-one and onto mapping of the plane to the plane.

Notice that a transformation of the plane is a function whose domain is the set of points in the plane, whose range is the same as the domain, and which has the property that different points have different images. From now on, we will refer to transformations of the plane simply as transformations.

If P is a point in the plane and T is some transformation, we say that $T(P)$ is the image of P under the transformation T. We frequently denote the image by P'. Thus $T(P) = P'$ and the point P is the preimage of P'. We also say that T **transforms** P to P'. As for any function, $T_1 = T_2$ if and only if $T_1(P) = T_2(P)$ for all points P in the plane. Because transformations are one-to-one and onto functions from the plane to the plane, the composition of two transformations is a transformation, and each transformation has an inverse.

Let I denote the identity function for the plane. That is, under I each point in the plane is transformed onto itself. We have the following definition:

Definition of the Identity Function for the Plane I is the identity function for the plane if $I(P) = P$, for all P in the plane.

Notice that T and I are names of functions, whereas $T(P)$ and $I(P)$ are values of the corresponding functions. If T_1 and T_2 are two transformations, then the **composition of T_1 with T_2** is a transformation written as $T_2 \circ T_1$. Notice that first T_1 acts on a point, and then T_2 acts on its image.

Definition of $T_2 \circ T_1$: The Composition of T_1 Followed by T_2 $(T_2 \circ T_1)(P) = T_2(T_1(P))$, for all P in the plane.

If T is a transformation, then the **inverse of T** (denoted by T^{-1}) is also a transformation. If $T(P) = Q$, then T^{-1} sends Q back to P. More formally, we have the following definition:

Definition of the Inverse T^{-1} of T $T^{-1}(Q) = P$ if and only if $Q = T(P)$.

We can picture the actions of T and T^{-1} in Figure 5.3, where the arrow at Q indicates that P is sent to Q by T and the arrow below it indicates that Q is sent back to P by T^{-1}. Notice that

$$(T^{-1} \circ T)(P) = T^{-1}(T(P)) = T^{-1}(Q) = P$$

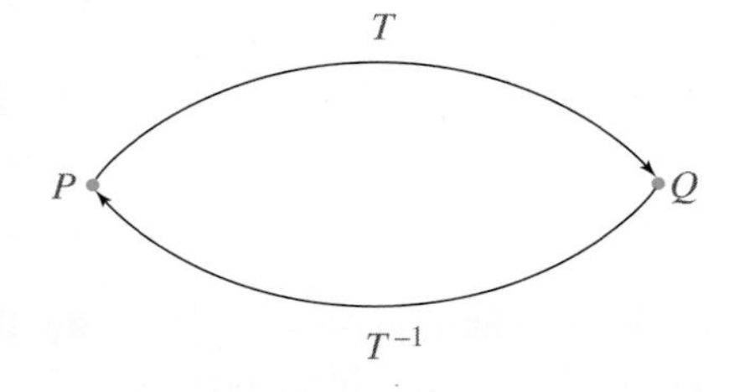

Figure 5.3

for every P in the plane. Hence $T^{-1} \circ T$ is the identity I. Similarly, $T \circ T^{-1}$ is the identity. Notice that because the range of T is the same as the domain (i.e., the plane), $T^{-1} \circ T = T \circ T^{-1} = I$.

Which properties does a reflection in a line have? It seems that the length of the image of a segment is always the same as the length of the segment. In other words, the distance between the images of any two points is the same as the distance between the original points. More succinctly, we say that reflection **preserves distance**. It also seems that the image of the segment $\overline{AB}$ is the segment $\overline{A'B'}$ and the image of the line $\overleftrightarrow{AB}$ is the line $\overleftrightarrow{A'B'}$. (The image of a geometric figure is defined as the set of all images of the points of the original figure.) In the problem set at the end of this section, we will investigate proofs of the fact that if A, B, and C are collinear (on the same line), their respective images A', B', and C' are also collinear.

Because a reflection preserves distance, the image of a triangle under a reflection is a congruent triangle. As we shall soon see, this is the case for any transformation that preserves distance. We will study such transformations and give them a special name: isometries.

homework

Definition of Isometry An isometry is a transformation that preserves distance. In other words, an isometry is a transformation T of the plane such that for every two points A and B, the distance between A and B equals the distance between their images $T(A)$ and $T(B)$.

Notice that $T(A)$, the image of A, can also be written as A'. Using this notation, a transformation is an isometry if and only if for every two points A and B we have $A'B' = AB$. If the distance between two points P and Q is denoted by $d(P,Q)$, then a transformation T of the plane is an isometry if for every two points A and B in the plane, $d(T(A), T(B)) = d(A, B)$. The word "isometry" is derived from the Greek words *isos* (meaning "equal") and *metron* (meaning "measure"), and is also commonly referred to as **rigid motion**.

Example 5.1 Given the nonperpendicular intersecting lines m and n in Figure 5.4, a mapping F from the plane to the plane is defined as follows: If P is on n, then $P' = P$. If P is not on n, then P' is the point for which $\overline{PP'}$ is parallel to m and bisected by n.

1. Is F a transformation of the plane?
2. Is F an isometry?

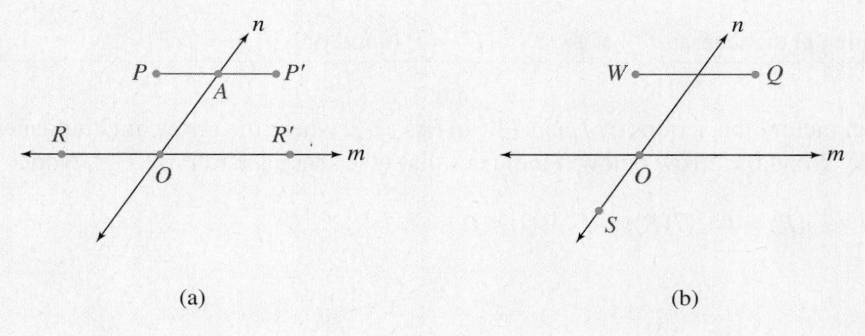

Figure 5.4

Solution

1. F is a transformation if it is onto and one-to-one. To show that F is onto, we need to demonstrate that every point on the plane has a preimage. In Figure 5.4b, the preimage of Q is W (and vice versa), and the preimage of S on n is S itself. Hence F is onto. It is

23. Let f be a function whose domain is $\mathbb{R}$ (the set of real numbers) and whose values are in $\mathbb{R}$. What kind of symmetry does each of the following functions have? Justify your answers.

a. $g(x) = \dfrac{f(x) + f(-x)}{2}$

b. $h(x) = \dfrac{f(x) - f(-x)}{2}$

24. **a.** Consider all lines $y = ax + b$. Which of these lines have the line $y = x$ as their line of symmetry?

★**b.** Find all curves $y = \dfrac{ax + b}{cx + d}$ whose line of symmetry is the line $y = x$. (*Hint:* You need to find all values of a, b, c, and d.)

25. **a.** Prove that a rotation R by 90° counterclockwise about the origin is given by $R(x, y) = (-y, x)$. Check that $R^2 = R \circ R = H_0$.

b. Find the equations of the images of the curves in Problem 21 under the rotation given in part (a) of this problem.

c. Use your result from part (a) to prove that the product of the slopes of two perpendicular lines, neither of which is vertical, is -1.

5.2 Congruence and Euclidean Constructions

Earlier, we defined two polygons as being congruent if there exists a one-to-one correspondence between their vertices such that the corresponding sides are congruent and the corresponding angles are congruent. Unfortunately, this definition does not generalize to figures that are not polygons, such as the two smooth curves in Figure 5.15.

Figure 5.15

Euclid's idea of congruence was based on the assumption that figures can be moved without changing their size or shape. Thus, if we could move one of the figures so that, when one is superimposed on the other, the figures coincide, then the figures would be congruent. As an example, trace Figure 5.15a on a tracing paper and try to move it by sliding and rotating this figure so that when superimposed on Figure 5.15b, it will coincide with the latter figure. You will find that this is impossible to accomplish. However, if you flip the tracing paper and trace Figure 5.15b on the other side of the paper, you should be able to move it so that it will coincide with Figure 5.15a. Because a formal concept of a function, and hence a transformation, was developed only in the nineteenth century, Euclid did not have the means to rigorously define his intuitively appealing approach to congruence. By contrast, our study of transformations—and in particular isometries—enables us to define congruence between any two figures.

Flipping the paper is equivalent to reflecting a figure in a line. Thus it seems that one figure can be made to coincide with the other by reflecting one of the figures in a line if necessary, and then translating or rotating it appropriately. Thus we could define figure S to be congruent with figure S' if there exists a sequence of isometries that maps S onto S'. Because any composition of isometries is an isometry, we could simply require the existence of an isometry that maps S onto S'. Also notice that if an isometry maps S onto S', then its inverse maps S' onto S. Thus, if S is congruent to S', then S' is congruent to S. We summarize this discussion in the following definition.

Definition of Congruence Two figures are congruent if there exists an isometry that maps one onto the other.

Now Solve This 5.7

In Figure 5.16, the ellipse E can be mapped onto the ellipse E' by a composition of two isometries. First use tracing paper to superimpose E onto E', and then write a composition of two familiar isometries that maps E onto E' (use the centers O and O' of the ellipses in your description of the isometries).

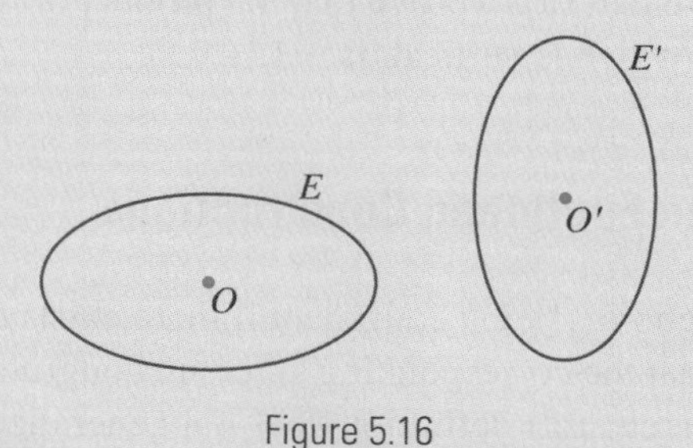

Figure 5.16

Applications of Isometries to Construction Problems

Transformations can be used to solve geometrically various maximum and minimum problems without the help of calculus. In fact, the transformational approach is often simpler and shorter. In Chapter 0 and Example 1.4, you may have worked on the problem of connecting two points on the same side of a line via a shortest path that goes through a point on the line. Because the approach in the solution of this problem is used in subsequent problems, we reintroduce the problem in the following example.

Example 5.3 Given points A and B on the same side of line ℓ (see Figure 5.17), find the point X on ℓ such that $AX + XB$ is minimum.

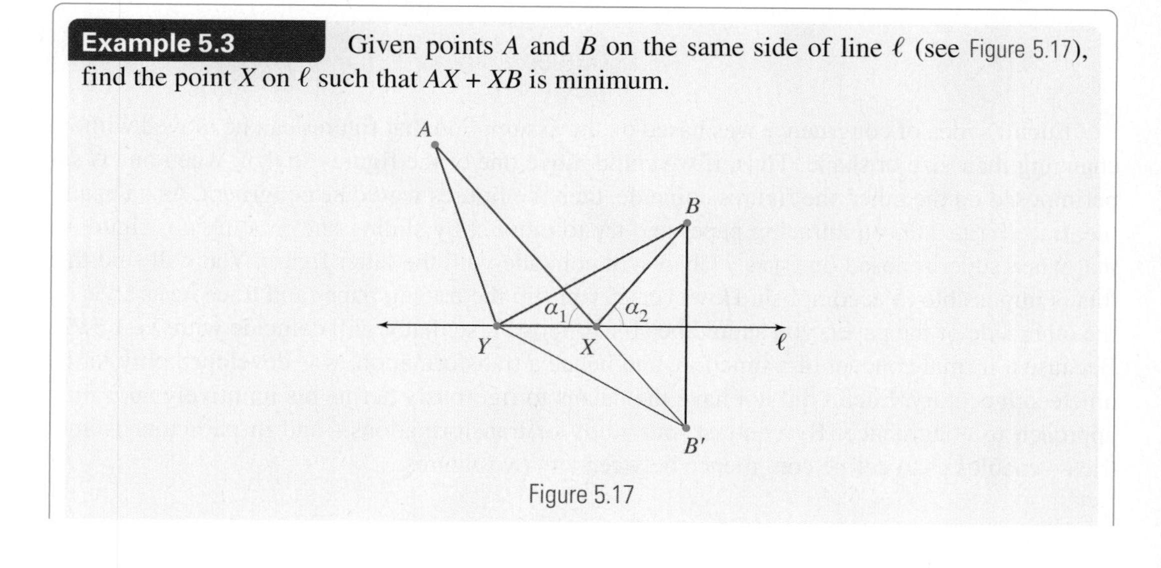

Figure 5.17

In Search of a Strategy A simpler, though trivial, question would be to look at the case where A and B are on opposite sides of ℓ, as in Figure 5.18. In this case, the line segment would intersect ℓ at X and the path $A-X-B$ would be the required path because any other path $A-Y-B$ would be longer (by the triangle inequality applied to $\triangle AYB$).

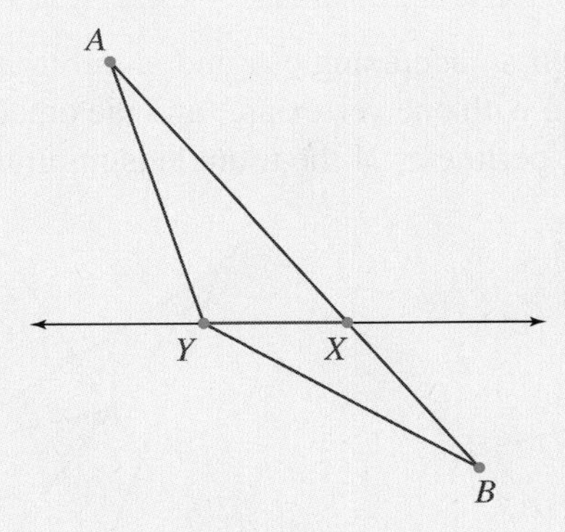

Figure 5.18

This simpler case suggests how to solve the original problem in Figure 5.17. We reflect B in ℓ and obtain its image B'. Because a reflection preserves distance, $XB = XB'$. Consequently, the problem is equivalent to finding a point X on ℓ such that the path $A-X-B'$ is minimal.

Solution

Connect A with B' (the image of B under reflection in ℓ). The intersection of $\overline{AB'}$ with ℓ is the required point X. Figure 5.17 shows why any other path is longer: $AY + YB = YB' + YB' > AB'$ by triangle inequality. Because $AB' = AX + XB$, it follows that $AY + YB' > AX + BX$.

In Figure 5.17, $\alpha_1 = \alpha_2$. This follows immediately from the fact that a reflection preserves the angle measure and that vertical angles are congruent. In fact, the proof of the converse statement—that is, if $\alpha_1 = \alpha_2$, then the path $A-X-B$ (where X is the point on ℓ for which $\alpha_1 = \alpha_2$) is the shortest—can be found by creating a figure similar to Figure 5.17. (This will be explored in Now Solve This 5.8.) This converse statement and its proof were discovered by Heron of Alexandria in the first century B.C.E. The Greeks knew that a ray of light r from a point A meeting a plane mirror m in a point X is reflected in the direction of a point B such that $\overline{AX}$ and $\overline{XB}$ form equal angles with the mirror (Figure 5.19). Heron proved that the path $A-X-B$ is the shortest possible path between A and B by way of the mirror.

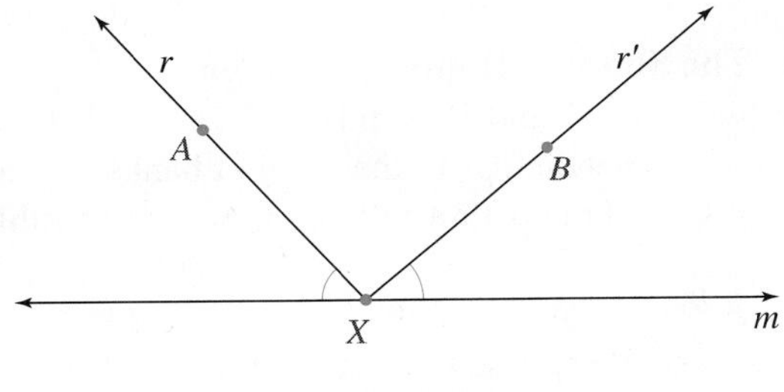

Figure 5.19

The shortest path problem in Example 5.3 can be generalized to include several lines, as shown in Example 5.4.

Now Solve This 5.8

Prove that if X in Figure 5.17 is a point on line ℓ such that $\alpha_1 = \alpha_2$, then the path $A-X-B$ is the shortest among all paths from A to a point on ℓ and then to B.

Example 5.4 Given an acute angle A and an arbitrary point P in the interior of the angle, construct a triangle with one vertex at P and the other two vertices on each of the sides of the angle so that the perimeter of the triangle is minimum. See Figure 5.20a.

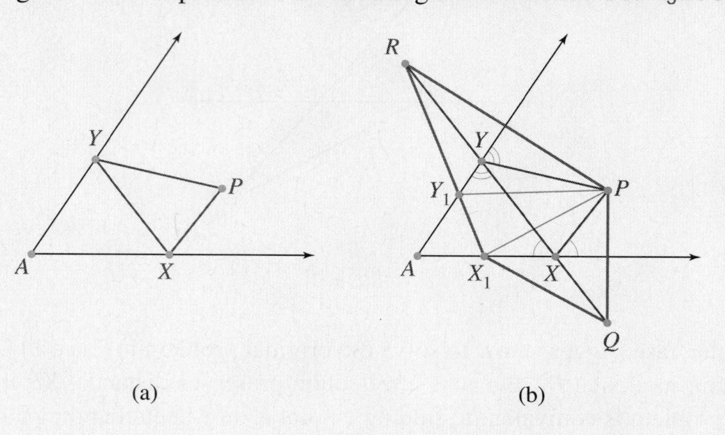

Figure 5.20

Solution

Following the ideas in Example 5.3, we reflect P in each of the sides of the angle and obtain the respective images R and Q as shown in Figure 5.20b. The segment intersects the sides of the angle in points Y and X. We claim that $\triangle PXY$ is the triangle with minimal perimeter. To prove that this is the case, consider any other triangle PX_1Y_1 in Figure 5.20b, The perimeter of that triangle is $RY_1 + Y_1X_1 + X_1Q$, while the perimeter of $\triangle PXY$ is $RY + YX + XQ$, which equals RQ. Because the segment connecting R and Q is the shortest path between R and Q, it follows that $RY_1 + Y_1X_1 + X_1Q > RQ$ and hence that the perimeter of $\triangle PY_1X_1$ is greater than the perimeter of $\triangle PYX$.

Remark Notice that our construction implies that the equally marked angles in Figure 5.20b are congruent. Consequently, if a ray of light emanates from a source P inside an angle whose sides act like mirrors, the light will be reflected from both sides of the angle and will return through P if the light is directed toward point Y (found as described in Example 5.4).

Example 5.5 **The Shortest Highway Problem**

A highway connecting two cities A and B as in Figure 5.21a needs to be built so that part of the highway is on a bridge perpendicular to the parallel banks b_1 and b_2 of a river. Where should the bridge be built so that the path $AXYP$ is as short as possible?

Solution

First we look at a special case. If the distance between b_1 and b_2 is 0, then the shortest path is the segment connecting A and B.

Returning to the original problem, we notice that regardless of where the bridge is located, its length is always the same: It is equal to the width of the river. Consequently, to make $AX + XY + YB$ (see Figure 5.21b) as short as possible, we need simply make $AX + BY$ as short as possible. Applying the "wishful thinking strategy," we wish that Y and X were the same point. Then the problem would be reduced to its special case. This can be achieved by moving $\overline{YB}$. Notice that Y can be moved to X by the translation τ_{YX}, which equals τ_{PQ} where PQ is the width of the river. The image of $\overline{YB}$ in Figure 5.21b under τ_{PQ} is $\overline{XB'}$. Consequently, $YB = XB'$ and therefore $AX + YB = AX + XB'$.

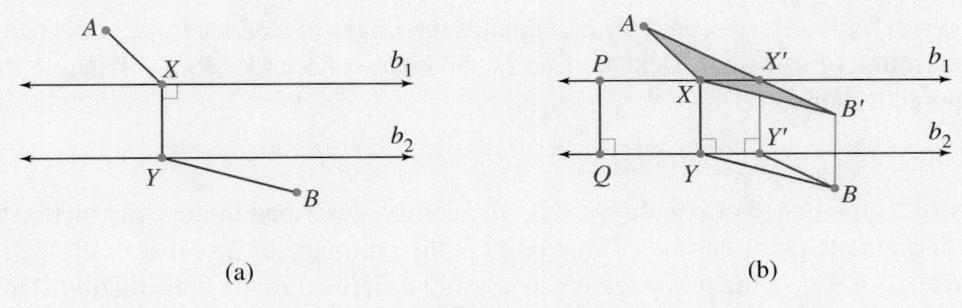

Figure 5.21

To solve the problem, we need to find the point X for which $AX + XB'$ is minimal. Thus we have reduced the problem to its special case. To find the shortest path, we connect A with B'. The point X' where $\overline{AB'}$ intersects b_1 is the location of X for which the highway is the shortest. The point Y' where the perpendicular to b_2 through X' intersects b_2 is the other end of the bridge. The proof that the highway along $\overline{AX'}$–$\overline{X'Y'}$–$\overline{Y'B}$ is the shortest possible path is embedded in the preceding discussion (consider the shaded $\triangle AXB'$). The formal writing of the proof is left for you as an exercise.

Other Construction Problems

All of our examples so far have involved solving minimum-type problems. The next example is somewhat different, however.

Example 5.6 Given three parallel lines a, b, and c and a point P on one of the lines, construct an equilateral triangle PQS that has one vertex on each of the lines.

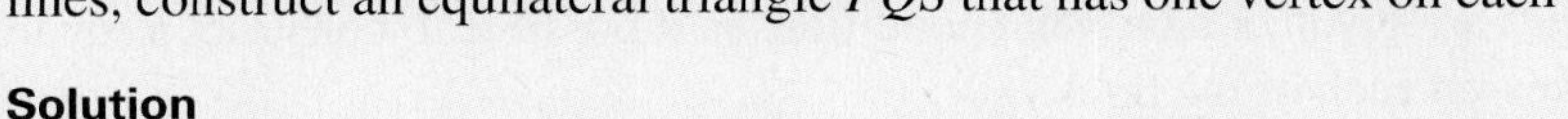

Solution

Suppose P is on b and $\triangle PQS$ is equilateral, as required and as shown in Figure 5.22. We need to find the locations of Q and S. What do we know about Q? Because $\triangle PQS$ is equilateral, if we rotate Q about P by 60° clockwise, we get the point S—that is, the image of Q under $R_{P,\frac{-\pi}{3}}$ is S.

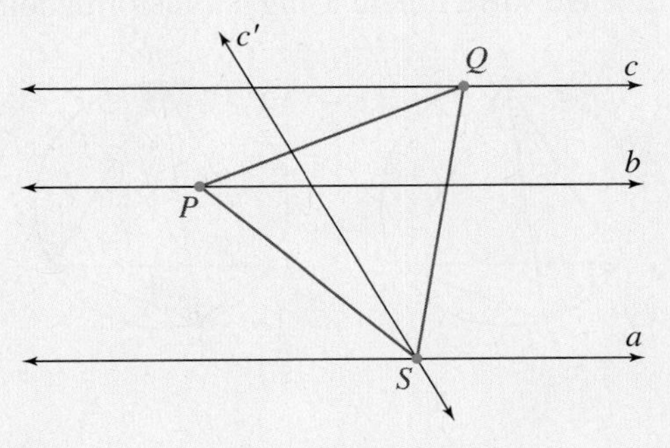

Figure 5.22

Thus, if we knew Q, we could find S. Using a "trial and error" approach, we could try many different points on c, rotate each about P by 60° clockwise, and see which one has its image on a. Of course, it would be better if we could rotate all the points on c about P by 60° clockwise. If c' is the image of c under the rotation, we can check which point on c' is also on a by finding the intersection of a with c': The point of intersection is S. This rotation is accomplished in Figure 5.22. Point Q can now be found in several ways, including by finding the image of S under the rotation $R_{P,\pi/3}$.

Construction

As shown in Figure 5.22, we construct c', which is the image of c under $R_{P,-\pi/3}$. We then find the intersection of c' with a. Next we find Q, the image of S under $R_{P,\pi/3}$. Triangle PQS is the required triangle.

Proof

Because c' is the image of c under $R_{P,-\pi/3}$, the point S described in the construction is the image of a unique point on line c. That point is the preimage of S_1—that is, $R^{-1}_{P,-\pi/3}$. Because $R^{-1}_{P,-\pi/3} = R_{P,\pi/3}$, the preimage is the point Q described in the construction. This implies that $\angle SPQ$ is 60° and $PS = PQ$. Consequently, $\triangle SPQ$ is an isosceles triangle with a vertex angle of 60°. Hence all the angles of $\triangle SPQ$ must be 60° and the triangle is equilateral as required. □

Now Solve This 5.9 includes some observations and questions related to Example 5.6 and its solution.

Now Solve This 5.9

1. In the solution, we assumed that the given point is on line b, which is between a and c. Will a similar construction work if the given point is on any of the parallel lines?
2. A moment of thought will probably convince you that there is actually another triangle that satisfies the conditions of the problem and has point P as a vertex. Can you find it?
3. It is possible to find an equilateral triangle with a vertex on each of any three given curves such as lines or circles. Under what conditions will a solution exist?
4. Suppose four lines are given. Under what conditions is it possible to construct a square that has one vertex on each of the lines?

Example 5.7 In Figure 5.23a $\triangle ABC$ is an equilateral triangle inscribed in a circle and P is any point on the minor arc AC. Prove that $PA + PC = PB$. (This problem was first introduced in Example 2.5. We solve it here using a transformational approach.)

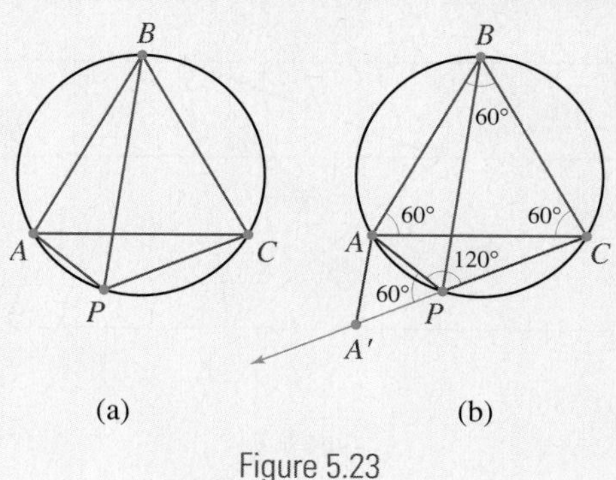

Figure 5.23

Proof

To obtain a single segment whose length is $PA + PC$, we extend the segment PC as shown in Figure 5.23b. Because $ABCP$ is a quadrilateral inscribed in a circle, the opposite angles of the quadrilateral are supplementary. Hence $m(\angle APC) = 120°$ and the supplementary angle $\angle APA'$ is 60°. Consequently, if we rotate point A by 60° about P, its image A' is on line PC and $PA = PA'$. Thus $PA + PC = PA' + PC = A'C$. It remains to prove that $A'C = PB$. To achieve this, we will show that the image of segment $A'C$ under an appropriate isometry is segment PB. Notice that under the rotation $R_{A,60°}$ the image of C is B. Thus we need simply show that the image of A' under that rotation is P. Because $\triangle AA'P$ is equilateral, this is the case and the theorem is proved. (Notice that the image of $\triangle ACA'$ is $\triangle ABP$.) □

Problem Set 5.2

In the following construction problems you should construct your own figures.

• **1.** In the following figure, $ABCD$ is a rectangular swimming pool. You are in a well-marked point P in the pool, and you need to swim to the bank BC, then to CD, then to DA, and finally back to the original point P. You want to minimize your total swimming distance around the path from P and back to P. The figure suggests how to accomplish this goal and how to justify that the path $PQST$ is the shortest. Examine the figure and answer the questions that follow.

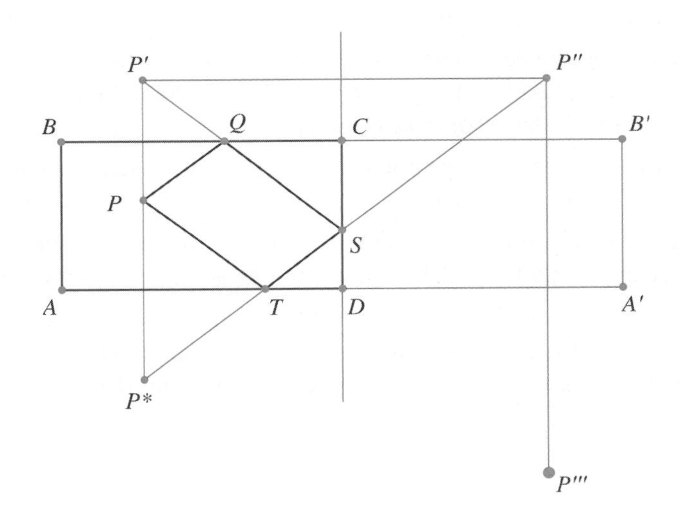

a. Describe how to construct the required path $PQST$.

• **b.** Prove that the path $PQST$ is shorter than any path from P to a point X on BC, then to a point Y on CD, then to a point Z on AD, and back to P.

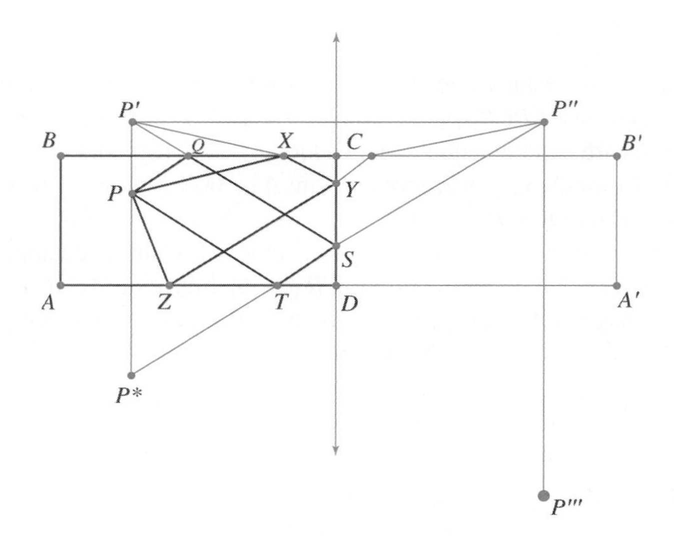

c. If P''' is the image of P'' under reflection in line AD, prove that the intersection of lines PP''' and AD is the point T.

•**d.** Prove that $PQST$ is a parallelogram.

e. Suppose the sides of the rectangle $ABCD$ have lengths a and b, where $BC = a$, $BA = b$, and $a > b$. Make a graph and a coordinate system so that A is at the origin, AD is on the x-axis, and AB is on the y-axis. Prove that $PQST$ is a rectangle if and only if P in the interior of the rectangle is at a distance $(a - b)$ from the side AB or the side CD.

f. Using the coordinate system you developed in part (e), let the coordinate of P be (x_0, y_0) and let P be $a - b$ units away from side AB. If $PQST$ is a rectangle, find the coordinates of Q, S, and T in terms of a, b, x_0, or y_0.

g. Show that there is only one point P in the interior of $ABCD$ such that the path $PQST$ is a square. Where is that point?

h. Suppose the swimming pool is in the form of a square. For which points P will $PQSR$ be a rectangle and for which points will it be a square? Justify your answer.

i. Prove that if P is at the distance $a - b$ from side AB of the rectangle $ABCD$, and a line through P that is perpendicular to side AD is then drawn, the boundary of the shaded region below is a square. Use this result to prove part (h).

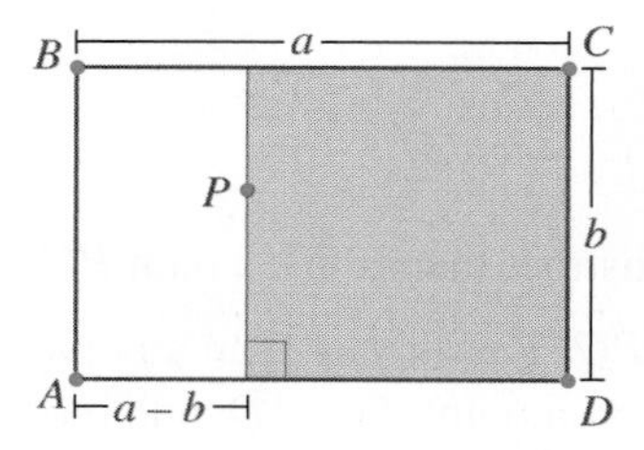

2. In the following figure, B is a ball and H is a hole on a miniature golf course. If you want the ball to bounce off all three walls, describe how to find the points P, Q, and S in the diagram. Assume that the angle of incidence equals the angle of reflection. Prove that the path $B-P-Q-S-H$ is the shortest path connecting B with H through points on all three walls.

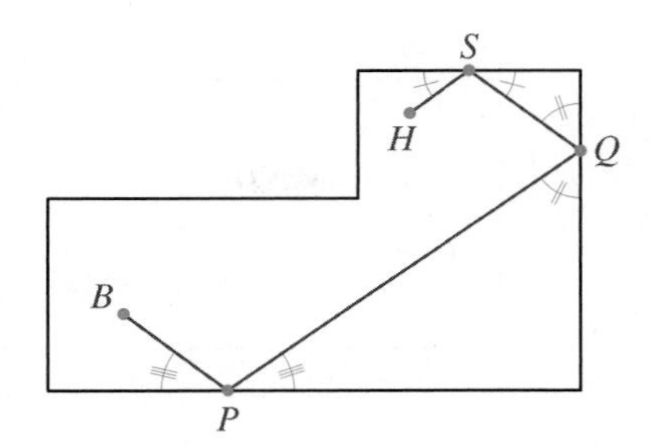

3. A kaleidoscope is in the shape of a prism and has a base in the form of an equilateral triangle whose side is a units. A beam of light is sent from a point P on the base of the prism at a 60° angle and is reflected from the mirrored sides, which are perpendicular to the base.

a. Construct the path that the beam of light follows.

b. Find, in terms of a, the length of the path that the beam of light follows from P to the moment it reaches P again.

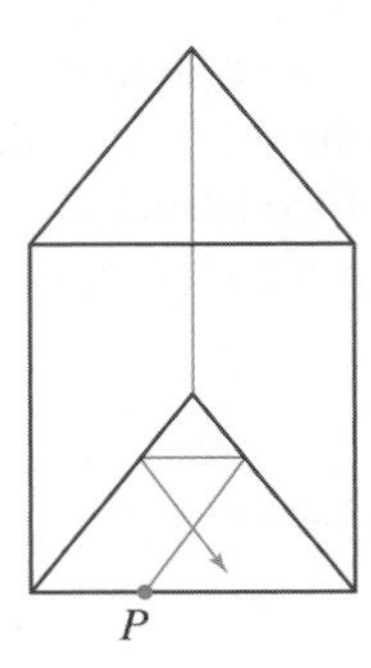

•**4.** Given a point P, a line ℓ, and a circle as shown, construct points X and Y (X on ℓ and Y on the circle) such that P is the midpoint of segment XY. Motivate your steps in the construction (that is, explain how you knew to take the major steps) and prove that your construction is correct.

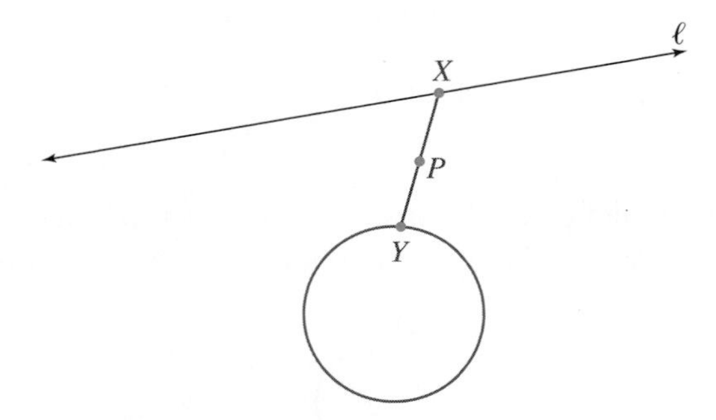

5. $ABCD$ is an a-unit by b-unit rectangular pool table. A ball positioned on side AB is hit toward side BC at a 45° angle. It bounces off the sides as shown in the figure and returns to its original position at P.

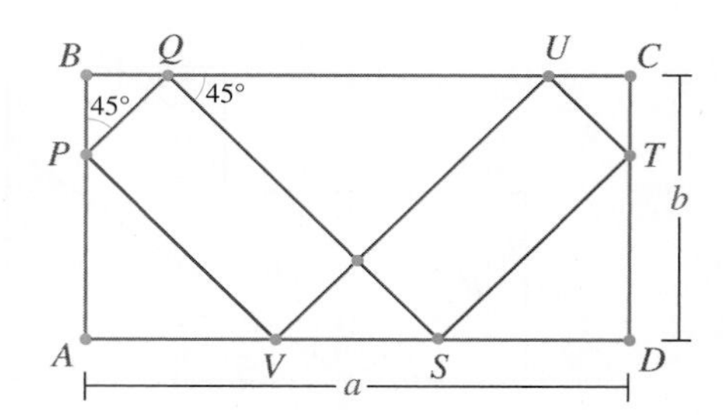

a. Find all the values of a and b for which this phenomenon happens for every point on side AB. [*Hint:* Let $P(0, h)$, where $0 < h < b$, and calculate the coordinates of points Q, U, T, S, and V in terms of h, a, and b.]

b. Suppose that P is at (k, h) in the interior of the rectangle. Use your result from part (a) to find a necessary and sufficient condition (relating k, h, a, and b) such that a ball sent from point P at a 45° angle toward side BC will follow a similar path to the one shown in the figure and return to P.

6. A highway connecting two cities A and B needs to be constructed such that a part of the highway is on a bridge perpendicular to the parallel banks a_1 and a_2 of a river and another part of the highway is on a second bridge perpendicular to the parallel banks b_1 and b_2 of a second river. Where should the bridges be built so that the highway is as short as possible? Describe the construction, construct the bridges and the highway, and prove that the highway you constructed is the shortest option.

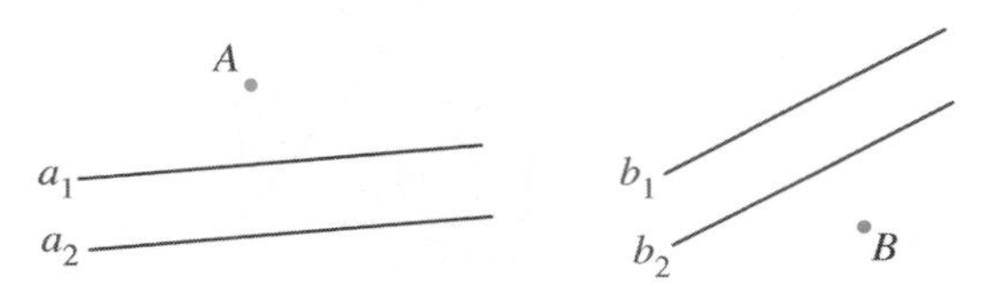

•**7.** A new road connecting streets A and B needs to be constructed. The road, marked XY in the figure, needs to be parallel to PQ and the same length as PQ. Construct the road. Motivate your construction and prove that it is correct.

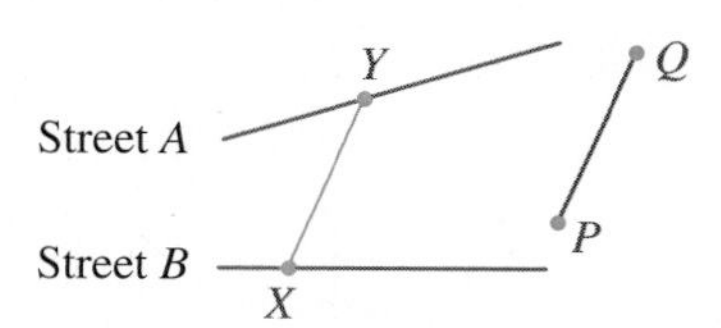

8. Given two circles and point P, construct a segment AB such that $AP = PB$, A is on circle O_1, and B is on circle O_2. How many solutions are there? Is there always a solution?

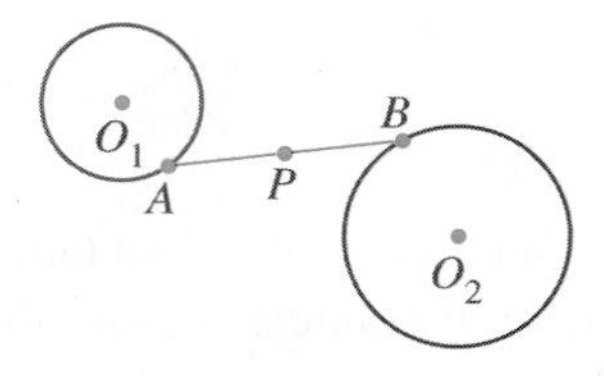

9. Construct an equilateral triangle such that there is exactly one vertex on each of the concentric circles shown.

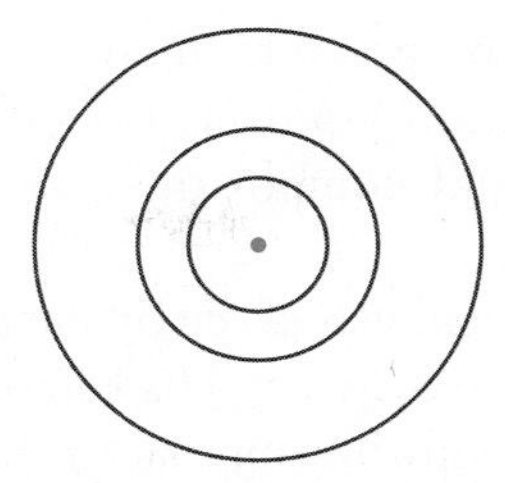

10. Construct two chords of equal length through points A and B that are perpendicular to each other. (*Hint:* If a line is rotated 90° about a point, the line and its image are perpendicular.)

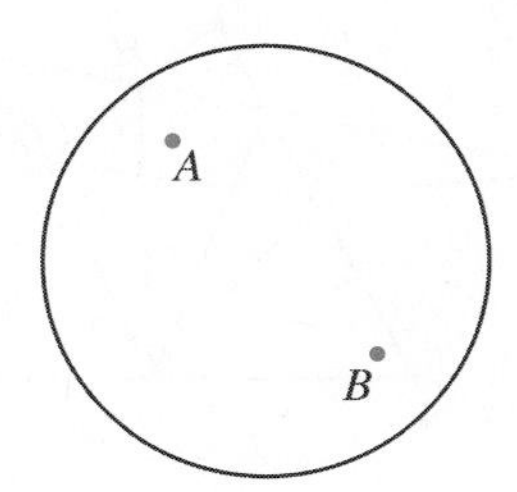

5.3 More on Extremal Problems

In the nineteenth century, several mathematicians investigated the solution of maximum and minimum problems by **synthetic methods**—that is, without using coordinates, algebra, or calculus. The Swiss mathematician Jacob Steiner (1796–1863) was one of the greatest synthetic geometers of all time. He solved, using a purely geometric method, maxima and minima problems that had previously required the use of calculus, including some problems that even required the use of calculus of variations (a higher branch of mathematics that is concerned with finding maxima or minima of definite integrals). Perhaps one of the most famous problems that Steiner investigated was the **isoperimetric theorem**: Of all plane figures with a given perimeter, the circle encloses the greatest area. The famous German mathematician H. A. Schwarz (1843–1921) gave a complete proof of the isoperimetric problem in three dimensions; he also solved synthetically the following problem.

Inscribed Triangle with Minimum Perimeter

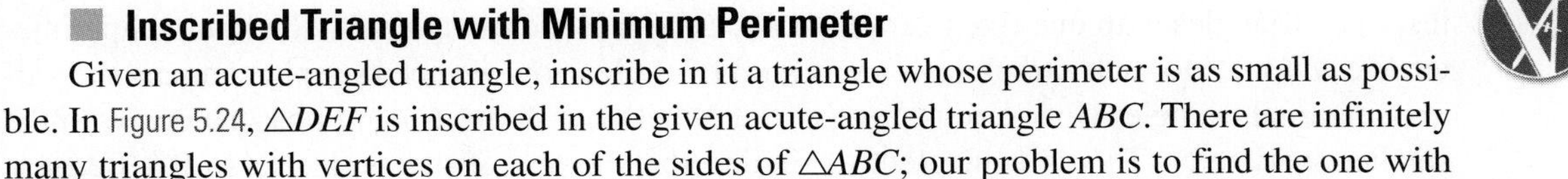

Given an acute-angled triangle, inscribe in it a triangle whose perimeter is as small as possible. In Figure 5.24, $\triangle DEF$ is inscribed in the given acute-angled triangle ABC. There are infinitely many triangles with vertices on each of the sides of $\triangle ABC$; our problem is to find the one with minimum perimeter.

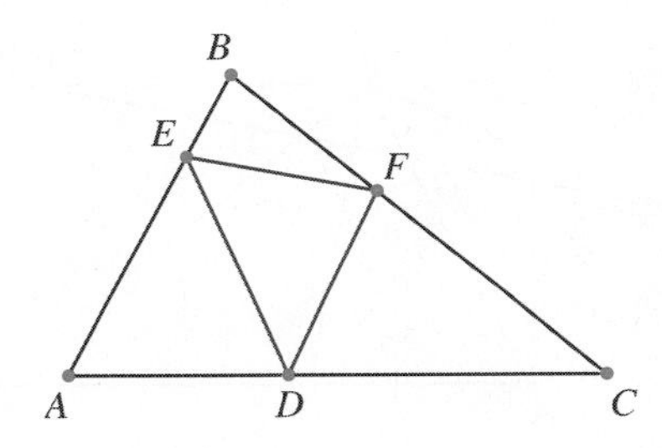

Figure 5.24

This problem was originally proposed by the Italian mathematician J. F. de'Toschi di Fagnano (1715–1797), who solved it in 1775 using calculus. Schwarz gave an ingenious synthetic solution that uses reflections and compositions of reflections. (See http://www.cut-the-knot.org/triangle/Fagnano.shtml.)

Another approach to the problem was given by the Hungarian mathematician Leopold Fejér (1880–1959), who was a student of Schwarz. His solution was praised by Schwarz as being especially simple and elegant; the following approach is based on that solution.

Investigation In Figure 5.25, $\triangle ABC$ is an acute-angled triangle.

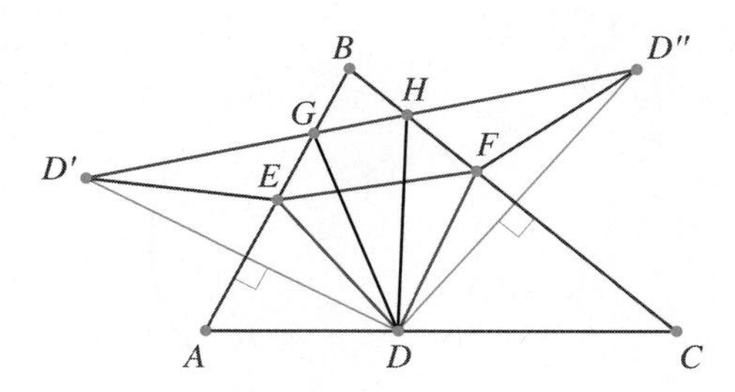

Figure 5.25

Let $\triangle DEF$ be an arbitrary inscribed triangle (with D on side AC, E on side AB, and F on side BC). To find which inscribed triangle will have the smallest perimeter, we will represent the perimeter of $\triangle DEF$ by the length of a path connecting two points (recall the minimum problems in Example 5.4). For that purpose, we reflect one of the vertices of $\triangle DEF$ in the two sides of $\triangle ABC$ that do not contain the vertex. We reflect D in side AB to obtain the image D' and in side BC to obtain the image D'', as shown in Figure 5.25. Because $ED = ED'$ and $FD = FD''$, the perimeter of $\triangle EDF$ equals $D'E + EF + FD''$, the length of the path $D'EFD''$. Certainly, the shortest path between D' and D'' is the segment $D'D''$. That segment intersects the sides of $\triangle ABC$ at points G and H. The perimeter of $\triangle DGH$ is $DG + GH + HD = D'G + GH + HD'' = D'D''$. Because the segment $D'D''$ is shorter than the length of the path $D'EFD''$, the perimeter of $\triangle DGH$ is less than the perimeter of $\triangle DEH$.

The trouble is that we don't know the position of D, and hence we don't know the positions of D' and D'' or the positions of G and H. But let us see what we have proved so far.

If we choose a point D (on side AC) and keep it fixed, then we have shown that among all inscribed triangles with one fixed vertex (at D), $\triangle DGH$ is the triangle with the minimal perimeter. We will refer to such a triangle as a **minimal triangle**. Now we need simply compare the various minimal triangles (for various positions of D between A and C) and pick the one with the smallest perimeter. This is equivalent to finding the shortest of the segments $D'D''$. Because the segment $D'D''$ is a side in $\triangle D'BD''$ (see Figure 5.26), we look more closely at that triangle.

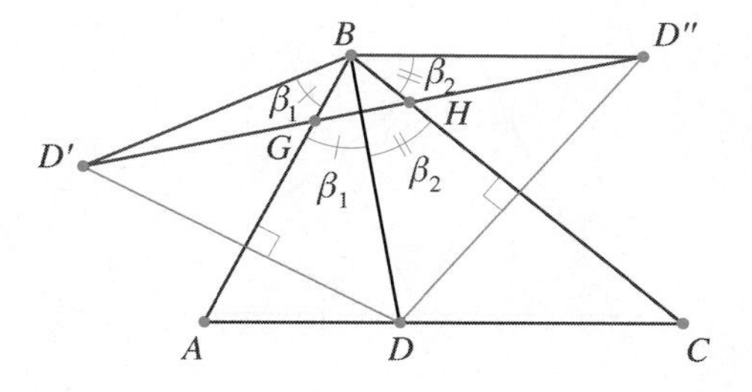

Figure 5.26

Notice that $BD' = BD$ (D' is the image of D under the reflection in side AB). Similarly, BD'' $= BD$. Thus $\triangle D'BD''$ is isosceles. Also, because a reflection preserves the angle measure, we have two pairs of equal-size angles with vertices at B—namely, two angles whose measure is designated by β_1 in Figure 5.26 and two angles with measure β_2.

Thus $m(\angle D'BD'') = 2\beta_1 + 2\beta_2 = 2(\beta_1 + \beta_2) = 2 \cdot m(\angle ABC)$. For any position of D on side AC, $m(\angle D'BD'')$ is always the same: It is twice the measure of $\angle B$ in $\triangle ABC$.

Let us get back now to our main goal: finding the location of D on side AC for which $D'D''$ is the smallest. Because the angle at B in $\triangle D'BD''$ is the same for all D, the segment $D'D''$ will be the shortest when the congruent sides BD' and BD'' are shortest. (This fact should be intuitively obvious and not hard to prove.) Because $BD' = BD'' = BD$, we need simply find the point D on side AC for which BD is smallest. This happens if and only if D is the foot U of the altitude from B to side AC, as shown in Figure 5.27.

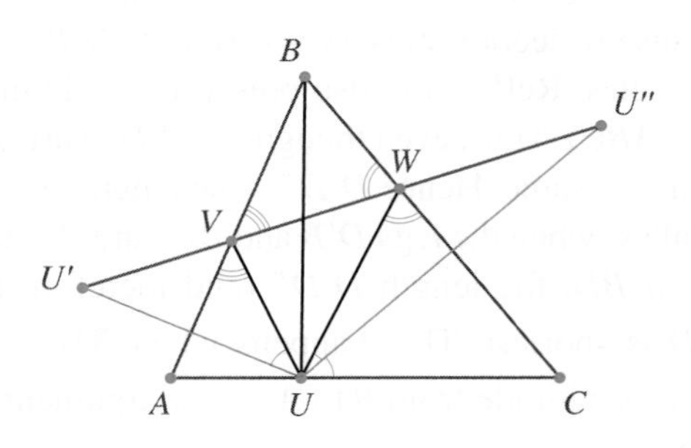

Figure 5.27

Thus the vertices of the required minimal triangle are point U and the points V and W where the line $U'U''$ intersects the sides AB and BC, respectively. (U' and U'' are the images of U under the reflections in sides AB and BC, respectively.) We have shown that the problem of finding an inscribed triangle with minimal perimeter has only one solution—that is, the minimal triangle UVW is unique. Had we started our solution by considering various triangles with vertices on side AB, we would have found that the minimal triangle must have as its vertex the foot of the altitude from A to side BC. Because the minimal triangle is unique, its vertices must be the feet of the three altitudes to the three sides of the original triangle. The triangle whose vertices are the feet of the three altitudes of a given triangle is called a **pedal triangle**.

The preceding investigation also proved that the minimal triangle and hence the pedal triangle share the property that at each vertex, the similarly marked angles in Figure 5.26 are congruent. (This follows immediately from the congruency of vertical angles and the fact that reflection preserves angle measure.) We will soon state this fact as a theorem. In the meantime, notice that the congruency of the previously mentioned angles implies that if three mirrors are placed on the sides of a given triangle and positioned perpendicular to the plane of the triangle, then whenever a beam of light emanates from any vertex of the pedal triangle toward another vertex of the pedal triangle, it will return to the original vertex from which it emanated. Using the principle that a beam of light travels via a path of shortest distance, it should be clear why the minimal triangle should have the property that at each vertex, the angles that its sides make with a side of the original triangle are equal.

We now state the theorems that follow from this investigation. The proofs are embedded in the investigation. In the proofs that follow, we omit the explorations and motivations we discussed earlier.

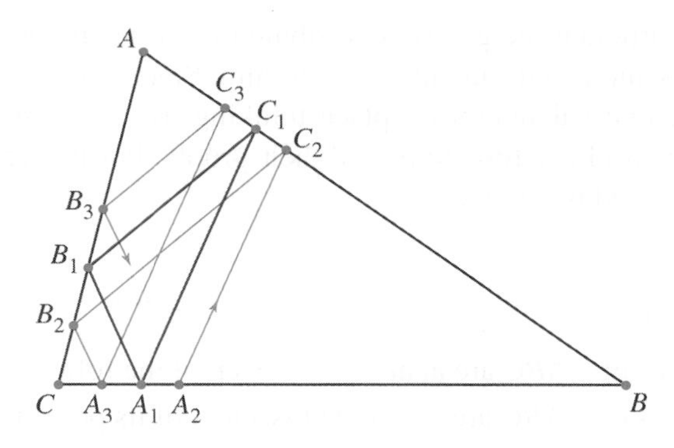

4. In $\triangle ABC$, each angle measures less than 120°. On each of the sides of the triangle, equilateral triangles have been constructed externally, as shown in the figure. Also, circles circumscribing the triangles have been constructed.

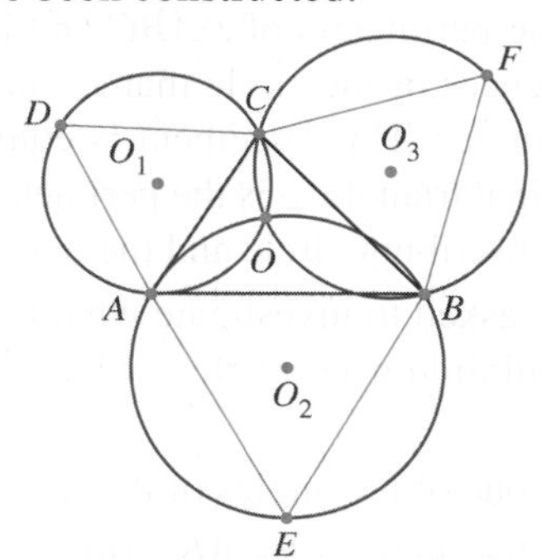

a. Prove that the three circles are concurrent (intersect in a single point). What is the significance of the point O at which the circles intersect? Justify your answer.

b. Use rotations to prove that the segments joining each vertex of $\triangle ABC$ with the remote vertex of the corresponding equilateral triangle constructed on the opposite side of $\triangle ABC$ are congruent.

c. Draw a triangle in which one of the angles is greater than 120°. What happens with the construction in part (a) for this triangle?

d. Is it possible to construct the Fermat's point for the triangle in part (c)? Why or why not?

e. Prove that the centers O_1, O_2, and O_3 of the three circles determine an equilateral triangle. (We will later prove this statement using a newly introduced size transformation; for now, you can prove it using similar triangles.)

•**5.** Given a convex quadrilateral, prove that the point determined by the intersection of the diagonals is the minimum distance point for the quadrilateral—that is, the point from which the sum of the distances to the vertices is minimal.

6. After completing Problem 5, determine whether the following construction of the minimum distance point for a convex quadrilateral is valid. Justify your answer.

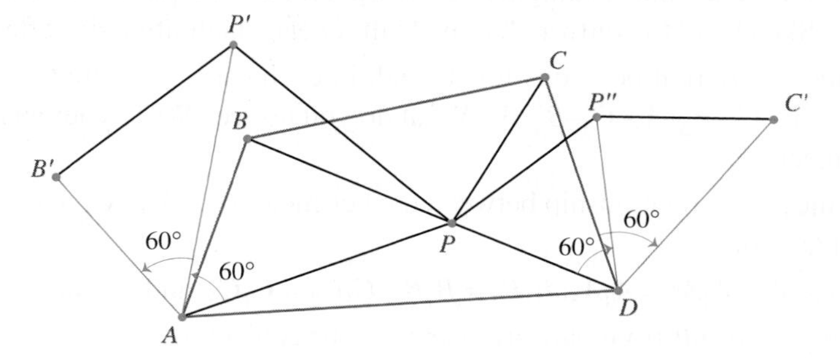

Given a convex quadrilateral $ABCD$, let P be a point in its interior, let B' be the image of B under a rotation with center A by 60° counterclockwise, and let P' be the image of P under the same rotation. Then $B'P' = BP$ and $AP = PP'$. Similarly, consider the rotation with center D by 60° clockwise. Let C' be the image of C and P'' be the image of P under the second rotation. Then $DP = PP''$ and $CP = C'P''$. Thus $BP + AP + PD + PC = B'P' + P'P + PP'' + P''C'$. Consequently, for every point P in the interior of $ABCD$, there exists a corresponding route connecting the fixed points B' and C' whose length is the sum of the distances from P to the four vertices of the quadrilateral. The shortest such route will be the one connecting B' to C' by a straight line. Thus P will be on the segment $B'C'$.

Next we make a similar construction for the other pair of opposite sides of the quadrilateral by connecting C'' with D' (not shown in the figure), where C'' is the image of C under a rotation with center B, 60° counterclockwise, and D' is the image of D under a rotation with center A, 60° clockwise. Because P must also be on the segment $C''D'$, it is determined by the intersection of the segments $B'C'$ and $C''D'$.

Does this construction result in the same point as the construction in Problem 5? Why or why not?

• **7.** Let $ABCD$ be a square with side a. Find, in terms of a, the length of the network connecting the vertices of the square shown in the figure. In other words, show that this network is shorter than the path determined by the intersection of the diagonals.

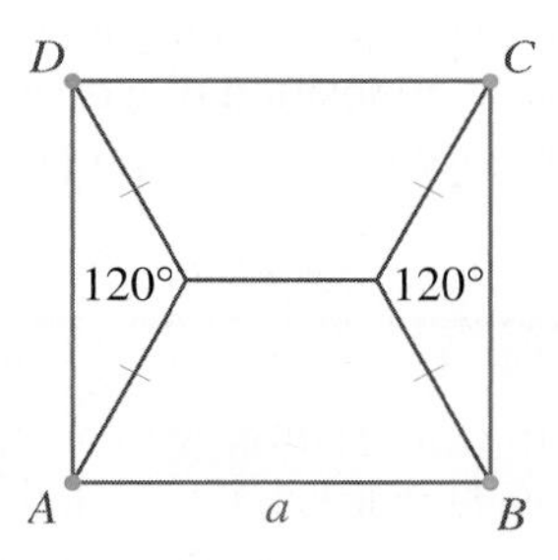

5.4 Similarity Transformation with Applications to Constructions

A similarity of the plane is a transformation S_k of the plane that multiplies all distances by the same positive constant k called the scale factor.[1] That is, for all points A and B in the plane, if $S_k(A) = A'$ and $S_k(B) = B'$ then $A'B' = k \cdot AB$.

Notice that every isometry is a similarity with scale factor 1, but not every similarity is an isometry. A special similarity is a **dilatation**, also called a **central similitude** or **homothety**.

A dilatation with center O and scale factor k is a transformation of the plane that, for $k > 0$, assigns to each point P in the plane a point P' on the ray $\overrightarrow{OP}$ such that $OP' = k \cdot OA$. If $k < 0$, then P' is on the ray opposite to $\overrightarrow{OP}$ (so that O is between P and P') and $OP' = |k|\, OA$. Point O is called the **center of the similitude**.

1. The terminology related to similarity transformations is not standardized. For example, some texts use the term "similarity ratio" instead of "scale factor."

Now Solve This 5.10

Justify each of the following statements:

1. A dilatation is a similarity transformation.
2. Under a dilatation:
   (a) Segment AB not belonging to a line through the center O is transformed onto the segment $A'B'$ parallel to segment AB.
   (b) A line through O is transformed onto itself, and a line not through O is transformed onto a parallel line.
   (c) Every triangle is transformed onto a similar triangle, and consequently angles are transformed to congruent angles.
   (d) A circle is transformed onto a circle. The image of the center of the circle is the center of the transformed circle, and the ratio of the radius of the transformed circle to the radius of the given circle is equal to the scale factor of the dilatation.
   (e) In a coordinate plane, if the center of the dilatation is at the origin, then the image of point (x, y) is (kx, ky)—that is, $(x, y) \to (kx, ky)$.

Using similarity transformations, we can define any two figures to be **similar** if there exists a similarity transformation that maps one figure onto the other. Two figures are called **centrally similar** if there exists a dilatation that maps one figure onto the other. In Chapter 6, we will prove that every similarity transformation is a composition of a dilatation followed by an isometry. You may want to try to prove this statement as well as another property of dilatations in Now Solve This 5.11.

Now Solve This 5.11

1. (a) Let S_k be a similarity transformation with scale factor k, and let $C_{1/k}$ be a dilatation about some point O. Prove that $S_k \circ C_{1/k}$ is an isometry and hence that any similarity transformation is a composition of a dilatation followed by an isometry. (Notice that point O can be chosen anywhere in the plane.)
   (b) Is it true that a similarity transformation is a composition of an isometry followed by a dilatation? Justify your answer.
2. Prove that the centers of dilatation of three triangles that are pairwise centrally similar are collinear. You may want to proceed as follows: In Figure 5.35, O_{12}, O_{13}, and O_{23} are the respective centers of the dilatations that take $\triangle P_1Q_1S_1$ onto $\triangle P_2Q_2S_2$, $\triangle P_1Q_1S_1$ onto $\triangle P_3Q_3S_3$, and $\triangle P_2Q_2S_2$ onto $\triangle P_3Q_3S_3$. Let ℓ be the line $O_{23}O_{13}$, and show that ℓ goes through O_{12}. For that purpose, show that the image of ℓ under the dilatation with center O_{13} is ℓ itself and that the image of ℓ under the dilatation with center O_{23} is ℓ itself.

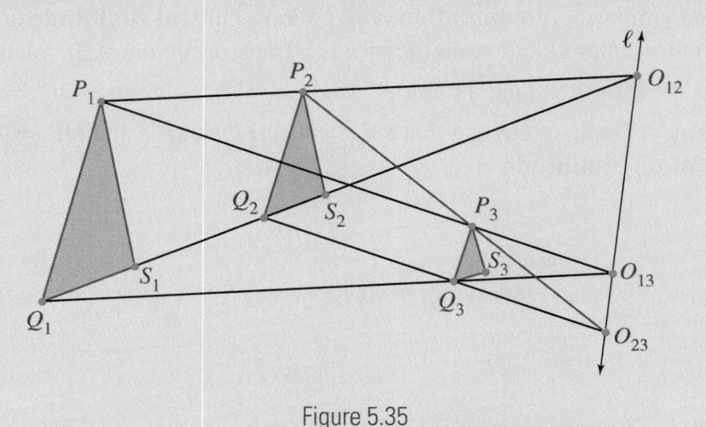

Figure 5.35

Similarity of Circles

Any two circles with different radii can be shown to be centrally similar. We first consider nonconcentric circles. Let P_1 and P_2 be the centers of two circles with radii r_1 and r_2, respectively, as shown in Figure 5.36. We want to find the dilatation that will transform the circle with center P_1 onto the circle with center P_2. That is, we want to find the center of the dilatation and the scale factor. For that purpose, we can construct any diameter through P_2 intersecting the circle at A_2 and B_2 as shown in Figure 5.36. The required dilatation should map the center P_2 onto P_1 and the radius A_2P_2 to a parallel radius A_1P_1. If the image of A_2 is to be A_1 (under the required dilatation), then the center of the dilatation is O_1, which is the point of intersection of the lines A_1A_2 and P_1P_2. However, if the image of B_2 is to be A_1, then the center of the dilatation is the point O_2. For each of the dilatations transforming circle P_1 onto circle P_2, the scale factor $k = \frac{A_2P_2}{A_1P_1} = \frac{r_2}{r_1}$.

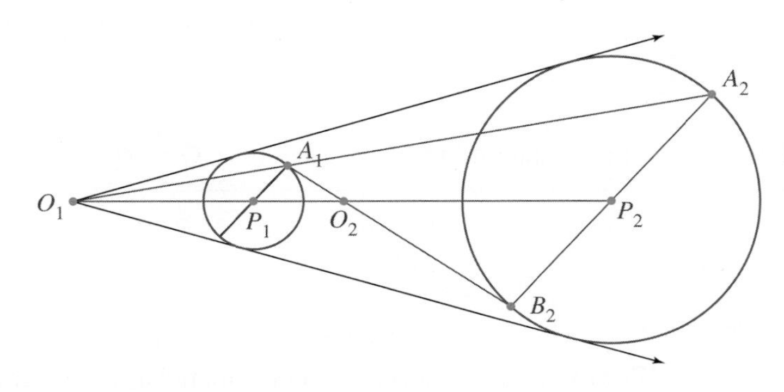

Figure 5.36

Construction of a Common Tangent

Figure 5.36 and our earlier discussion suggest a new way to construct a common tangent to two circles with different radii (when one circle is not in the interior of the other). We first find O_1, the center of dilatation; we then construct the tangent to one of the circles through O_1 (an easy task). That tangent must also be the tangent to the other circle because one point of tangency must be the image or preimage of the other point of tangency under the dilatation. The internal tangent can be constructed in a similar way.

Compare this construction of a common tangent to the one presented in Chapter 2. Which seems to be easier? Why? Next we prove a celebrated theorem attributable to Leonhard Euler.

Theorem 5.3

Euler Line

The centroid G (the intersection of the medians), the orthocenter H (the intersection of the altitudes), and the circumcent O (the center of the circumscribing circle) of a triangle are collinear. (This line is called the **Euler line**.*) Moreover* $OG = \frac{1}{3}OH$.

Proof

In Figure 5.37, A', B', and C' are the midpoints of the sides of $\triangle ABC$ opposite the vertices A, B, and C, respectively. The point O is the orthocenter of $\triangle A'B'C'$. Under dilatation with center G and dilatation factor $-\frac{1}{2}$, the image of $\triangle ABC$ is $\triangle A'B'C'$. Because under the dilatation the orthocenters must correspond, the image of H is O. By definition of a dilatation O, G, and H are collinear. Notice also that the medians of $\triangle ABC$ bisect the sides of $\triangle A'B'C'$. Therefore the centroids of the triangles are the same point, G. (This should be also clear because the dilatation with center G takes G to itself.) If we name the dilatation f, we have

$$f(G) = G$$
$$f(H) = O$$

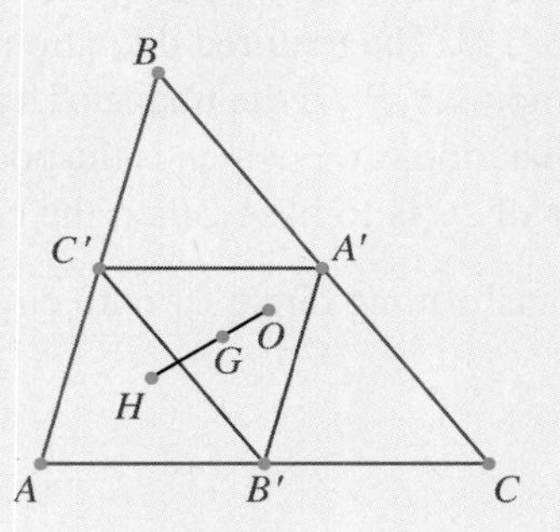

Figure 5.37

Since the absolute value of the dilatation factor is $\frac{1}{2}$, we have $OG = \frac{1}{2}HG$. □

Historical Note: Leonhard Euler

Leonhard Euler (1707–1783) was the most prolific mathematician of all time. Euler was born in Basel, Switzerland, and finished his studies at the University of Basel at age 15.

Euler published more than 800 books and articles. His collected works are published in 75 substantial volumes. Euler's name is attached to concepts and theorems in almost all branches of mathematics. Among them are *Euler's characteristic*, *Euler circle*, *Euler circuit*, *Euler's phi function*, *Euler's formula*, and *Euler multiplier*. Euler introduced the notation $f(x)$ (in 1734), e for the base of the natural logarithm (in 1727), i for $\sqrt{-1}$ (in 1777), π for the ratio of circumference to the diameter of a circle, Σ for summation (in 1755), R and r for the radii of circumscribing and inscribed circles, and $\sin x$ and $\cos x$ for values of the sine and cosine functions, among many other notations.

Euler lost the sight in one of his eyes at age 31; soon after, he became totally blind. He produced about half of his work while being blind. Euler had 13 children and many grandchildren. He claimed that he did some of his best work while holding a baby, with other children playing around his feet.

Constructions Using Dilatation

The definition and properties of dilatation can simplify many construction problems, as we shall see next.

Construction 5.1

Construct a circle through a point in the interior of a given angle that is not on the angle bisector of the angle, and is tangent to the sides of the angle. In the following solution we combine the three stages of investigation, construction, and proof into one.

In Figure 5.38, $\angle A$ and the point P_1 in the interior of the angle are given. We need to construct a circle through P_1 tangent to the sides of the angle. We imagine that such a circle is already constructed. The center of the desired circle must be on the angle bisector of $\angle A$. We can easily construct a different circle with center O at some point on the angle bisector of $\angle A$, tangent to the sides of $\angle A$ (the radius of this circle is OH, the distance from O to a side of the angle, as shown in Figure 5.38). The desired circle through P_1 and center O_1 can be regarded as the image of the circle O under a dilatation with center of similitude at A. To find the point on the circle O that is mapped to P_1, we connect A with P_1. The point P that is the preimage of P_1 under the dilatation must be on $\overrightarrow{AP_1}$ as well as on the circle O. Two such points, P and P^*, exist as shown in Figure 5.38.

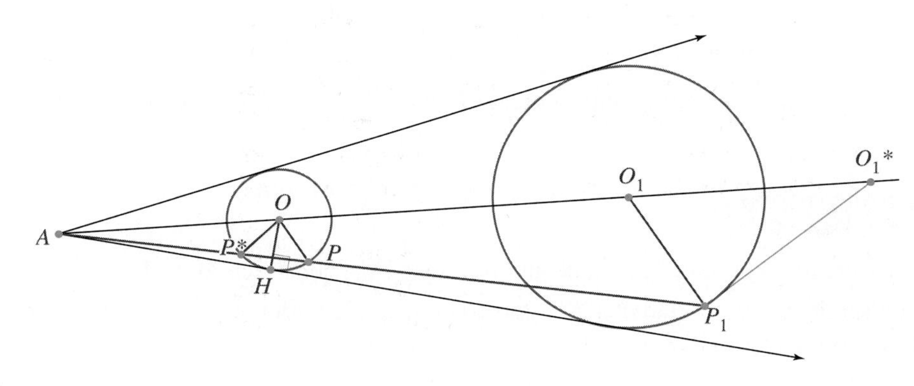

Figure 5.38

We know that under dilatation $\overline{O_1P_1}$, the image of $\overline{OP}$ and $\overline{OP}$ itself are parallel radii. Thus we can find the center of the desired circle by drawing a line through P_1 parallel to line OP. That line intersects the angle bisector at O_1, which is the center of the required circle (whose radius is O_1P_1).

Another circle also satisfies the requirements of this problem. If we draw a line parallel to line OP^* through P_1, the point where that line intersects the angle bisector at O_1^* is the center of the second circle that satisfies the requirements of the problem (that circle is not shown in Figure 5.38).

The next construction was introduced in Chapter 4. Here we solve it using a similarity transformation. □

Construction 5.2

Construct a square inscribed in a given triangle such that one of the sides of the square is on a side of the triangle.

We imagine that the required square $E'F'G'H'$ in Figure 5.39a is already constructed and that it is the image of another square $EFGH$ under a dilatation with center A. We can easily construct the square $EFGH$ with F on line AB and G as well as H on line AC. Because E' is on $\overline{BC}$ as well as on $\overrightarrow{AE}$, we can find it by identifying the point of intersection of $\overrightarrow{AE}$ with side BC. Dropping the perpendicular from E' to side AC, we obtain the vertex H'. The vertex F' is the point of intersection of the line through E' parallel to line AC with side AB. The perpendicular from F' to line AC determines the fourth vertex, G'. Because under dilatation the sides of $EFGH$ are mapped to segments parallel (or on the same line) to the corresponding sides of $EFGH$ and each segment is enlarged by the same factor, $E'F'G'H'$ is, indeed, a square.

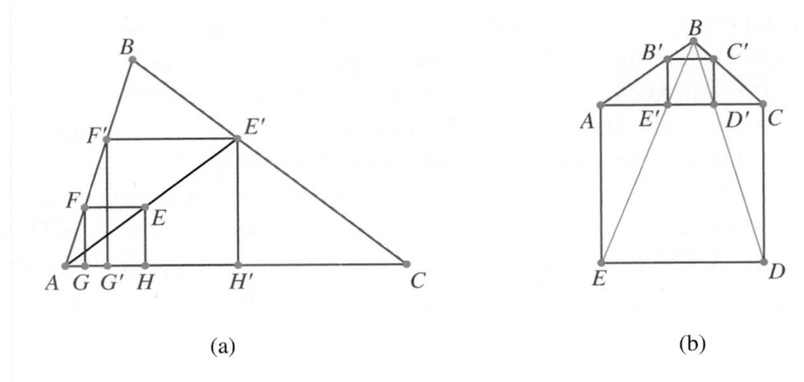

Figure 5.39

Figure 5.39b shows a dilatation with center at B that achieves the same construction.

Problem Set 5.4

In your answers to all problems, include an investigation, a construction, and a proof.

1. In a semicircle with diameter AB, construct a square with two vertices on the semicircle and two vertices on the diameter.

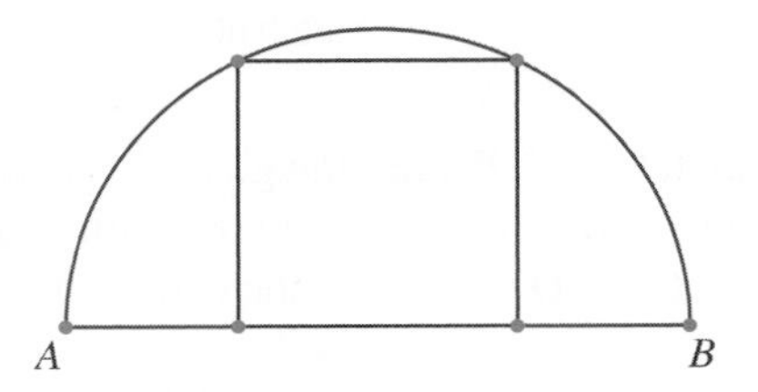

2. a. In a given circle, inscribe a square $ABCD$ and a smaller square $EFGH$ with two vertices on the arc BC and the other two vertices on side BC.
   • b. If the side of the larger square is s, find the side of the smaller square in terms of s.

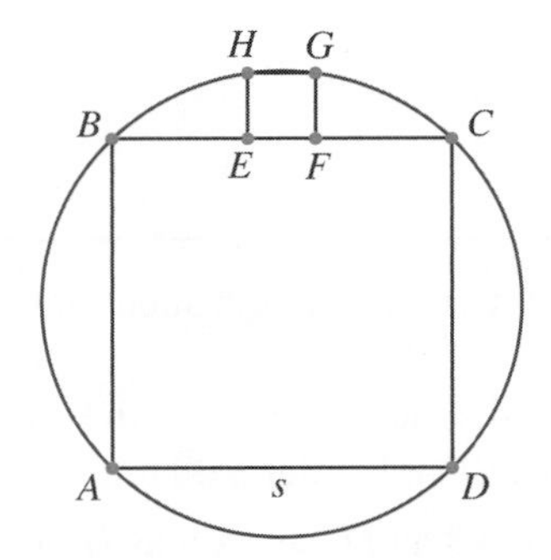

3. Given three lines a, b, and c intersecting at point O, construct a circle tangent to line a, having its center on b, and cutting off on line c a chord of length d.

4. Inscribe a rectangle in which one side is twice as long as the other side in each of the following:
   a. A triangle
   b. A semicircle

5. Two circles with centers O_1 and O_2, respectively, and a point P are given. Construct a circle with center X (not known) such that the line through the points A and B, which are the points of tangency (of each of the given circles with the unknown circle), contains the point P.

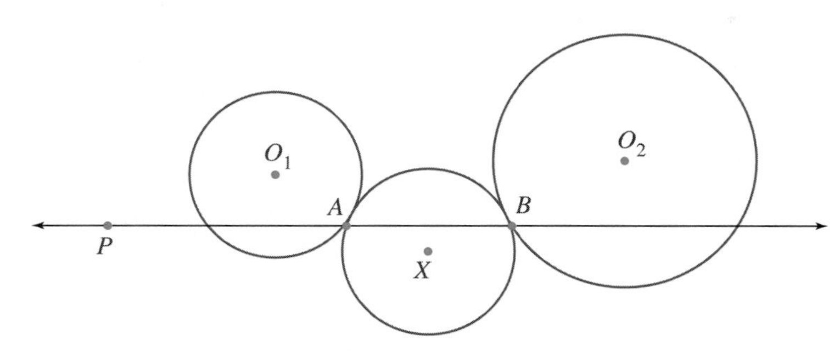

6. Through a point in the interior of a given circle, construct a chord that is trisected by the point.

7. In a given circle, inscribe a triangle similar to a given triangle.

8. Are all parabolas similar to each other? Justify your answer.

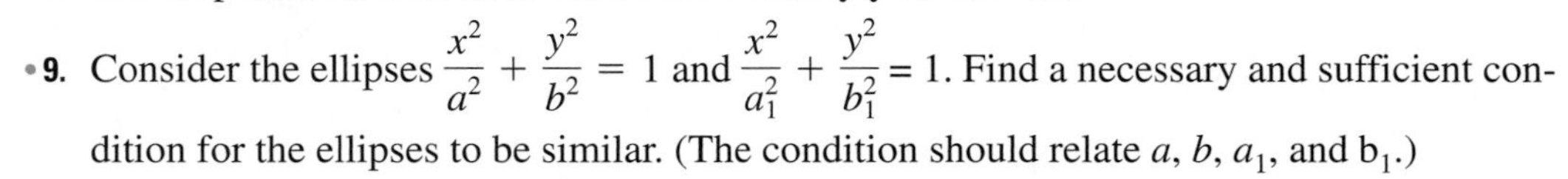

•**9.** Consider the ellipses $\frac{x^2}{a^2} + \frac{y^2}{b^2} = 1$ and $\frac{x^2}{a_1^2} + \frac{y^2}{b_1^2} = 1$. Find a necessary and sufficient condition for the ellipses to be similar. (The condition should relate a, b, a_1, and b_1.)

10. Are all hyperbolas $xy = c$, where c is any real number, similar to each other? Justify your answer.

6 Composition of Transformations

The pursuit of truth and beauty is a sphere of activity in which we are permitted to remain children all our lives.

—Albert Einstein (1879–1955),
American physicist and Nobel laureate

Introduction

In Chapter 5, we saw that a composition of two transformations is a transformation and that a composition of two isometries is an isometry. We have encountered four types of isometries: reflections, translations, rotations, and glide reflections. Are these the only isometries, or do others exist?

Because a composition of two isometries is an isometry, it is natural to ask whether a composition of any of the isometries we have encountered so far results in one of those isometries or, perhaps, in a new isometry. By studying all of the possible compositions of reflections, translations, and rotations, we will rediscover an isometry that is not one of these three, and hence is a new type of isometry: glide reflection, which was first introduced in Chapter 5. We will then prove that there are no other isometries. At the same time, we will discover that reflections are the building blocks for the plane isometries in the sense that every isometry is a composition of at most three reflections. The information about the results of various compositions of isometries will enable us to solve some geometrical problems in simple and efficient ways. (The Treasure Island Problem introduced in Chapter 0 is one such problem. Knowing the result of composition of two rotations about two different points will not only give us a surprisingly simple solution and proof, but will also show how similar "treasure island" problems can be created.)

In geometry, crucial roles are played not only by individual transformations but also by sets of transformations satisfying prescribed conditions called **transformation groups**. (The concept of a group will be precisely defined later.) The concept of a group is a unifying concept for all of mathematics, and in geometry it will enable us to measure and investigate symmetry of figures.

In 1827, the German astronomer and mathematician A. F. Möbius gave a meaning to the composition of isometries by extending the idea of transformation to the whole plane. This extension enabled him to introduce the concept of a *group* to geometry and to define notions such as congruence and similarity as properties that were invariant (unchanged) under certain groups of transformations of the plane. These concepts did not gain wide acceptance among mathematicians until 45 years later, when another German mathematician, Felix Klein, extended Möbius's ideas in his inaugural lecture in honor of his appointment to a chair at Erlanger University. In that lecture, which became known as the Erlanger Programme, Klein introduced a unifying principle for classifying Euclidean and various non-Euclidean geometries: He defined **geometry** as the study of those properties of a given set that remain invariant under some group of transformations. Klein's approach has had an enormous impact on the development of modern mathematics.

Historical Note

August Ferdinand Möbius (1790–1868) was a student of Gauss and became a professor of astronomy at Leipzig in 1815. In a paper presented to the French Academy and discovered only after his death, Möbius describes the properties of one-sided surfaces, including the now-famous Möbius strip. Around 1850, Möbius pioneered the study of transformations in the geometry of complex numbers. A transformation named after him has important applications both in pure and applied mathematics.

At age 23, **Felix Klein** (1849–1925) became a professor at the University of Erlanger, where he delivered his famous Erlanger Programme. In 1886, Klein accepted a position at Göttingen University, which, under his leadership, became a world center for mathematical research. Klein was a gifted teacher and personally supervised 48 doctoral dissertations. In 1895, the English mathematician Grace Chisholm, who, as a woman, was not permitted to attend graduate school in England, was awarded a Ph.D. in mathematics under Klein. Her doctorate was the first in any field to be awarded to a woman in Germany. In 1897, the American mathematician Mary Winston also earned her doctorate under Klein. She was the first American woman to obtain a Ph.D. in mathematics from a European university.

6.1 In Search of New Isometries

Chapter 5 introduced three basic types of isometries: reflections, translations, and rotations (we are assuming here that the glide reflection has not yet been discovered). To find other possible isometries, we will look at various compositions of the isometries we have encountered so far. For that purpose, it is useful to know which qualities determine an isometry. Is there a unique isometry that maps a known point A to point A'? The reflection M_ℓ, where ℓ is the perpendicular bisector of the segment AA', is one such isometry because $M_\ell(A) = A'$. But the translation $\tau_{AA'}$ (the translation from A to A') also maps A to A'. Moreover, any rotation with center O on the line ℓ, which is the perpendicular bisector of AA' (see Figure 6.1), and by $\angle AOA'$ in an appropriate direction (counterclockwise in Figure 6.1) will map A onto A'.

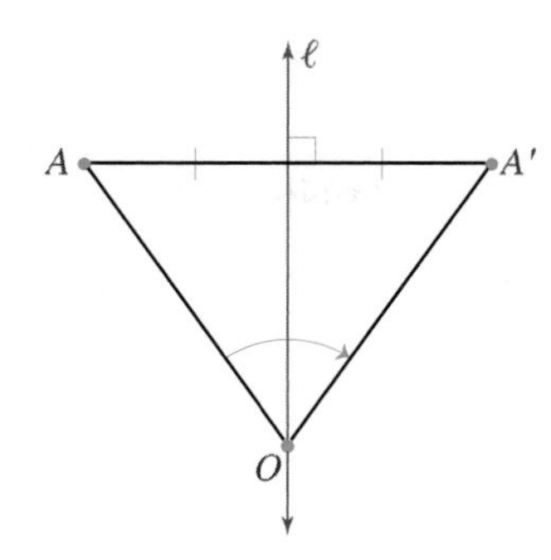

Figure 6.1

Is there a unique isometry that maps any two points A, B to any other two points A', B'?[1] Because any isometry preserves distance, we must have $A'B' = AB$. We will show that at least two different isometries map A to A' and B to B'. In Figure 6.2, A is mapped to A' by the reflection M_k in the perpendicular bisector k of $\overline{AA'}$. Under M_k, the point B is mapped to B^*. Next B^* is mapped to B' by the reflection M_ℓ in the perpendicular bisector ℓ of $\overline{B^*B'}$. Thus the composition $M_\ell \circ M_k$ maps B to B'. [Notice that $(M_\ell \circ M_k)(B) = M_\ell(M_k(B)) = M_\ell(B^*) = B'$.] We will show that $M_\ell \circ M_k$ maps A to A'.

First A is mapped by M_k to A'. Careful drawing as in Figure 6.2 suggests that A' is on ℓ.

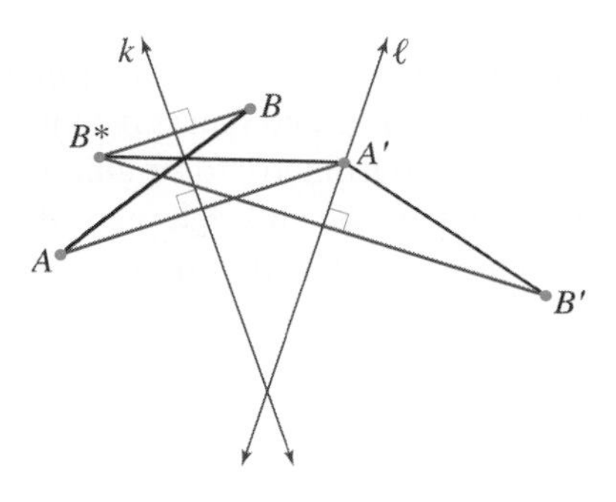

Figure 6.2

If that were the case, then we would have $(M_\ell \circ M_k)(A) = M_\ell(M_k(A)) = M_\ell(A') = A'$. Because ℓ is the perpendicular bisector of $\overline{B^*B'}$, to prove that A' is on ℓ we need simply show that A' is equidistant from the endpoints of $\overline{B^*B'}$—that is, that $B^*A' = A'B'$. To show that this is the case, notice that $\overline{B^*A'}$ is the image of $\overline{BA}$ under M_k and hence $B^*A' = BA$. It was given that $BA = B'A'$, however, so $B^*A' = B'A'$ and, therefore, A' is on ℓ. Consequently, $M_\ell \circ M_k$ maps A to A' and B to B'.

To show that another isometry maps A to A' and B to B', we use the fact that an isometry maps a triangle onto a congruent triangle. We pick any point C not collinear with A and B, as in Figure 6.3, and obtain its image C' under the isometry $M_\ell \circ M_k$ that mapped the points A, B to A', B', respectively. For convenience, we denote this isometry by T. We will express the new isometry in terms of T. Notice that reflecting $\triangle A'B'C'$ in line $A'B'$ (denoted by n in Figure 6.3) results in $\triangle A'B'C''$ congruent to $\triangle A'B'C'$. Because A' and B' are fixed points under M_n, the transformation $M_n \circ T$ still maps A to A' and B to B'. However, $M_n \circ T \neq T$ because T maps C to C' and $M_n \circ T$ maps C to C''. (Recall that two functions with the same domain are equal if and only if

1. When we say that isometry maps A, B to A', B', we mean that it maps A to A' and B to B', unless otherwise specified.

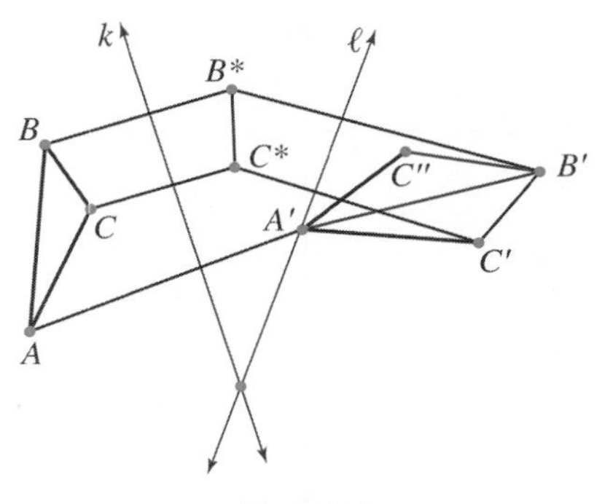

Figure 6.3

they give the same output for every input.) Thus we have found two different isometries that map A to A' and B to B'.

Are these two isometries the only isometries that map A, B onto A', B'? You will soon be able to answer this question.

Necessary Conditions for Three Points and Their Images to Determine a Unique Isometry

We just discovered that two points and their given images do not determine a unique isometry. You may now be wondering if three points and their given images determine a unique isometry, or if you need even more points for an isometry to be unique. Given the three points A, B, C and the points A', B', C', is there a unique isometry that maps A to A', B to B', and C to C'? As we are looking for an isometry, distances must be preserved:

$$\begin{aligned} A'B' &= AB \\ B'C' &= BC \\ A'C' &= AC \end{aligned} \tag{6.1}$$

Thus the given points must satisfy condition 6.1. If the points A, B, C are collinear, then under every isometry the images of these points will also be collinear. Thus A', B', C' must be collinear and, if B is between A and C, B' must be between A' and C' as in Figure 6.4.

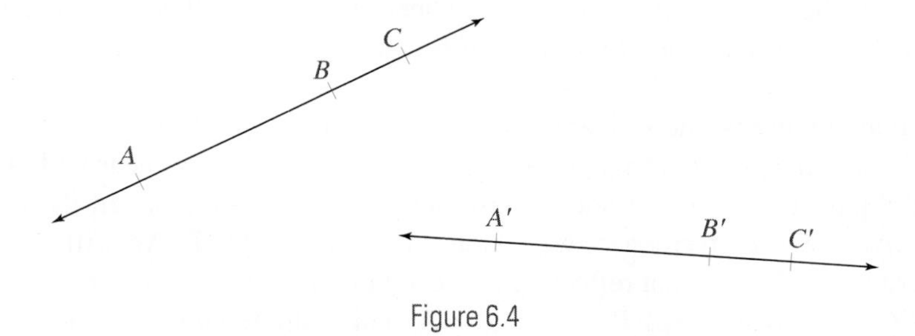

Figure 6.4

Let us focus for a moment only on A, B and A', B'. We have seen that two different isometries map A, B onto A', B'. We can show that these isometries—and, in fact, any isometry that maps A, B onto A', B'—will automatically map C onto C' as long as C and C' satisfy Equation 6.1. (You will explore the proof of this fact in Problem 3 of Problem Set 6.1.) Thus the two different isometries that map A, B onto A', B' will also map C to C'. Consequently, there is no unique isometry that maps A, B, C onto A', B', C' (satisfying Equation 6.1 when the points are collinear). Therefore, if we want the isometry to be unique, the three points must be noncollinear. The next theorem shows that this condition is sufficient.

Theorem 6.1

The First Fundamental Theorem of Isometries

If A, B, C are three noncollinear points, and A', B', C' are also noncollinear such that $A'B' = AB$, $B'C' = BC$, and $A'C' = AC$, then there exists a unique isometry that maps A to A', B to B', and C to C'. This isometry can be expressed as a composition of at most three reflections.

Proof

Part I: Existence We need to prove that there exists an isometry that maps the noncollinear points A, B, C onto the points A', B', C' and that there is only one such isometry. If $A \neq A'$, $B \neq B'$, and $C \neq C'$, the isometry $M_\ell \circ M_k$, described in Figure 6.2, will map A to A' and B to B'. Assume that the image of C under $M_\ell \circ M_k$ is not C'. Before reading on, try to find the line m such that $M_m \circ (M_\ell \circ M_k)$ will map $\triangle ABC$ onto $\triangle A'B'C'$.

In Figure 6.5, points A, B are mapped to points A', B' by $M_\ell \circ M_k$ (this was done in Figure 6.2). Now, however, we also have the points C and C', and we need to find an isometry that will not only map A, B onto A', B' but also map C onto C'. Let's see what $M_\ell \circ M_k$ does to C.

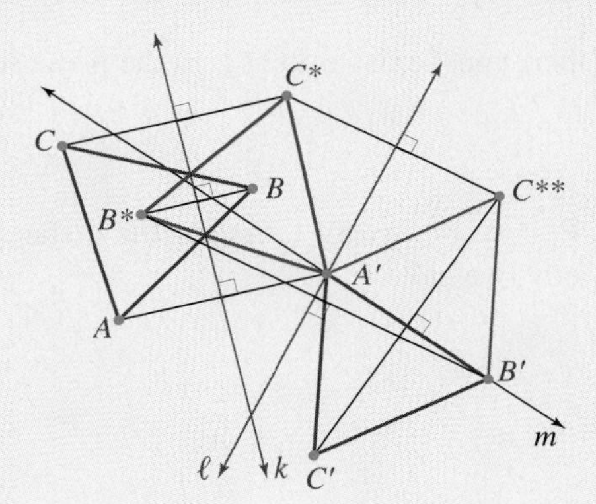

Figure 6.5

In Figure 6.5, $(M_\ell \circ M_k)(C) = M_\ell(C^*) = C^{**}$. If $C^{**} = C'$, we are done, and $M_\ell \circ M_k$ is the required isometry. Suppose that $C^{**} \neq C'$. Figure 6.5 suggests that the line $A'B'$ (denoted by m) is such that the reflection of C^{**} in it is C'. If this were true, it would follow that $M_\ell \circ M_k$ (which maps $\triangle ABC$ onto $\triangle A'B'C^{**}$) followed by M_m will map $\triangle ABC$ onto $\triangle A'B'C'$ (since M_m will not change A' and B' and will map C^{**} to C'). To show that $M_m(C^{**}) = C'$, we need simply prove that A' and B' are equidistant from C' and C^{**}. We first show that $A'C' = A'C^{**}$. From the hypothesis,

$$A'C' = AC \tag{6.2}$$

When we consider the reflection in ℓ, we see that the image of $\overline{A'C^*}$ is $\overline{A'C^{**}}$ and hence that

$$A'C^{**} = A'C^* \tag{6.3}$$

Finally, reflection in k implies

$$A'C^* = AC \tag{6.4}$$

The last two equations imply

$$A'C^{**} = AC$$

This, along with Equation 6.2, implies

$$A'C' = A'C^{**}$$

The last equation tells us that A' is equidistant from the endpoints of the segment $\overline{C'C^{**}}$. In the same way, we can show that B' is also equidistant from the endpoints of $\overline{C'C^{**}}$. Thus the line m through A' and B' is the perpendicular bisector of $\overline{C'C^{**}}$. Consequently, $M_m \circ M_\ell \circ M_k$ maps A, B, C, onto A', B', C', respectively.

In the preceding argument, we assumed that $A \neq A'$, $B \neq B'$, and $C \neq C'$. You will investigate the cases when one or more of the points are the same as their corresponding images in the problem set at the end of this section.

Part II: Uniqueness We need to prove that the isometry we have found is unique—that is, that no other isometry will accomplish the required mapping. Suppose F and G are two isometries such that $F(A) = G(A) = A'$, $F(B) = G(B) = B'$, and $F(C) = G(C) = C'$, where A, B, C are noncollinear. We need to prove that $F = G$ [i.e., $F(P) = G(P)$ for all points P in the plane].

Suppose $F \neq G$. Then there must exist a point P in the plane such that

$$F(P) \neq G(P)$$

Let $F(P) = P'$ and $G(P) = P''$ (see Figure 6.6). Consider the distance AP. Because $F(A) = A'$, $F(P) = P'$, and F is an isometry, we have

$$A'P' = AP \tag{6.5}$$

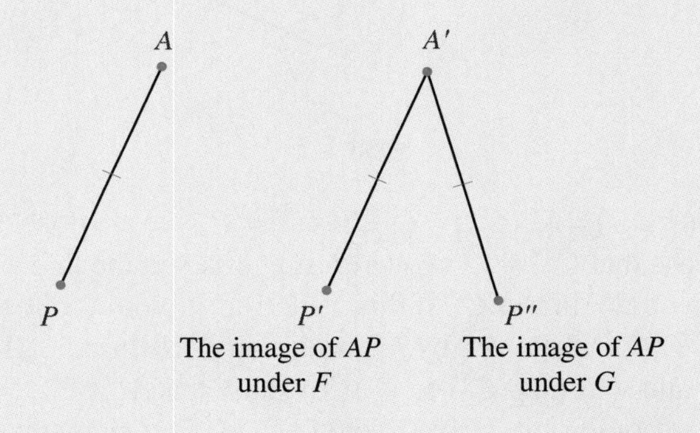

Figure 6.6

We apply now the same reasoning to the isometry G. Because $G(A) = A'$ and $G(P) = P''$, we get

$$A'P'' = AP \tag{6.6}$$

From Equations 6.5 and 6.6, we get $A'P' = A'P''$, which in turn implies that A' is on the perpendicular bisector of $\overline{P'P''}$. If we apply the same reasoning to B rather than A, we find that B' must also be on the perpendicular bisector of $\overline{P'P''}$. Likewise, C' must be on the perpendicular bisector of $\overline{P'P''}$. Consequently, A', B', C' are on the same line (the perpendicular bisector of $\overline{P'P''}$) and, therefore, are collinear. But if A', B', C' are collinear, then A, B,

C must also be collinear, because an isometry preserves collinearity. This contradicts the fact that A, B, C are not collinear and hence $F = G$. Consequently, the isometry we found in Part I is unique. □

We state the result just proved separately from Theorem 6.1 as follows:

Corollary 6.1 *Two isometries that have the same values (images) at three noncollinear points are equal.*

Corollary 6.1 and Theorem 6.1 imply:

Theorem 6.2

The Second Fundamental Theorem of Isometries

Every isometry is equal to a composition of at most three reflections.

Proof

(Try to write your own proof before reading on.) Let F be an isometry. Pick any three noncollinear points A, B, C. Let A', B', C' be their corresponding images under F. Because F is an isometry and A, B, C are noncollinear, their corresponding images A', B', C' are noncollinear and $A'B' = AB$, $A'C' = AC$, and $B'C' = BC$. By the first part of Theorem 6.1, there exists a succession of at most three reflections that maps the points A, B, C onto A', B', C'. By the uniqueness part of Theorem 6.1 (or Corollary 6.1), F equals the composition of these reflections. □

Theorem 6.2 tells us that all the possible isometries including rotations and translations are either reflections, the composition of two reflections, or the composition of three reflections. Thus, to find all possible isometries, we need simply investigate the compositions of two or three reflections.

We will continue our investigation of isometries by looking at the composition of two reflections in two intersecting lines.

Composition of Two Reflections in Two Intersecting Lines

In Figure 6.7, the lines k and ℓ intersect at O. We reflect figure A in line k and obtain its image A'. In turn, we reflect A' in line ℓ to obtain figure A''. It looks as if we could obtain A'' directly from A by rotating A about the point O.

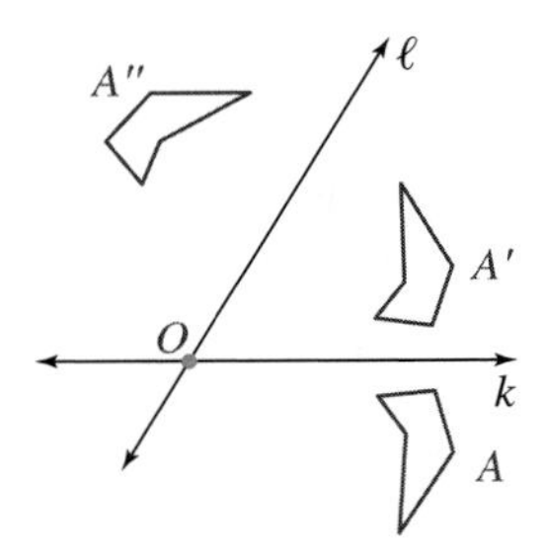

Figure 6.7

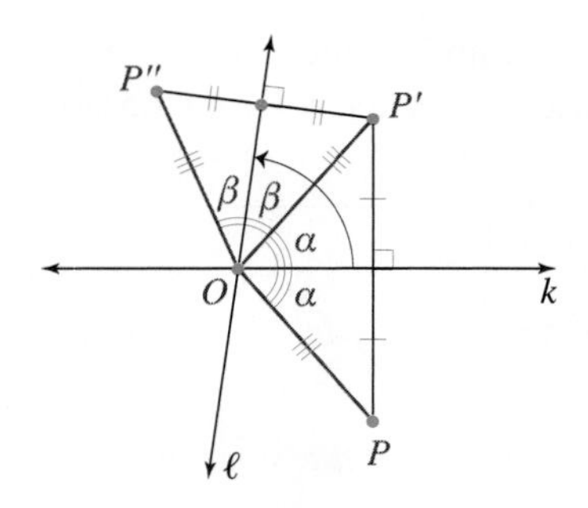

Figure 6.8

Thus it seems that a composition of two reflections in two intersecting lines is a rotation about the point O, where the lines intersect. To prove that this is the case, let P be any point in the plane and P' be its image under reflection in k. Then let P'' be the image of P' under reflection in ℓ. We need to show that for every point P in the plane, $OP = OP' = OP''$, and that the measure of $\angle P''OP$ is always the same. What do we know about P, P', and P''?

In Figure 6.8, line k is the perpendicular bisector of $\overline{PP'}$. Because O is on k, $OP' = OP$. Similarly, $OP' = OP''$ and hence $OP = OP''$. Also, the pairs of angles marked in the same way in Figure 6.8 are congruent. (The measure of the angles in the first pair is designated by α and in the second by β.) We see from Figure 6.8 that

$$m(\angle POP'') = 2\alpha + 2\beta = 2(\alpha + \beta) = 2\theta$$

where θ is the measure of the angle between lines k and ℓ, as shown in Figure 6.8. Notice that θ is a fixed angle. Hence it seems that

> P'' can be obtained from P by rotation with center O by an angle whose measure is twice the angle between the lines of reflection in the direction from line k to line ℓ.

Upon closer examination of the preceding argument, you might see that we have not proved the key assertion for all points P in the plane. For example, if P is situated as shown in Figure 6.9, the argument showing that $m(\angle POP'') = 2\theta$ is somewhat different.

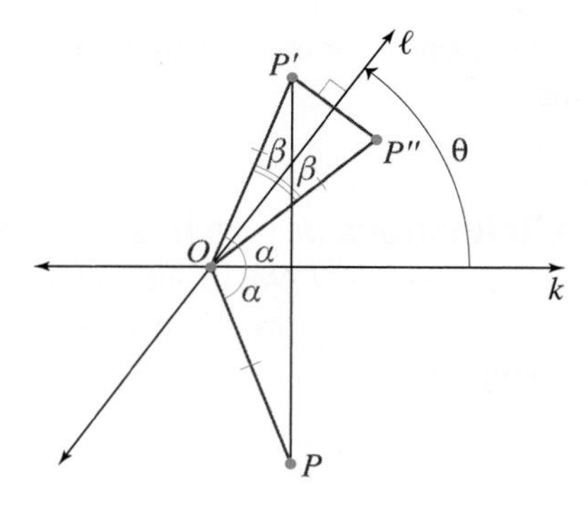

Figure 6.9

From Figure 6.9, we have

$$m(\angle POP'') = 2\alpha - 2\beta = 2(\alpha - \beta) = 2\theta$$

To continue proving in this manner that $m(\angle POP'') = 2\theta$ for all points in the plane, we would need to consider all possible positions of P in relation to the lines k and ℓ. There are only a small number of cases to consider, so we could complete our argument on a case-by-case basis, although it would certainly be tedious. There is also the danger that we might miss a case. Fortunately for us, Theorem 6.1 (or Corollary 6.1) comes to rescue, as we can see in the following proof.

A Shorter Proof

In addition to point P in Figure 6.8, we can choose two other points, Q and R, such that the images of these points behave like the images of P and such that P, Q, and R are noncollinear. (We can choose Q and R to be close to line k in the half plane determined by line k and P.) $M_\ell \circ M_k$, the reflection in line k followed by the reflection in line ℓ, is an isometry (as a composition of two isometries) and, therefore, maps the noncollinear points P, Q, R to the noncollinear points P'', Q'', R''. We have seen from our previous explanation that the rotation $R_{O,2\theta}$ (where θ is the angle between the lines k and ℓ and the rotation is in the direction from line k to line ℓ) also maps P, Q, R onto P'', Q'', R''. But Theorem 6.1 (or Corollary 6.1) assures us that the isometry that maps P, Q, R onto P'', Q'', R'' is unique. Consequently,

$$M_\ell \circ M_k = R_{O,2\theta} \qquad \square$$

We have just proved the following theorem:

Theorem 6.3

If k and ℓ are two lines that intersect at O, then the composition $M_\ell \circ M_k$ (the reflection in line k followed by the reflection in line ℓ) is a rotation about O, with the angle of rotation being twice the directed angle formed by the intersecting lines in the direction from k to ℓ.

Remark Theorem 6.3 tells us that $M_\ell \circ M_k = R_{O,2\theta}$ where θ is the directed angle formed by lines k and ℓ in the direction from k to ℓ. But the angle whose measure is δ (see Figure 6.10) is another angle between the lines k and ℓ that is in the direction from k to ℓ.

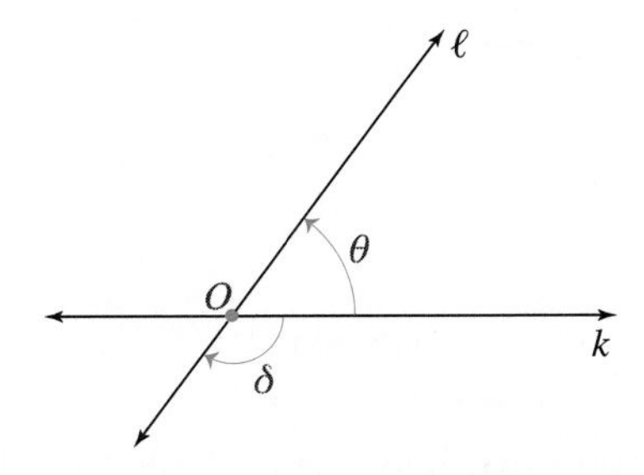

Figure 6.10

Thus we could also conclude from Theorem 6.3 that $M_\ell \circ M_k = R_{O,2\delta}$, and we had better have $R_{O,2\theta} = R_{O,2\delta}$. To see that this is the case, notice that

$$|\theta| + |\delta| = \pi$$

If $\theta > 0$ and $\delta < 0$ as in Figure 6.10, then $|\theta| = \theta$ and $|\delta| = -\delta$. Hence

$$\theta - \delta = \pi$$

This implies

$$2\theta - 2\delta = 2\pi$$

$$2\theta = 2\pi + 2\delta$$

Consequently,

$$R_{O,2\theta} = R_{O,2\pi+2\delta}$$

Because rotating a point about O by $2\delta + 2\pi$ is the same as rotating the point by 2δ, $R_{O,2\pi+2\delta} = R_{O,2\delta}$ and hence $R_{O,2\theta} = R_{O,2\delta}$.

Let's take a closer look at Theorem 6.3. It says that a reflection in line k followed by a reflection in line ℓ is a rotation about the point O (where the lines intersect) by twice the angle θ (the angle between the lines in the direction from k to ℓ). Of course, there are infinitely many different pairs of lines through O such that the angle between them is θ and in the direction from k to ℓ. Figure 6.11 shows a pair of such lines that are different from the pair k and ℓ. Notice how $M_\ell \circ M_k = M_n \circ M_m$. In fact, one of the lines through O—say, line m—could be chosen at will. The other line—here, line n—is the line that makes angle θ with line m in the same direction as the direction from k to ℓ. This discussion along with Theorem 6.3 implies the following corollary:

Corollary 6.2 *Any rotation $R_{O,\alpha}$ (the rotation about a point O through a directed angle α) can be expressed in infinitely many ways as the composition of two reflections. The lines of reflection can be any pair of lines through O such that the directed angle between the lines is $\alpha/2$.*

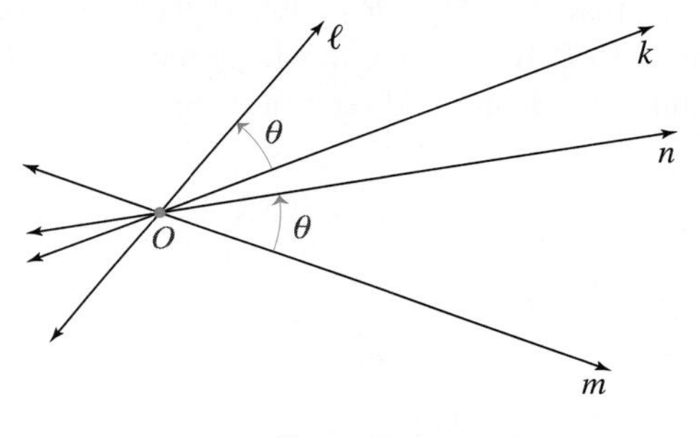

Figure 6.11

Example 6.1 Are there two intersecting lines for which the composition of reflections in these lines is commutative? If so, find all such lines.

Solution

Let k and ℓ be two lines intersecting at O. We want to find all the lines for which

$$M_\ell \circ M_k = M_k \circ M_\ell \tag{6.7}$$

That is, every point in the plane has the same image under each composition.

On the one hand, by Theorem 6.3, $M_\ell \circ M_k = R_{O,2\theta}$, where the rotation is about the directed angle in the direction from line k to line ℓ. On the other hand, $M_k \circ M_\ell$ is a rotation about O by the same angle (in absolute value) but in the direction from line ℓ to line k (that is, in the opposite direction). Thus $M_k \circ M_\ell = R_{O,-2\theta}$. Consequently, Equation 6.7 is true if and only if

$$R_{O,2\theta} = R_{O,-2\theta} \tag{6.8}$$

For which θ does this equation hold? We know that $R_{O,\pi} = R_{O,-\pi}$. Consequently, if $2\theta = \pi$ (that is, $\theta = \pi/2$), then Equation 6.8 holds and therefore Equation 6.7 is true. Because θ is the angle between the lines, we see that if the lines are perpendicular, then the composition of reflections in the lines is commutative.

To show that if the lines are not perpendicular, then the composition is not commutative, we can find all θ for which Equation 6.8 is true. Because $R_{O,-2\theta}$ is the inverse of $R_{O,2\theta}$, each of the following is equivalent to Equation 6.8:

$$R_{O,2\theta} \circ R_{O,-2\theta} = R_{O,-2\theta} \circ R_{O,2\theta} \tag{6.9}$$

$$R_{O,4\theta} = I$$

Equation 6.9 is true if and only if $4\theta = 2\pi k$, where k is an integer. If $k = 0$, then $\theta = 0$, which is impossible because the lines are distinct. If $k = 1$, then $4\theta = 2\pi$ and $\theta = \pi/2$, and hence the lines are perpendicular. Because the smaller of the two angles between two intersecting lines is less than or equal to $\pi/2$, there is no need to consider any other values of k.

We could have shortened this solution by proceeding immediately from Equation 6.8 to Equation 6.9.

Now Solve This 6.1

Show that for any two lines k and ℓ (intersecting or parallel), the following three equations are equivalent:

$$M_\ell \circ M_k = M_k \circ M_\ell$$

$$(M_\ell \circ M_k)^2 = I$$

$$(M_\ell \circ M_k)^{-1} = M_\ell \circ M_k$$

Recall that if T is a transformation, we write $T \circ T$ as T^2.

Composition of Two Reflections in Two Parallel Lines

In Figure 6.12, lines k and ℓ are parallel. To find what kind of isometry $M_\ell \circ M_\ell$ (the reflection in line k followed by the reflection in line ℓ) is, we pick any three convenient noncollinear points P, Q, and S so that their reflections in k fall between the two lines. (This is always possible.) P'' and Q'' are then as shown in Figure 6.12. Looking at Figure 6.12, it seems that $\triangle P''Q''S''$ can be obtained from $\triangle PQS$ by a translation. To see that this is, indeed, the case, we will look at P, P', and P''.

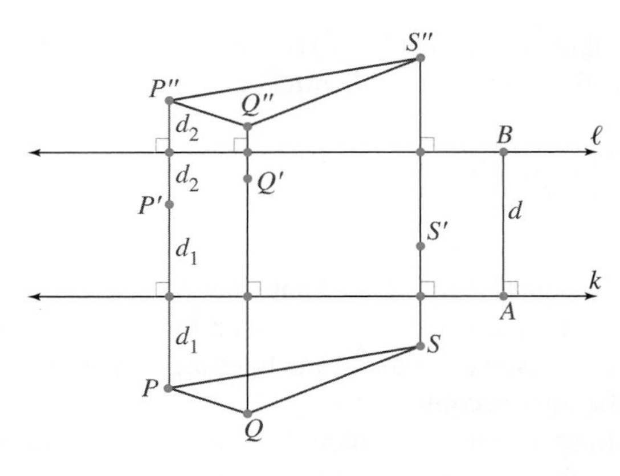

Figure 6.12

Let d be the distance between the lines. Referring to the notation in Figure 6.12, we have

$$PP'' = 2d_1 + 2d_2 = 2(d_1 + d_2) = 2d$$

Because the line PP'' is perpendicular to the given lines, we can obtain P'' from P by a translation in the direction from line k to line ℓ perpendicular to these lines. Notice that the same is true for Q'' and Q as well as for S'' and S. If A is on k, B is on ℓ, and $\overline{AB}$ is perpendicular to the lines, we have

$$(M_\ell \circ M_k)(P) = T_{2AB}(P)$$
$$(M_\ell \circ M_k)(Q) = T_{2AB}(Q)$$
$$(M_\ell \circ M_k)(S) = T_{2AB}(S)$$

In other words, $M_\ell \circ M_k$ and T_{2AB} have the same values (images) at three noncollinear points. But, by Corollary 6.1, if two isometries have the same values at three noncollinear points, then they have the same values at all points, because they are equal. We can write this fact as $M_\ell \circ M_k = T_{2AB}$. We state this result in the following theorem:

Theorem 6.4

The composition of two reflections in two parallel lines is a translation by twice the distance between the lines in the direction from the first to the second.

Like Corollary 6.2 to Theorem 6.3, which stated that a rotation can be written in infinitely many ways as a composition of two reflections in two intersecting lines, we have an analogous corollary to Theorem 6.4.

Corollary 6.3 *A translation T_{PQ} can be expressed in infinitely many ways as a composition of two reflections in two lines, each of which is perpendicular to the line PQ such that the distance between the lines is $\frac{PQ}{2}$.*

Notice that we can choose one of the lines of reflection in Corollary 6.3 arbitrarily. For example, given T_{PQ}, if we choose line k to be perpendicular to line PQ at P, as in Figure 6.13, then line ℓ must be the line parallel to k through M, the midpoint of $\overline{PQ}$. Then T_{PQ} =

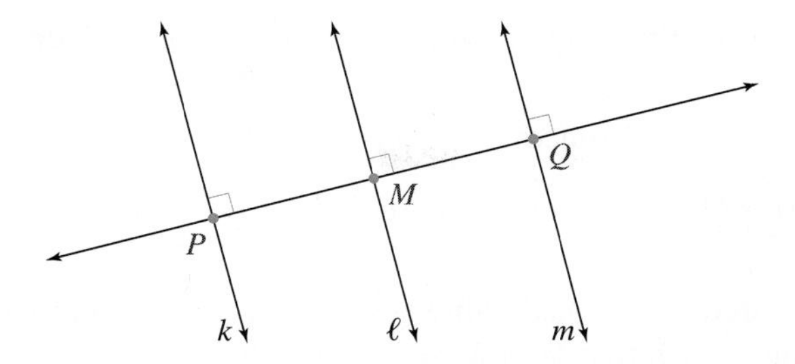

Figure 6.13

$M_\ell \circ M_k$. If we choose line ℓ to be our first line of reflection, then the second line of reflection must be line m through Q, which is parallel to lines k and ℓ. We have $T_{PQ} = M_\ell \circ M_k = M_m \circ M_\ell$.

Remark We have seen that if two lines k and ℓ are parallel, then $M_\ell \circ M_k$ is the translation by twice the distance between the lines in the direction from k to ℓ. Similarly, $M_k \circ M_\ell$ is the translation by twice the distance between the lines in the direction from ℓ to k. Thus, if $M_\ell \circ M_k = T_{PQ}$, then $M_k \circ M_\ell = T_{QP} = T_{PQ}^{-1}$. Consequently, $M_\ell \circ M_k$ and $M_k \circ M_\ell$ are inverses of each other.

By Theorem 6.2, we know that we can express any isometry as a composition of at most three reflections. So far, we have investigated compositions of two reflections. To account for all possible isometries, we next consider composition of three reflections. Three lines could be concurrent, all parallel, two parallel and the third line intersecting each of the other two lines, or in a configuration in which any two lines intersect in exactly one point.

Composition of Reflections in Three Concurrent Lines

Suppose lines k, ℓ, and m intersect at O as shown in Figure 6.14. Our goal is to express $M_m \circ M_\ell \circ M_k$, if possible, as a single familiar isometry. Because each reflection reverses orientation, three reflections will also reverse orientation. Among reflections, translations, and rotations, only a reflection reverses orientation. (Although glide reflection was introduced in Chapter 5, we are not assuming here a familiarity with this isometry.) Thus, if $M_m \circ M_\ell \circ M_k$ is one of these isometries, it must be a reflection. To prove that $M_m \circ M_\ell \circ M_k$ is a single reflection, we use Theorem 6.3 and Corollary 6.2, which tell us that we can express a composition of any two reflections in infinitely many ways as a composition of two reflections. For example, $M_m \circ M_\ell = M_y \circ M_x$, where either line x or line y could be chosen at will. Therefore we have

$$M_m \circ M_\ell \circ M_k = (M_m \circ M_\ell) \circ M_k = (M_y \circ M_x) \circ M_k = M_y \circ (M_x \circ M_k)$$

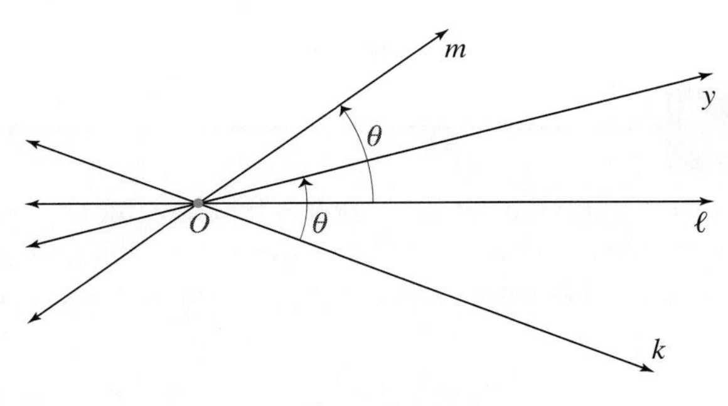

Figure 6.14

The result will be especially simple if we choose x to be line k. Then:

$$M_x \circ M_k = M_k \circ M_k = I \quad \text{and hence}$$

$$M_m \circ M_\ell \circ M_k = M_y \circ (M_x \circ M_k) = M_y \circ I = M_y$$

Thus a composition of three reflections in three concurrent lines is a reflection. We can identify that reflection by finding y such that $M_m \circ M_\ell = M_y \circ M_k$. In Figure 6.14, y is the line for which the directed angle from k to y is the same as the directed angle from ℓ to m (angle θ in Figure 6.14). In a completely analogous way, we can show that a composition of three reflections in three parallel lines is a reflection. The details are left to you as an exercise.

Composition of Reflections in Three Lines That Are Neither Concurrent Nor All Parallel

Perhaps the simplest case is when two lines are parallel and the third line is perpendicular to them, as shown in Figure 6.15. Consider

$$M_m \circ M_\ell \circ M_k \text{ where } k \parallel \ell \text{ and } m \perp k \tag{6.10}$$

Is this composition of three reflections a familiar isometry? Notice that $M_m \circ M_\ell \circ M_k$ reverses orientation and hence cannot be a rotation or a translation. Is it a reflection? A reflection has infinitely many fixed points (the points on the line of reflection). The composition of the three reflections in Equation 6.10, however, does not seem to have any fixed points. (The justification of this fact will be explored in Now Solve This 6.2.) Thus the transformation in Equation 6.10 is not a rotation, a translation, or a reflection in a single line. Consequently, it is a new isometry. Because $k \parallel \ell$, $M_\ell \circ M_k$ is a translation. Referring to Figure 6.15, $M_\ell \circ M_k = T_{2AB}$. Thus $M_m \circ M_\ell \circ M_k = M_m \circ T_{2AB}$. Because m is perpendicular to k and ℓ, m must be parallel to $\overline{AB}$. Consequently, $M_m \circ T_{2AB}$ is a translation followed by a reflection in a line parallel to the direction of the translation. We call such a composition of a translation followed by a reflection a **glide reflection**.

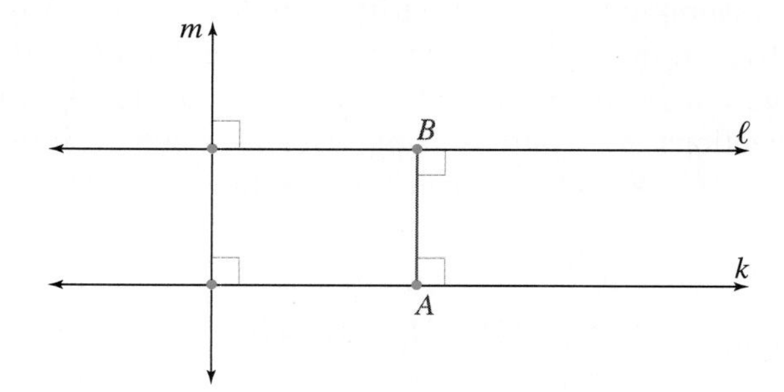

Figure 6.15

Now Solve This 6.2

1. Prove that a glide reflection has no fixed points by showing that any point in one half plane determined by line m in Figure 6.15 is transformed by the glide reflection to a point in the other half plane. Also show that every point on m is transformed to a different point on m.
2. In Figure 6.16, P_2 is the image of P_1 under reflection in m, $PP_1 = 2AB$, and $PP_1P_2P_3$ is a rectangle. Use the figure to show that $M_m \circ T_{2AB} = T_{2AB} \circ M_m$.

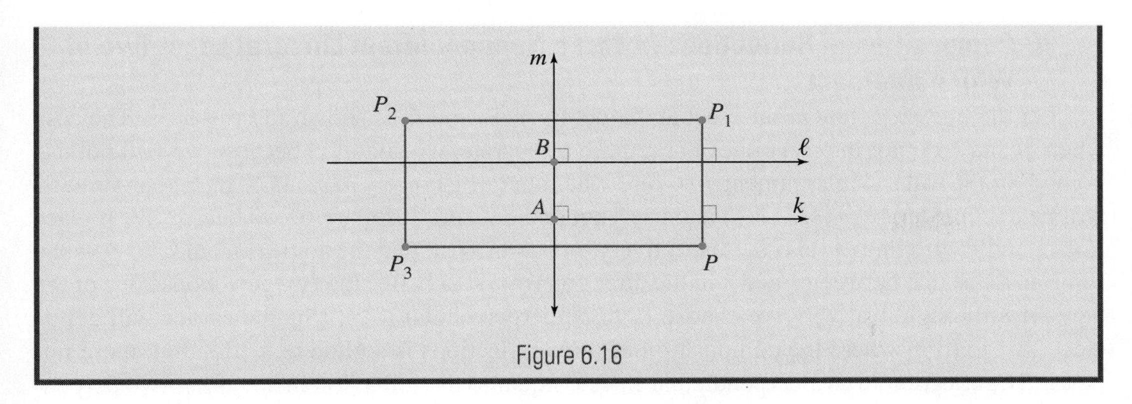

Figure 6.16

In part 2 of Now Solve This 6.2, you saw that in a glide reflection, instead of performing the translation (glide) first and then the reflection, we obtain the same result if we reverse the order and perform the reflection first and then the translation. We can also prove this fact using the result in Example 6.1—namely, if k and ℓ are perpendicular, then $M_\ell \circ M_k = M_k \circ M_\ell$. This is done in Example 6.2 using the algebra of the composition of functions.

Example 6.2 If $k \parallel \ell$ and m is perpendicular to k and ℓ, use Example 6.1 to prove that $M_m \circ (M_\ell \circ M_k) = (M_\ell \circ M_k) \circ M_m$.

Solution

Our strategy is to gradually "move" M_m all the way to the right.

$$\begin{aligned} M_m \circ (M_\ell \circ M_k) &= (M_m \circ M_\ell) \circ M_k \\ &= (M_\ell \circ M_m) \circ M_k \quad \text{because } m \perp \ell \\ &= M_\ell \circ (M_m \circ M_k) \\ &= M_\ell \circ (M_k \circ M_m) \quad \text{because } k \perp \ell \\ &= (M_\ell \circ M_k) \circ M_m \end{aligned}$$

Thus $M_m \circ (M_\ell \circ M_k) = (M_\ell \circ M_k) \circ M_m$.

We summarize the result of Now Solve This 6.2 and Example 6.2 in the following theorem:

Theorem 6.5

If τ_{PQ} is a translation defined by the vector PQ, and m is parallel to line PQ, then $M_m \circ \tau_{PQ} = \tau_{PQ} \circ M_m$.

Now Solve This 6.3

The reflections M_k, M_ℓ, and M_m of Example 6.2 can be composed in 3! (i.e., six) different ways. In Example 6.2, we saw that $M_m \circ M_\ell \circ M_k = M_\ell \circ M_k \circ M_m$. There are four other composition arrangements. Write them and show that each is a glide reflection equal either to $M_m \circ M_\ell \circ M_k$ or to its inverse.

Composition of Reflections in Three Nonconcurrent Lines, at Least Two of Which Intersect

Let the lines k, ℓ, and m intersect at the points A, B, and C as shown in Figure 6.17. (The case when m and ℓ do not intersect does not require a separate investigation because we will not use point C in the following argument.) To find what kind of isometry $M_m \circ M_\ell \circ M_k$ is, our strategy will be to consider $M_m \circ (M_\ell \circ M_k)$ and use Theorem 6.3 and Corollary 6.2 as before. We replace lines k and ℓ with lines k_1 and ℓ_1, respectively, in Figure 6.17 so that the angle α from k_1 to ℓ_1 is the same angle as that between k and ℓ in the direction from k to ℓ. Because we can choose one of the new lines through A at will, we choose ℓ_1 to be perpendicular to m. (Such a choice will get us closer to the form where the composition of three reflections is a glide reflection but, more importantly, reflection in two perpendicular lines is equal to the reflection in any other two perpendicular lines through the same intersection point.) Next we choose k_1 (see Figure 6.17) through A so that the angle from k_1 to ℓ_1 is also α. We have

$$\begin{aligned} M_m \circ (M_\ell \circ M_k) &= M_m \circ (M_{\ell_1} \circ M_{k_1}) \\ &= (M_m \circ M_{\ell_1}) \circ M_{k_1} \quad \text{(worthwhile because } m \perp \ell_1) \end{aligned} \tag{6.11}$$

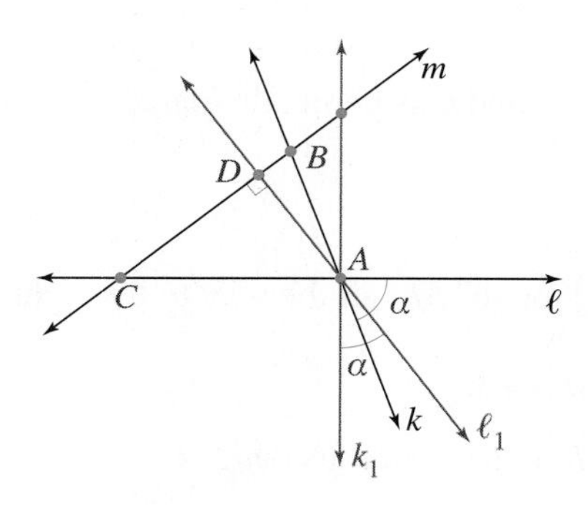

Figure 6.17

Because $\ell_1 \perp m$, we can replace ℓ_1 and m with any two perpendicular lines through D, the point where ℓ_1 and m intersect. As shown in Figure 6.18, ℓ_1 is replaced with ℓ_2 so that $\ell_2 \parallel k_1$ (because we want to obtain the glide for a glide reflection). We replace m with m_1 so that $m_1 \perp \ell_2$.

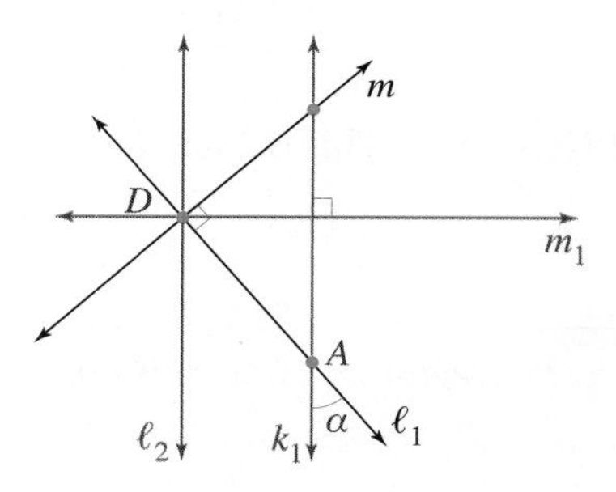

Figure 6.18

We get $M_m \circ M_{\ell_1} = M_{m_1} \circ M_{\ell_2}$. This, along with Equation 6.11, implies

$$\begin{aligned} M_m \circ (M_\ell \circ M_k) &= M_m \circ (M_{\ell_1} \circ M_{k_1}) \\ &= (M_{m_1} \circ M_{\ell_2}) \circ M_{k_1} \\ &= M_{m_1} \circ (M_{\ell_2} \circ M_{k_1}) \end{aligned}$$

Because $\ell_2 \parallel k_1$ and m_1 is perpendicular to these lines, the resulting isometry and thus the original $M_m \circ M_\ell \circ M_k$ is a glide reflection. We have thus proved the following theorem:

Theorem 6.6

A composition of reflections in three nonconcurrent lines, at least two of which intersect, is a glide reflection.

Problem Set 6.1

1. Construct any two congruent segments AB and $A'B'$, and then find two different isometries, each as a composition of reflections and each mapping A, B onto A', B'. Construct the lines of reflections, and explain why the isometries are different.
2. Consider two perpendicular congruent segments AB and $A'B'$ such that $A = A'$. Describe two different isometries, each being a reflection or composition of reflections and each mapping A, B onto A', B'.

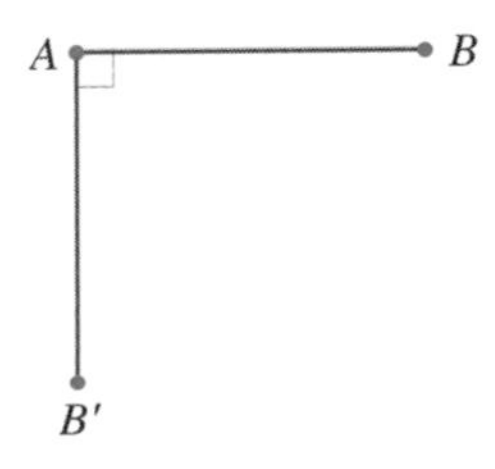

3. In this problem, points A, B, C are on line k and points A', B', C' are on line ℓ. In addition, $A'B' = AB$, $A'C' = AC$, and $B'C' = BC$.

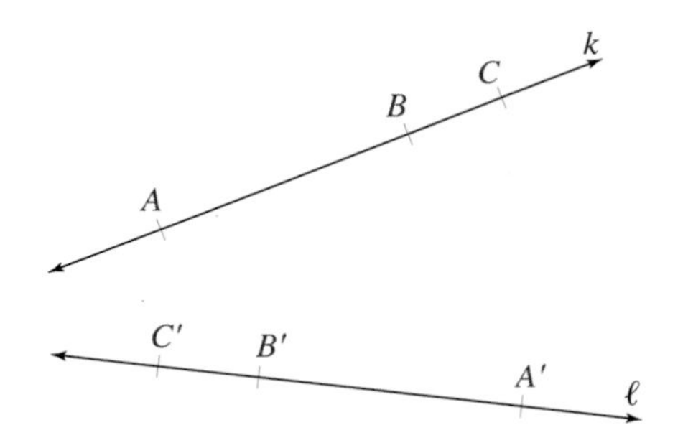

   a. Find two different isometries that map A, B onto A', B'. Check experimentally (using a compass and a straight edge or Sketchpad) that each isometry maps C onto C'.
   b. Prove that every isometry that maps A, B onto A', B' also maps C to C'. [*Hint:* Let F be an isometry that maps A to A' and B to B', and let $F(C) = C^*$. You need to show that $C^* = C'$. Because every isometry preserves betweenness and B is between A and C, B' must be between C^* and A'. Consequently, C' and C^* must be on the ray $B'C'$. Show that this implies that $C^* = C'$.]

•4. a. Construct two congruent triangles ABC and $A'B'C'$ so that the orientation $A'–B'–C'$ is the same as the orientation $A–B–C$. Find an isometry in the form of a composition of reflections that maps A, B, C onto A', B', C'. How many reflections are needed?
   b. Repeat part (a) but with the opposite orientations. (If $A–B–C$ is clockwise, $A'–B'–C'$ is counterclockwise.)

•**5.** **a.** Suppose triangles ABC and $A'B'C'$ are congruent and $A = A'$. What is the smallest number of reflections whose composition will map A, B, C onto A', B', C'? Justify your answer.

b. Suppose triangles ABC and $A'B'C'$ are congruent with $A = A'$ and $B = B'$. Will one reflection map triangle ABC onto triangle $A'B'C'$? Justify your answer.

6. Suppose that F is an isometry, and A, B, and C are three noncollinear points for which $F(A) = A$, $F(B) = B$, and $F(C) = C$. What kind of isometry must F be? Prove your answer in two different ways. One way should use Corollary 6.1 and the other should start as follows:

Suppose there exists a point P such that $F(P) = P' \neq P$. Then each of the points A, B, and C must be equidistant from P and P' because . . .

7. Let point P be in the interior of angle θ formed by the lines k and ℓ as shown. Construct P' and P'' defined by $(M_\ell \circ M_k)(P) = M_\ell(P') = P''$. Show that in this case, $m(\angle POP'') = 2\theta$ and, consequently, that if P is in the interior of θ, P'' can be obtained from P by a rotation $R_{O,2\theta}$.

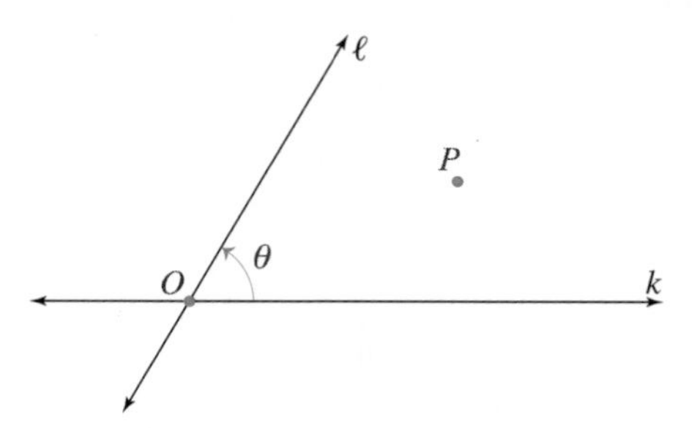

8. Lines k and ℓ intersect at O.

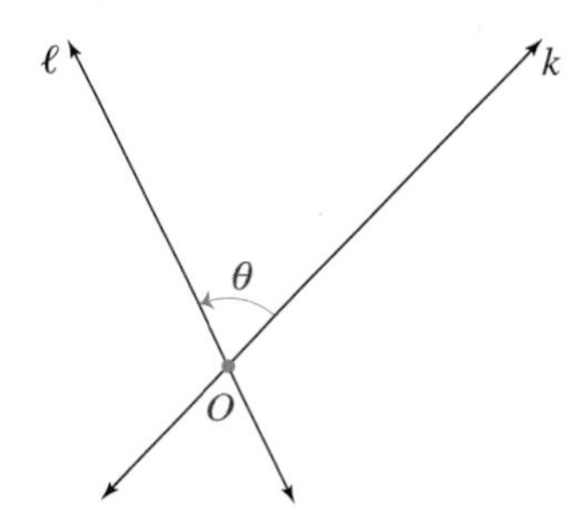

•**a.** Construct line x such that

$$M_\ell \circ M_k = M_x \circ M_\ell$$

Is line x unique? Justify your answer.

b. Given the lines in part (a), construct and describe line y through O such that

$$(M_\ell \circ M_k) \circ (M_\ell \circ M_k) = M_y \circ M_k$$

9. **a.** In the accompanying figure, $x \perp y$ and k makes a 45° angle with line x. Prove that $M_y \circ M_k \circ M_x$ can be expressed as a reflection in a single line. Find that line.

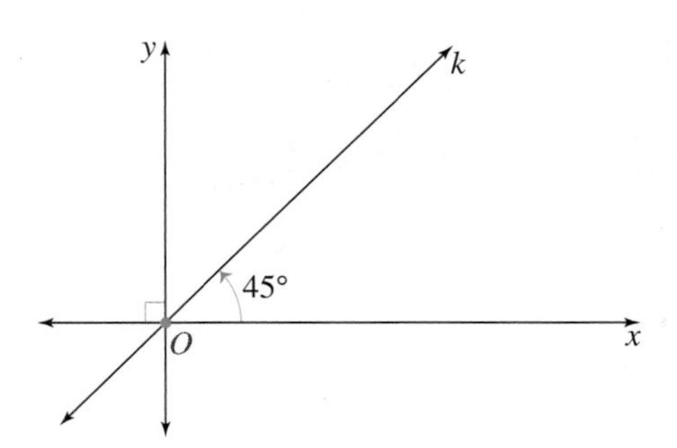

b. The accompanying figure suggests a generalization of the result in part (a). State this generalization and justify it.

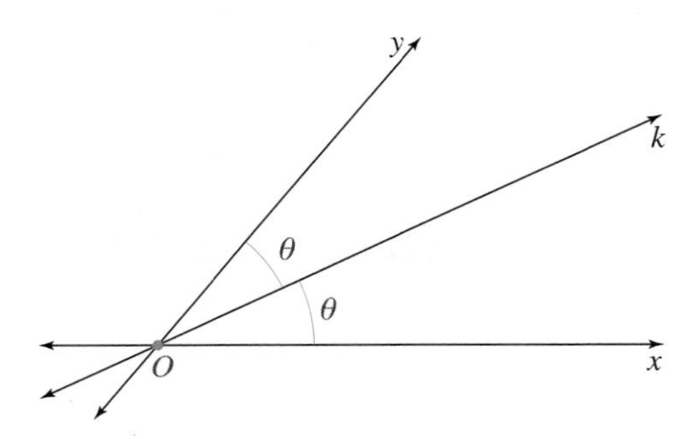

•**10.** What kind of isometry is the composition of three reflections in three parallel lines? Prove your answer.

11. What kind of isometry is a composition of a glide reflection with itself? Prove your answer in two different ways:

a. Geometrically by looking at the effect of a glide reflection applied twice on a point P

b. Algebraically by using the fact that $M_k \circ T_{AB} = T_{AB} \circ M_k$, where AB is parallel to the line of reflection k

12. For lines k, ℓ, and m (where k is parallel to ℓ) in the accompanying figure, prove that $M_m \circ M_\ell \circ M_k$ is a glide reflection. Determine the translation vector and the line of reflection in the glide reflection.

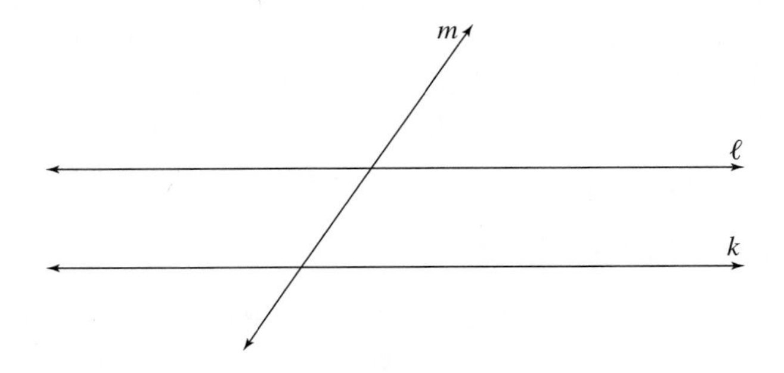

•**13.** Suppose the lines k, x, and y form an isosceles right triangle as shown in the figure, where A is at (1, 0) and B at (0, 1). Write $M_k \circ M_y \circ M_x$ in a standard form $M_\ell \circ T_{OP}$, where O is at the origin and line OP is parallel to ℓ. Find the equation of ℓ and the coordinates of P.

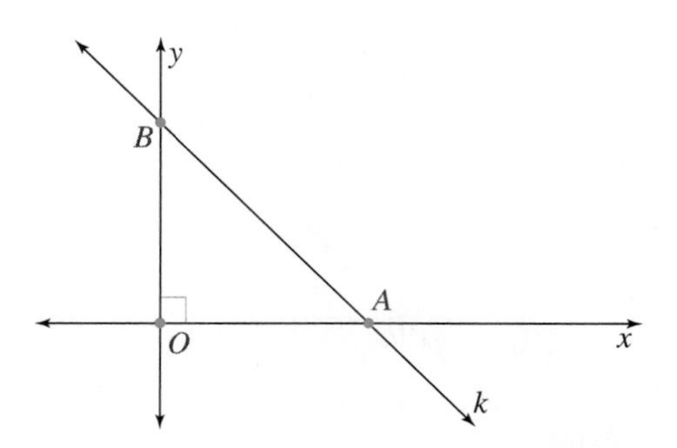

14. The lines a, b, and c form an equilateral triangle.

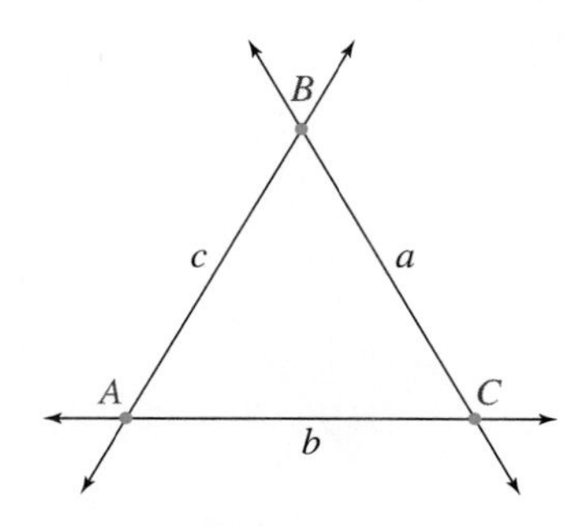

a. Prove that

$$M_c \circ M_b \circ M_a = M_n \circ T_{\frac{3}{2}CA},$$

where n is the line through the midpoints of the two sides of the triangle, which meet at B, and $T_{\frac{3}{2}CA}$ is the translation defined by the vector $\frac{3}{2}\overrightarrow{CA}$ (the translation in the direction from C to A by the distance $\frac{3}{2}CA$.)

b. Keeping in mind the notation in part (a), use an analogy to write $M_a \circ M_b \circ M_c$ as a glide reflection in the standard form. Check that your answer, when composed with the right side of the equation in part (a), gives the identity. [It should, because $M_a \circ M_b \circ M_c$ is the inverse of $M_c \circ M_b \circ M_a$, the left side of the equation in part (a).]

15. Investigate the types of isometries resulting from composition of reflections in

a. n concurrent lines

b. n parallel lines

16. In a coordinate system a reflection M_x in the x-axis can be described as $M(x, y) = (x, -y)$. Similarly describe a reflection in the line $y = b$ using the following approaches.

a. Let M_b be the reflection in the line $y = b$ and τ be the translation τ_{OP}, where O is the origin and $P(0, b)$. Explain why $M_b = \tau \circ M_x \circ \tau^{-1}$ and use this fact to find $M_b(x, y)$.

b. Use the midpoint formula $x_N = \dfrac{x_A + x_B}{2}$, $y_N = \dfrac{y_A + y_B}{2}$ for the coordinates of the midpoint N of the segment AB.

17. Prove that the composition of a translation followed by a reflection in a line is a glide reflection, and identify the line of reflection, and the glide, using the following:

a. Prove that $M_\ell \circ \tau_{AB}$, where $\overline{AB}$ is perpendicular to ℓ, is a reflection. Identify the line of reflection.

b. Let τ_{CD} be a translation from C to D and ℓ be a line not parallel to $\overline{CD}$. Write τ_{CD} as a composition of two translations, one parallel to ℓ and the other perpendicular to ℓ.

c. Use parts (a) and (b) to prove that $M_\ell \circ \tau_{CD}$ is a glide reflection.

• **18.** Let k and ℓ be two lines intersecting at O and θ be the angle by which ℓ needs to be rotated about O so that its image is line k. If R_θ is the rotation about O by θ prove that $M_\ell = R_{-\theta} \circ M_k \circ R_\theta$.

6.2 Composition of Rotations, the Treasure Island Problem, and Other Treasures

The result of composition of two rotations about the same point is quite obvious: If we rotate a point about O by angle α and then rotate its image about O by angle β, the result is the same as if we had applied a single rotation about O by angle $\alpha + \beta$. In fact, the order of the rotations is immaterial. We write this relationship as follows:

$$R_{O,\beta} \circ R_{O,\alpha} = R_{O,\alpha} \circ R_{O,\beta} = R_{O,\alpha+\beta}$$

What if the rotations are about different points. Is the result always a rotation and, if so, about what point? We state our question more formally:

What kind of isometry is a composition of two rotations about two different points?

Figure 6.19 demonstrates what happens to a point P under the composition of two rotations: one with center A and positive angle α, and the other with center B and positive angle β.

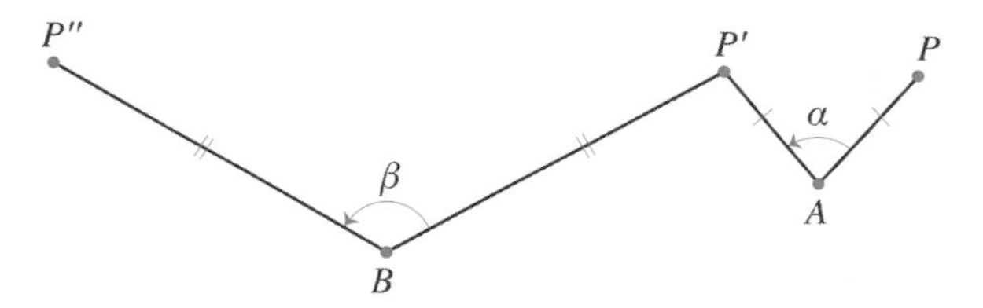

Figure 6.19

Because $R_{A,\alpha}(P) = P'$ and $R_{B,\beta}(P') = P''$, we have

$$\begin{aligned}(R_{B,\beta} \circ R_{A,\alpha})(P) &= R_{B,\beta}(R_{A,\alpha}(P)) \\ &= R_{B,\beta}(P') \\ &= P''\end{aligned}$$

We want to find a single isometry that will have the same effect as $R_{B,\beta} \circ R_{A,\alpha}$. Notice that composition of rotations preserves orientation, whereas the only isometries that preserve orientation are rotations and translations. Thus $R_{B,\beta} \circ R_{A,\alpha}$ is either a rotation or a translation. To find out what kind of isometry $R_{B,\beta} \circ R_{A,\alpha}$ is, we write each rotation as a composition of reflections. We know from Theorem 6.3 that a rotation can be written as a composition of two reflections in infinitely many ways with two conditions. The lines of reflection must go through the center of rotation, and the angle formed by the lines must be equal to half the angle of the rotation. Let $R_{A,\alpha} = M_\ell \circ M_k$ and $R_{B,\beta} = M_n \circ M_m$, where the lines k, ℓ, m, and n are still to be determined. We can choose one of the lines in each pair in any way we want. Before we decide, though, let's see what do we want to accomplish. We have

$$R_{B,\beta} \circ R_{A,\alpha} = (M_\ell \circ M_k) \circ (M_n \circ M_m) \tag{6.12}$$

To make the right side of Equation 6.12 as simple as possible, it would be advantageous to make $M_k = M_n$; that is, we choose $k = n$ (then $M_k \circ M_k = I$). But line n goes through A and line k goes through B (see Figure 6.19), so $k = n$ if and only if the line goes through A and B. This line of reflection, along with lines m and ℓ, is shown in Figure 6.20.

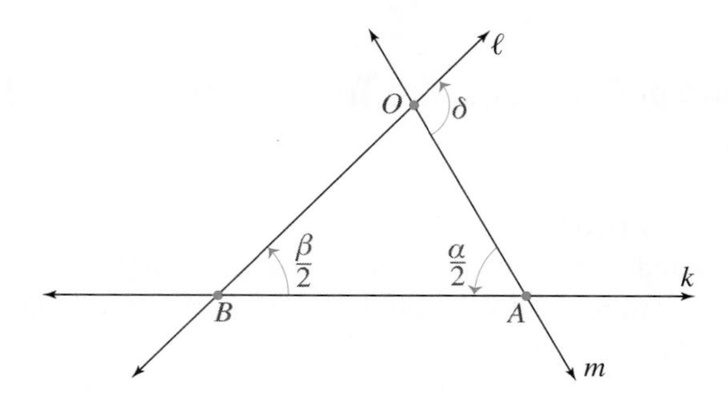

Figure 6.20

Because α and β were chosen to be positive, we use Equation 6.12 for $n = k$ and the associative property of composition of isometries to obtain

$$\begin{aligned} R_{B,\beta} \circ R_{A,\alpha} &= (M_\ell \circ M_k) \circ (M_k \circ M_m) \\ &= M_\ell \circ (M_k \circ M_k) \circ M_m \\ &= M_\ell \circ I \circ M_m \\ &= M_\ell \circ M_m \end{aligned}$$

Consequently,

$$R_{B,\beta} \circ R_{a,\alpha} = M_\ell \circ M_m \tag{6.13}$$

If lines m and ℓ intersect (in point O in Figure 6.20) and the angle from m to ℓ is δ, then $M_\ell \circ M_m$ is a rotation with center O by angle 2δ. How is 2δ related to the given angles α and β? Notice that δ is an exterior angle in $\triangle OAB$, so $\delta = \alpha/2 + \beta/2$ and, therefore, $2\delta = \alpha + \beta$. Thus, if the lines m and ℓ intersect, then $R_{B,\beta} \circ R_{A,\alpha}$ is a rotation with center O (as found in Figure 6.20) by the angle $\alpha + \beta$.

Lines ℓ and m in Figure 6.20 intersect if and only if a triangle is formed. Because the sum of the interior angles in a triangle is 180°, the lines in Figure 6.20 will not form a triangle if $\alpha/2 + \beta/2 = 180°$ (that is, if $\alpha + \beta = 360°$). Thus, if $\alpha + \beta \neq 360°$, the composition of the two rotations is a rotation. If $\alpha + \beta = 360°$, lines m and ℓ are parallel and Equation 6.13 implies that the composition of the rotations is a translation.

Now Solve This 6.4

Express each of the following as a single rotation or translation.

1. $R_{B,90°} \circ R_{A,-120°}$
2. $R_{B,90°} \circ R_{A,-90°}$

Based on the preceding discussion, we have the following theorem:

Theorem 6.7

If $R_{A,\alpha}$ and $R_{B,\beta}$ are two rotations with centers at A and B and angles of rotation α and β, respectively (α and β may be positive or negative), then $R_{B,\beta} \circ R_{A,\alpha}$ is a rotation by the angle $\alpha + \beta$ if and only if $\alpha + \beta$ is not a multiple of 360°; it is a translation otherwise.

Problem 6.1: The Treasure Island Problem We introduced the Treasure Island Problem in Chapter 0, where we suggested that you investigate it with Sketchpad in Problem 1 of Problem Set 0; we later suggested a proof in Now Solve This 1.12 and Problem 37 in Problem Set 1.3. We restate the problem here for easy reference.

Among his great-grandfather's papers, Marco found a parchment describing the location of a hidden pirate treasure buried on a deserted island. The island contained a coconut tree, a banana tree, and a gallows (Γ) where traitors were hanged. A reproduction of the map appears in Figure 6.21. It was accompanied by the following directions:

> Walk from the gallows to the coconut tree, counting the number of steps. At the coconut tree, turn 90° and go to the right. Walk the same distance, and put a spike in the ground. Return to the gallows and walk to the banana tree, counting your steps. At the banana tree, turn 90° and go to the left. Walk the same number of steps, and put another spike in the ground. The treasure is halfway between the spikes.

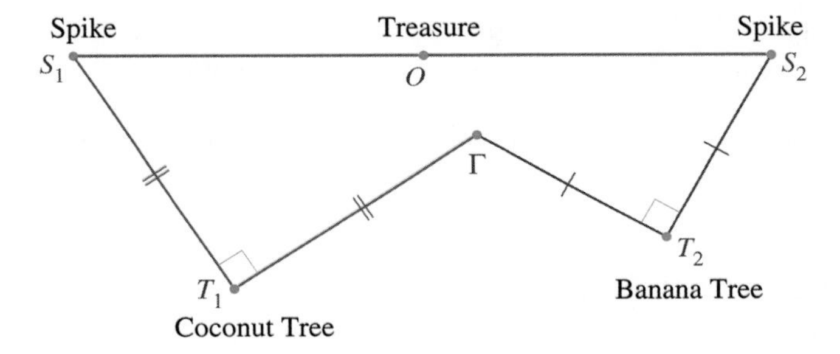

Figure 6.21

Marco found the island and the two trees but no trace of the gallows or the spikes, which had probably rotted. In desperation, he began to dig at random but soon gave up because the island was too large. Devise a plan for finding the treasure and prove your solution.

Solution

This problem brings to mind composition of rotations. No matter where Γ is, we can obtain S_2 from S_1 by rotating S_1 by $-90°$ about T_1 and then rotating its image Γ by $-90°$ about the center T_2. We can write this idea as follows:

$$(R_{T_2,-90} \circ R_{T_1,-90})(S_1) = S_2$$

Based on the proof of Theorem 6.7 and Figure 6.20, we know that

$$\begin{aligned} R_{T_2,-90} \circ R_{T_1,-90} &= (M_m \circ M_k) \circ (M_k \circ M_\ell) \\ &= M_m \circ M_\ell \\ &= R_{O,180°} \end{aligned}$$

Thus, no matter where Γ is, we can obtain the corresponding spike S_2 from S_1 by a half-turn about a fixed point O. Thus O is the midpoint of the segment S_1S_2 resulting from any location of Γ and, therefore, is the location of the treasure.

To find the treasure, we could start with any location for Γ, find the corresponding spikes, and then construct the midpoint O of the segment S_1S_2. Alternatively, we could construct the triangle T_1T_2O, as shown in Figure 6.22. Yet another way to find O is to notice that O is the vertex of an isosceles right triangle and hence on the perpendicular bisector of the segment T_1T_2 at a distance of $\frac{1}{2}T_1T_2$ from the line T_1T_2.

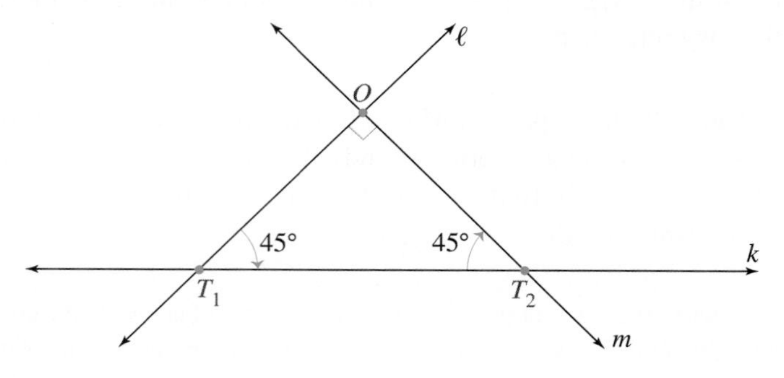

Figure 6.22

Now Solve This 6.5

In the Treasure Island Problem, assume that we are supposed to turn at T_1 by angle α (rather than 90°) and at T_2 by angle β. For which α and β will the solution be the same as in the original problem? Why?

Composition of Half-Turns

Rotations by 180° about a given point deserve separate attention because they possess special properties. A rotation by 180° about a point A is called a half-turn about A and is denoted by H_A, as mentioned earlier. By Theorem 6.7, a composition of two half-turns is a translation. For better understanding, however, we will independently verify that this is true.

Problem 6.2 Show that $H_B \circ H_A$ is a translation, and identify that translation.

Solution

In Figure 6.23, A and B are two arbitrary points. By Theorem 6.3, H_A can be written as a composition of two reflections in any pair of perpendicular lines through A, and H_B can be written as a composition of two reflections. As in the composition of two rotations (Figure 6.20), we use the lines shown in Figure 6.23 as the lines of reflection.

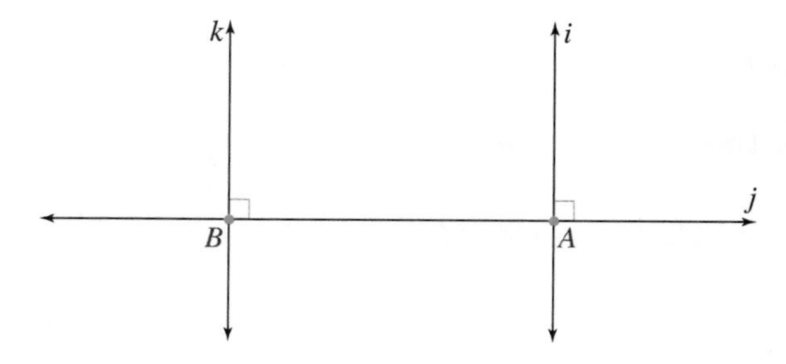

Figure 6.23

Thus

$$H_B \circ H_A = M_k \circ M_j \circ M_j \circ M_i$$
$$= M_k \circ M_i$$
$$= T_{2AB}$$

Problem 6.3 What kind of isometry is a composition of three half-turns $H_C \circ H_B \circ H_A$?

Solution

We have just seen that $H_B \circ H_A = \tau_{2AB}$, a translation in the direction from A to B by twice the distance AB. We also know that $\tau_{2AB} = M_q \circ M_p$, where lines p and q are any two lines perpendicular to line AB, with a distance of AB between each other. Lines p and q must also be such that the direction from p to q is the same as from A to B. In addition, we can express H_C as a composition of two reflections in any two perpendicular lines through C (see the remark following Corollary 6.2). If r and s are such lines through C, we have

$$H_C \circ H_B \circ H_A = (M_s \circ M_r) \circ (M_q \circ M_p)$$

To simplify this composition of four reflections, we choose the lines r and q such that $r = q$. Then $M_r \circ M_q = I$. But where are those lines? Because r goes through C, and q must be perpendicular to line AB, the line is uniquely determined as the line through C perpendicular to line AB. This location of q determines p as shown in Figure 6.24.

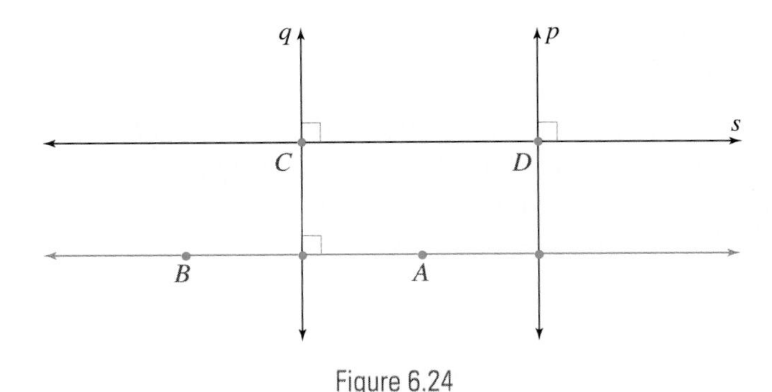

Figure 6.24

Because s must be perpendicular to q through C, it is uniquely determined as shown in Figure 6.24. Let the point where s and q intersect be D. Then

$$H_C \circ H_B \circ H_A = (M_s \circ M_q) \circ (M_q \circ M_p)$$
$$= M_s \circ M_p$$
$$= H_D$$

Thus a composition of three half-turns is a half-turn. If A, B, and C are noncollinear, the center point of the half-turn D is determined as the fourth vertex of the parallelogram with vertices A, B, and C where the fourth vertex is such that the direction from D to C is the same as from A to B.

Now Solve This 6.6

Abby found a description of a buried treasure on a deserted island. There were originally three landmarks on the island: points A, B, and C. The pirates put a spike P in the ground and then a spike Q in the ground so that A was the midpoint of segment PQ. Then they put a spike R in the ground so that B was the midpoint of $\overline{RQ}$. Then they put a fourth spike S in the ground so that C was the midpoint of RS. Finally, they connected S to P and buried the treasure at the midpoint M of the segment PS. (See Figure 6.25.)

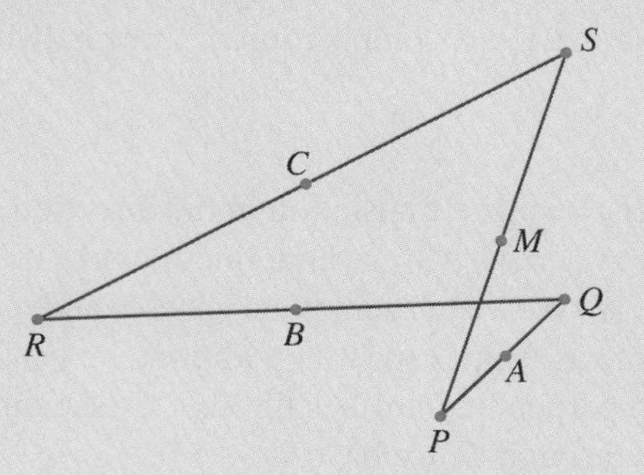

Figure 6.25

Abby arrived at the island and found no trace of the four spikes, but the landmarks A, B, and C were there. How did she find the treasure?

Now Solve This 6.7

1. Let F be a transformation of the plane. We denote $F \circ F$ by F^2. Prove that for arbitrary three points A, B, and C we have

$$(H_C \circ H_B \circ H_A)^2 = I$$

2. In Figure 6.26, P_1 is the image of P under H_A, P_2 is the image of P_1 under H_B, and P_3 is the image of P_2 under H_C. We then find P_4, the image of P_3 under H_A; then P_5, the image of P_4 under H_B (not shown in the figure); and finally P_6, the image of P_5 under H_C. How is P_6 related to P? Prove your answer.

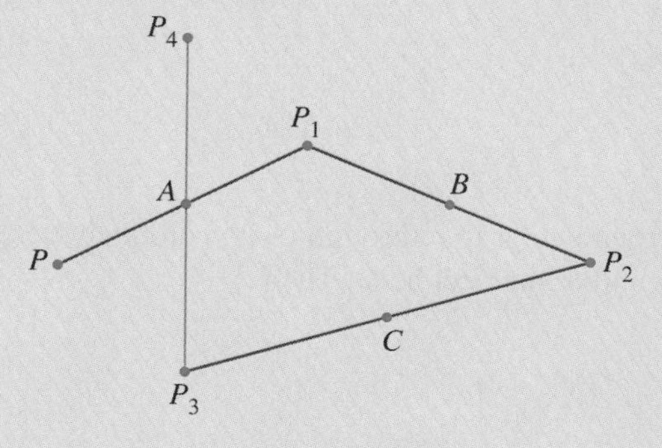

Figure 6.26

3. Write a "buried treasure" problem based on what you found in part 1.

Problem 6.4 **Squares on the Sides of a Quadrilateral**

1. Use Sketchpad or pencil-and-paper constructions to experimentally verify for several different quadrilaterals an amazing property of quadrilaterals. (The same problem was introduced in Chapter 0; see Figure 0.9.) Let $ABCD$ be a quadrilateral as shown in Figure 6.27.

Outside the quadrilateral, construct squares on a pair of opposite sides of the quadrilateral. Let Q and S be the centers of the squares (where the diagonals intersect). Similarly, construct such squares on the pair of the other opposite sides, and let T and P be the centers of these squares. No matter which quadrilateral you start with, the segments QS and TP are congruent and perpendicular.

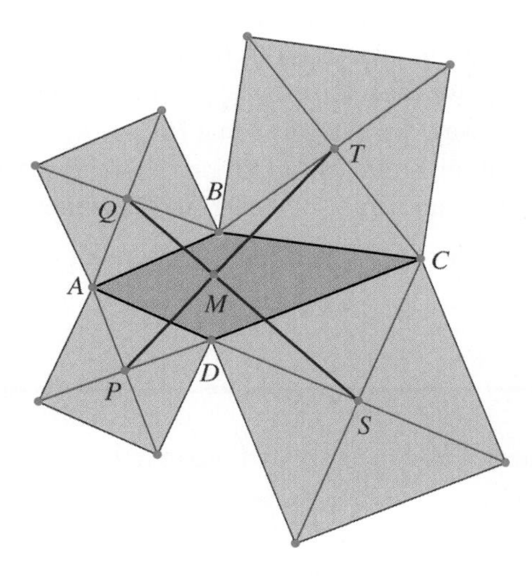

Figure 6.27

2. Prove that $\overline{QS}$ and $\overline{TP}$ are congruent and perpendicular.

Solution to Part 2

The assertion would follow if we could show that segment TP is the image of segment QS under an isometry. Because we want to show that the segments are perpendicular, the most likely isometry to do the job is a rotation by 90°. Recall that the image of a line under a rotation by an angle θ is a line that makes angle θ (or $180° - \theta$) with the original line. Under such rotation, the image of Q could be T and the image of S could be P. Figure 6.28 shows the point O we wish we had, so that $OQ = OT$, $OS = OP$, and both $\angle QOT$ and $\angle SOP$ measures 90°. If we could find such a point O, the image of the segment QS under a rotation with center O by 90° would be segment TP, and our proof would be completed.

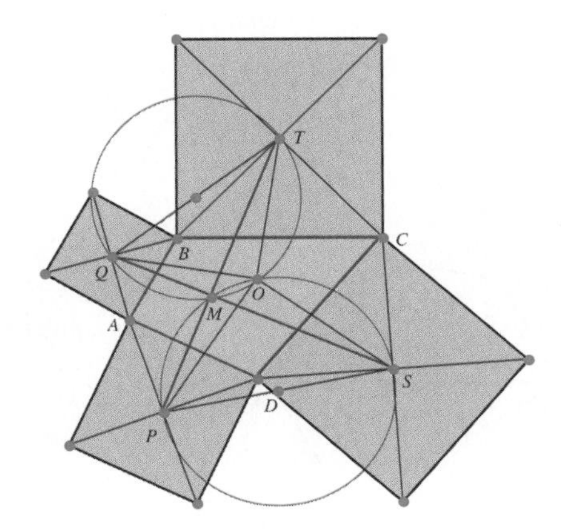

Figure 6.28

If you focus on points Q, T, A, B, C, and O, the corresponding part of Figure 6.28 may remind you of the composition of two rotations about points Q and T—and, in particular, the rotations

in the Treasure Island Problem (Q and T correspond to the trees, B to the gallows, A and C to the spikes, and the midpoint of $\overline{AC}$ to the treasure). By Theorem 6.7 (or the Treasure Island Problem), the composition $R_{T,90°} \circ R_{Q,90°}$ is a rotation by 180° about a point that is the vertex of an isosceles triangle with the base $\overline{QT}$ and base angles of 45°. This is the point O we wished we had in Figure 6.28. As in the Treasure Island Problem, O is the midpoint of segment AC. For another way to see why, we find the image of A under $R_{T,90°} \circ R_{Q,90°}$. Because the first rotation takes A to B and the second rotation takes B to C, the image of A under $R_{T,90°} \circ R_{Q,90°}$ is C. But because $R_{T,90°} \circ R_{Q,90°} = R_{O,180°}$, we must have $R_{O,180°}(A) = C$. Hence O is the midpoint of $\overline{AC}$.

In a completely analogous way, we can show that $R_{P,90°} \circ R_{S,90°}$ is a rotation by 180° about a point O_1, which is the vertex of an isosceles right triangle with base PS. We would like to show that $O_1 = O$. This is the case because $R_{S,90°}$ takes C to D, while $R_{P,90°}$ takes D to A. Because $R_{P,90°} \circ R_{S,90°} = R_{O_1,180°}$, $R_{O_1,180°}(C) = A$. Thus O_1 is the midpoint of $\overline{AC}$. Hence $O_1 = O$, and the proof is complete.

Now Solve This 6.8

1. State part 2 of Problem 6.4 as a theorem and prove it. (Write a concise proof without investigation and motivation.)
2. To discover a theorem similar to the one in part 1 of this exercise but for triangles, sketch figures analogous to Figure 6.27 when B gets closer and closer to C. (Your squares with base BC should get smaller and smaller.) State the theorem that follows for triangles.

Napoleon's Theorem

The French conqueror and emperor Napoleon I (1769–1821) is credited with the following theorem:

Theorem 6.8

In Figure 6.29, ABC is an arbitrary triangle. Equilateral triangles have been constructed on the sides of the triangle as shown. If P, Q, and S are the centers of the equilateral triangles (that is, the centers of the circumscribing as well as inscribed circles of the equilateral triangles), then $\triangle PQS$ is equilateral.

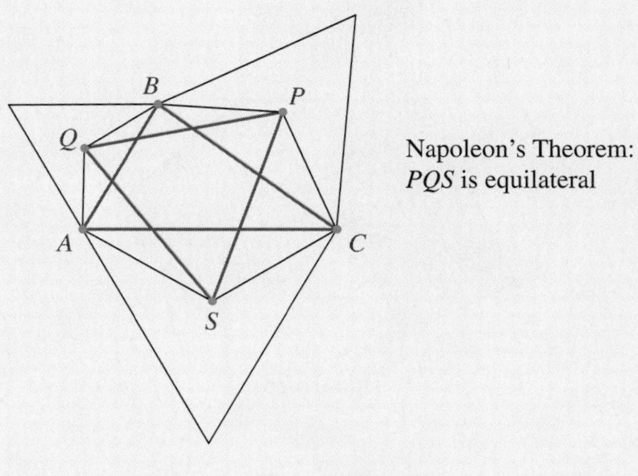

Figure 6.29

Examine the proof in Problem 6.4 and then try to prove Napoleon's Theorem before reading on.

Proof

To prove that $\triangle PQS$ is equilateral, it will suffice to show that one of the vertices of this triangle is the image of another vertex under a rotation by 60° in an appropriate direction about a third vertex. We will show that Q is the image of P under a rotation about S by 60° (counterclockwise). Because P, Q, and S are the centers of equilateral triangles, the triangles AQB, BPC, and CSA are isosceles triangles with angles of 120° at the vertices Q, P, and S. Analogous to what we did in Problem 6.4, we will consider $R_{P,120^\circ} \circ R_{Q,120^\circ}$. By Theorem 6.3,

$$R_{P,120^\circ} \circ R_{Q,120^\circ} = R_{O,240^\circ} \quad (6.14)$$

where O is a point determined by lines i, j, and k as shown in Figure 6.30.

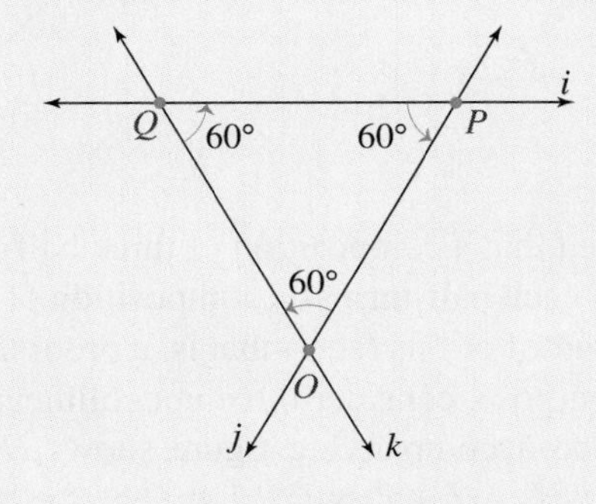

Figure 6.30

For a better understanding, it is useful to prove again Theorem 6.3 for the given rotations by decomposing each rotation as a composition of reflections.

$$R_{P,120^\circ} \circ R_{Q,120^\circ} = (M_j \circ M_i) \circ (M_i \circ M_k) = M_j \circ M_k = R_{O,240^\circ}$$

We now show that point O is actually point S of Figure 6.29. One way to do so is to show that

$$R_{P,120^\circ} \circ R_{Q,120^\circ} = R_{S,240^\circ} \quad (6.15)$$

This would imply that $R_{O,240^\circ} = R_{S,240^\circ}$, and hence that $O = S$. We compose each side of Equation 6.15 with $R_{S,-240^\circ}$ and obtain the equivalent equation

$$R_{S,-240^\circ} \circ R_{P,120^\circ} \circ R_{Q,120^\circ} = I \quad (6.16)$$

Because $R_{S,-240^\circ} = R_{S,120^\circ}$, the last equation is equivalent to

$$R_{S,120^\circ} \circ R_{P,120^\circ} \circ R_{Q,120^\circ} = I \quad (6.17)$$

To prove Equation 6.17, we focus on $R_{S,120^\circ} \circ R_{P,120^\circ} \circ R_{Q,120^\circ}$. Because the angles all add up to 360°, this composition of rotations is, by Theorem 6.7, a translation. It is the identity I if and only if the image of some point is the point itself. (A translation that fixes a point must be the identity.) Referring to Figure 6.29, we find that $R_{Q,120^\circ}$ takes A to B, $R_{P,120^\circ}$ takes B to C, and $R_{S,120^\circ}$ takes C to A. Thus the image of A under the composition of the three rotations in Equation 6.17 is A and Equation 6.17 is true. As already mentioned, Equation 6.17 is equivalent to Equation 6.15, which along with Equation 6.14 implies that $R_{O,240^\circ} = R_{S,240^\circ}$, which in turn implies that $O = S$, as in Figure 6.30. □

Problem Set 6.2

1. Using an approach similar to the one we used in proving Theorem 6.7, prove that $R_{B,\pi/2} \circ R_{A,\pi}$ is a rotation. Find the center of the rotation and the angle of the rotation.

2. In the accompanying figure, lines h and k are parallel, and line AB is perpendicular to both lines (A is on h and B is on k). $H_A(P) = P'$ and $H_B(P') = P''$. Write a "buried treasure" problem based on the figure and then solve it.

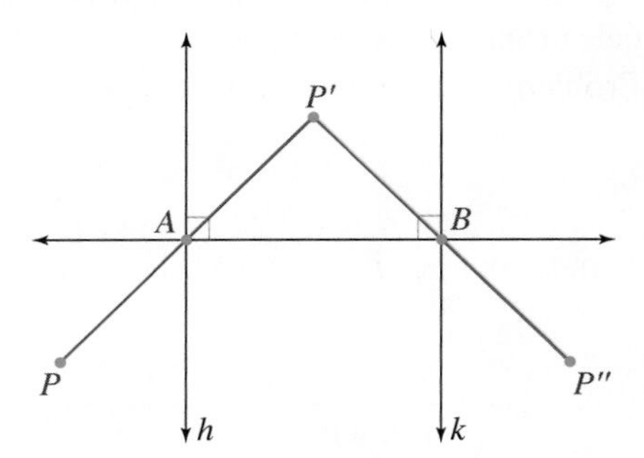

•**3.** In Section 6.2, we showed that a composition of three half-turns is a half-turn. Our proof was based on expressing each half-turn as a composition of two reflections. It is also possible to give a synthetic proof of this fact—that is, a proof that uses only classical Euclidean geometry. For that purpose, consider three noncollinear points A, B, and C, and let P be an arbitrary point. The accompanying figure shows $H_A(P) = P'$, $H_B(P') = P''$, and $H_C(P'') = P'''$. Thus $(H_C \circ H_B \circ H_A)(P) = P'''$. Let O be the midpoint of the segment PP'''. We want to show that for all points P in the plane, the midpoint of PP''' is the same point O. Use the properties of midsegments to prove this fact.

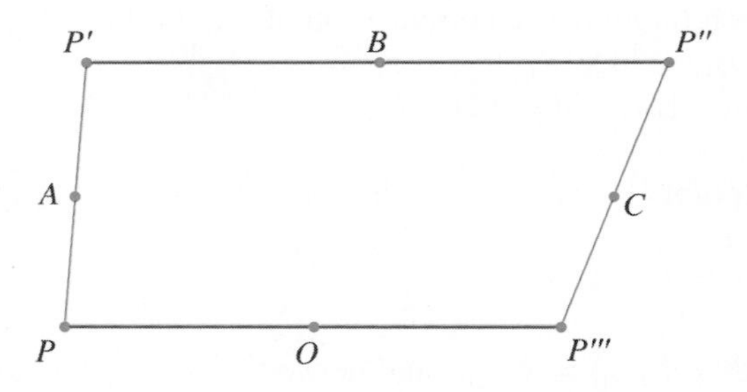

•**4.** In the accompanying figure, squares with centers O_1 and O_2 were constructed on sides AB and BC of an arbitrary $\triangle ABC$. On side AC, an arbitrary point D was chosen, and squares with centers O_3 and O_4 were constructed on $\overline{DC}$ and $\overline{AD}$ as sides. Prove that the segments O_1O_3 and O_2O_4 are congruent and perpendicular.

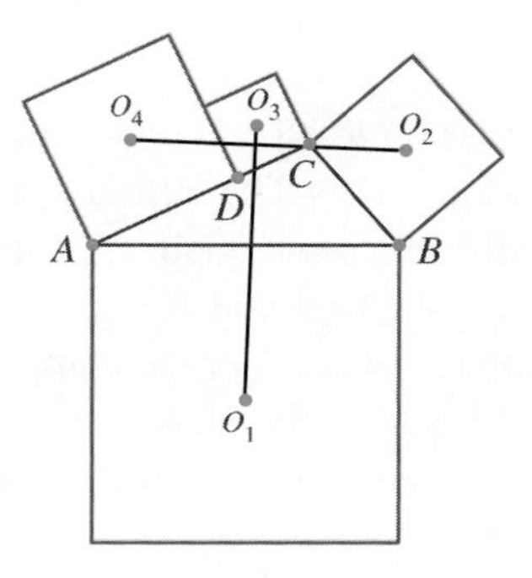

5. In the accompanying figure, squares were constructed on the sides of a concave quadrilateral. If the centers of the squares are O_1, O_2, O_3, and O_4, is it true that segments O_1O_3 and O_2O_4 are congruent and perpendicular? Justify your answer.

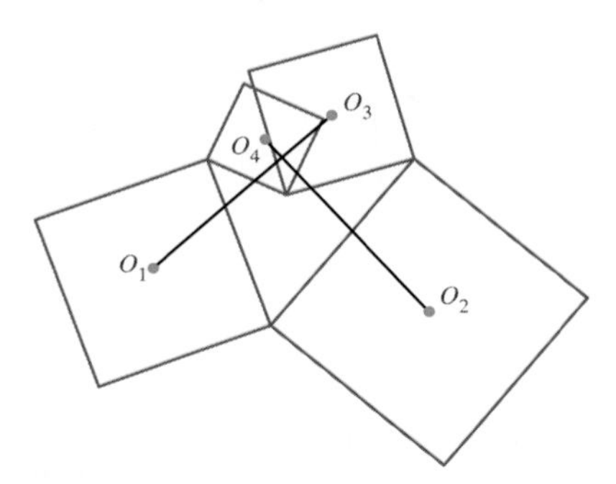

6. In the accompanying figure, *ABCD* is a convex quadrilateral. Squares are constructed on its sides with centers O_1, O_2, O_3, and O_4. Prove that $O_1O_2O_3O_4$ is a square.

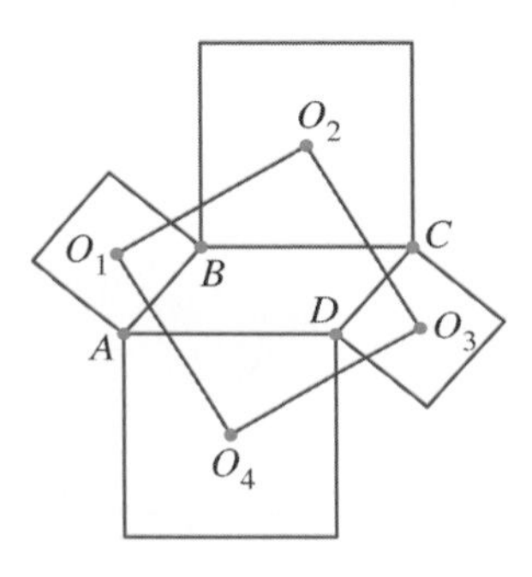

7. **a.** Use Sketchpad or manual instruments to conjecture whether Napoleon's Theorem remains true when equilateral triangles are constructed on the sides of an arbitrary triangle *ABC* in such a way that each equilateral triangle is on the same half-plane determined by the line containing a side of $\triangle ABC$ and the third vertex of that triangle. (See the accompanying figure.)

b. Justify your conjecture in part (a).

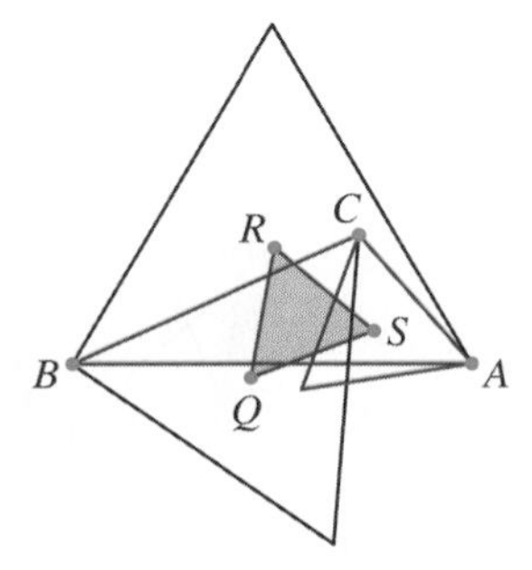

8. *ABCD* is an arbitrary quadrilateral. Equilateral triangles have been constructed on its sides so that one pair of triangles on opposite sides of the quadrilateral is constructed exterior to the quadrilateral ($\triangle GDC$ and $\triangle BEF$ in the figure), while the other pair of equilateral triangles is constructed so that their interiors intersect the interior of *ABCD*.

a. Use any method to check that the vertices *E*, *F*, *G*, and *H* of the equilateral triangles form a parallelogram.

b. Prove that $EFGH$ is a parallelogram.

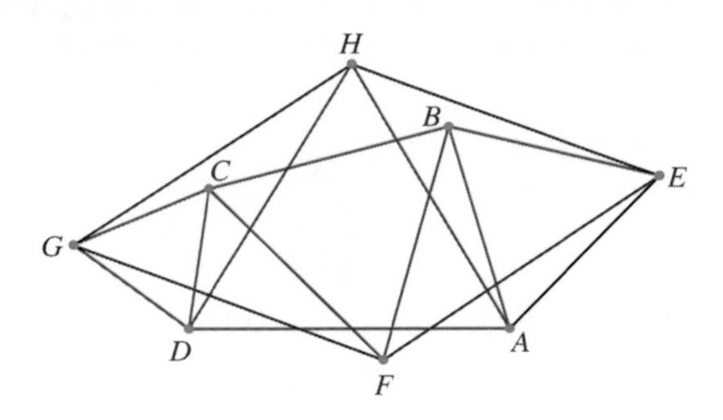

9. On two sides of an arbitrary triangle ABC, two equilateral triangles have been constructed outwardly as shown in the accompanying figure. M, P, and Q are the midpoints of sides AB, DC, and CE, respectively.

a. Use any method to check experimentally that PQM is equilateral.

b. Prove the assertion in part (a).

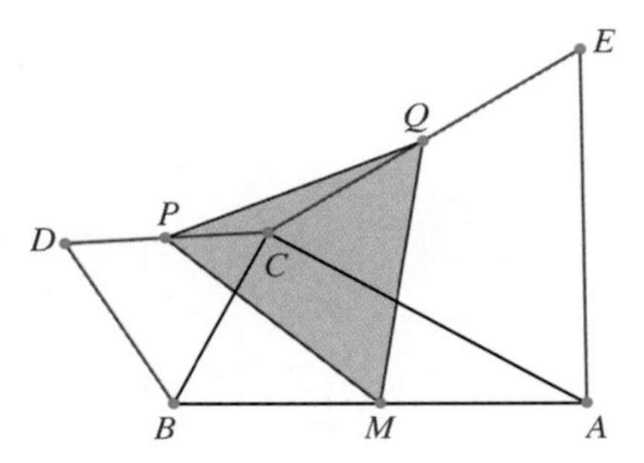

10. On the sides AC and BC of an arbitrary triangle ABC, squares $BCDE$ and $ACFG$ have been constructed outwardly as shown. P and Q are centers of the squares, and N and M are the midpoints of $\overline{EG}$ and $\overline{DF}$, respectively.

a. Experiment with different triangles ABC to conjecture the most that can be said about triangles ABN and PQM.

b. Prove your conjecture in part (a).

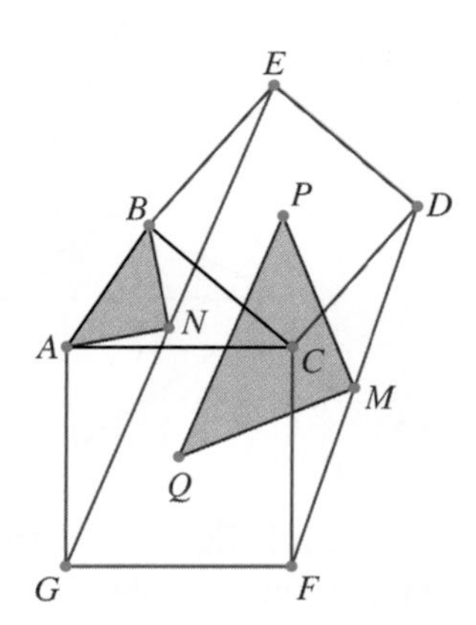

The following problems involve the concept of a **transformation group**. A set S of transformations from the plane onto the plane is a *group* under composition of transformations if the following hold:

a. **Closure:** If T_1 and T_2 are two transformations in S, then $T_2 \circ T_1$ is in S.

b. **Associative property:** For every three transformations T_1, T_2, and T_3,

$$(T_3 \circ T_2) \circ T_1 = T_3 \circ (T_2 \circ T_1)$$

c. **The identity transformation** I [$I(P) = P$ for all P in the plane], satisfying $I \circ T = T \circ I = T$ for all T in S, is in S.

d. If T is in S, so is T^{-1}.

If S_1 is a subset of S and is itself a group, we say that S_1 is a **subgroup** of S.

Notice that the set E of all the isometries of the plane is a subgroup of the group of all transformations of the plane.

11. Let T_1, T_2, and T_3 be any three transformations of the plane. Prove the *associative property* (stated above) by showing that for all points P in the plane

$$((T_3 \circ T_2) \circ T_1)(P) = (T_3 \circ (T_2 \circ T_1))(P)$$

• **12.** Which of the following are groups and which are not? Justify your answers.

a. The set of all translations

b. The set of all direct isometries (translations and rotations)

c. The set of all indirect isometries (reflections and glide reflections)

d. The set of all half-turns and all translations

13. Prove or disprove:

a. The group N of all translations is a **normal subgroup** of E; that is, if τ is any translation, then for all f in E (that is, for all isometries f), $f^{-1} \circ \tau \circ f$ is a translation.

b. The group N is a normal subgroup of all direct isometries; that is, if τ is any translation, then for all f that is a direct isometry, $f^{-1} \circ \tau \circ f$ is a translation.

14. Recall that a *symmetry* of a figure is an isometry that maps the figure onto itself, and answer the following.

• **a.** Explain why the set of all symmetries of any figure is a group.

b. The symmetry group of a rectangle that is not a square is the set $\{I, H, V, R\}$ where H is the reflection in a line through the midpoints of two opposite sides, V is the reflection in the line through the midpoints of the other pair of opposite sides, and R is a rotation about the center of the rectangle (the intersection of the diagonals) by 180°. Prove that $H^2 = V^2 = R^2 = I$, $H \circ V = V \circ H = R$, which can be displayed in the following *Cayley table*:

$\circ$	I	R	H	V
I	I	R	H	V
R	R	I	V	H
H	H	V	I	R
V	V	H	R	I

c. List the symmetries of the group of symmetries of an equilateral triangle and construct a Cayley table for the group.

15. a. Describe the symmetry group of a circle.

b. Use the concept of a group to define what it means to say that one figure is more symmetrical than another.

c. Why is a circle more symmetrical than any regular n-gon?

16. a. Prove that the set of all dilatations (see Section 5.4) with a common center of similatude constitutes a group.

b. Show that the set of all possible dilatations in a plane (with all points of the plane as similatude centers) does not constitute a group.

More Recent Discoveries 7

Everybody knows that mathematics is about miracles, only mathematicians have a name for them: theorems.

—Roger Howe, invited MAA address, Baltimore, Maryland, January 9, 1998

Introduction

In this chapter we prove a few of the more recent—from the nineteenth and twentieth centuries—geometrical results, some of which were previously introduced in Chapter 0. In Section 7.2, we discuss complex numbers and their use in proving geometrical theorems, including the Treasure Island Problem. We show how the most important trigonometric formulas can be derived from a property of a product of complex numbers.

7.1 The Nine-Point Circle and Other Results

The Nine-Point Circle was introduced (without proof) in Chapter 0 to arouse readers' curiosity. We will now prove the theorem. For convenience, we will review both the historical background of this problem and the theorem.

The nineteenth century experienced a renewed interest in classical Euclidean geometry. Probably the most spectacular discovery was the **Nine-Point Circle**, which was investigated simultaneously by the French mathematicians Charles Jules Brianchion (1785–1864) and Jean-Victor Poncelet (1788–1867), who published their findings jointly in 1821. The theorem is, however, commonly attributed to the German mathematician and high school teacher Karl Wilhem Feurbach (1800–1834), who independently discovered the theorem and published it with some related results in 1822.

Theorem 7.1

The Nine-Point Circle

With any triangle ABC, nine particular points can be associated with it, as shown in Figure 7.1. *The first three points—M_1, M_2, and M_3—are the midpoints of the three sides of the triangle. The next three points—N_1, N_2, and N_3—are the midpoints of the segments joining the vertices A, B, and C with the point H, the point of intersection of the three altitudes of the triangle. The final three points—F_1, F_2, and F_3—are the points of intersection of each altitude with each corresponding side (these points are known as the "feet" of the altitudes). All nine points lie on one circle called the Nine-Point Circle.*

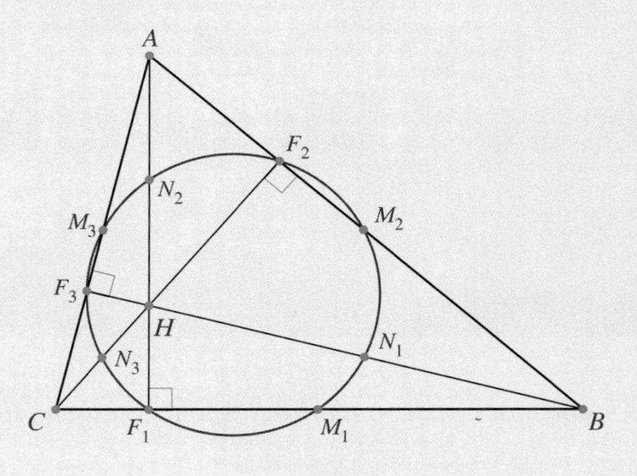

Figure 7.1

Proof of the Nine-Point Circle Theorem

In the proof, we will repeatedly use the Midsegment Theorem from Chapter 1. It states that the segment joining the midpoints of two sides of a triangle is parallel to the third side and half as long as the third side.

We first show that the midpoints M_1, M_2, and M_3 of the sides of $\triangle ABC$ and F_1, the foot of the altitude from A to the side BC, lie on one circle. For that purpose, consider Figure 7.2. We have $\overline{M_2M_3} \parallel \overline{BC}$, $\overline{M_1M_2} \parallel \overline{AC}$, and $M_1M_2 = \frac{1}{2}AC$. Because segment F_1M_3 is a median to the hypotenuse in the right triangle AF_1C, $F_1M_3 = \frac{1}{2}AC$. Thus $M_1M_2 = F_1M_3$ and, therefore, $M_1M_2M_3F_1$ is an isosceles trapezoid and hence cyclic. Consequently, M_1, M_2, M_3, and F_1 lie on a single circle. In the same way, we can show that M_1, M_2, M_3, and F_2 as well as M_1, M_2, M_3, and F_3 lie on the same circle. Thus we have six points on the same circle.

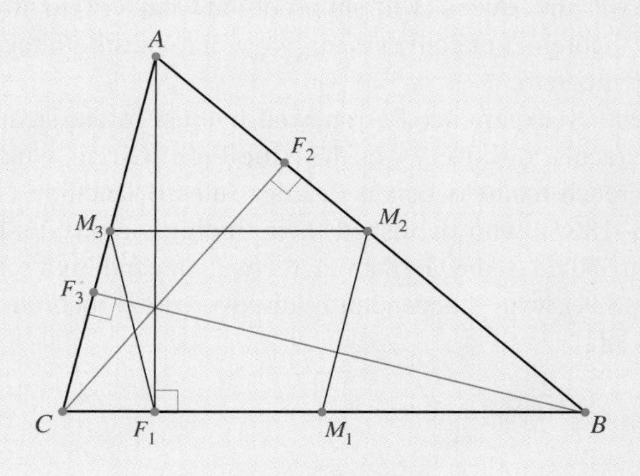

Figure 7.2

Next we focus on the seventh point N_2, the midpoint of $\overline{AH}$ in Figure 7.3. Notice that $\overline{M_3N_2} \parallel \overline{CF_2}$ and $\overline{M_1M_3} \parallel \overline{AB}$. Because $\overline{CF_2}$ is perpendicular to $\overline{AB}$, $\overline{M_3N_2}$ is perpendicular to $\overline{M_1M_3}$. Thus F_1 and M_3 are on the circle whose diameter is $\overline{M_1N_2}$. Because this circle passes through M_1, M_3, and F_1, it is the same circle that went through the six points M_1, M_2, M_3, F_1, F_2, and F_3. In an analogous way, we can show that the points N_1 and N_3 lie on the same circle. □

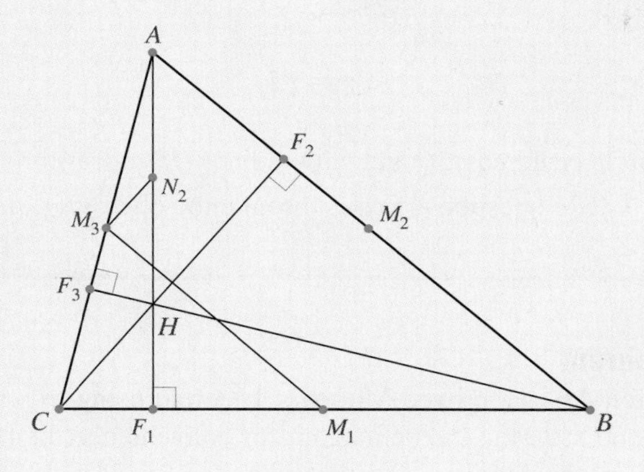

Figure 7.3

Now Solve This 7.1

The following questions will lead you to a different proof of the Nine-Point Circle Theorem. Notice that the result in question 2 has its own merit as an interesting property.

1. Let O be the center of the circle that circumscribes $\triangle ABC$. Use Figure 7.4, where H is the point of intersection of the altitudes. Prove that $AH = 2HF_1$ by showing that $ALCH$ is a parallelogram, $LC = AH$, and $LC = 2(OM_1)$.
2. Referring to Figure 7.4, prove that $HF_1 = F_1P$.

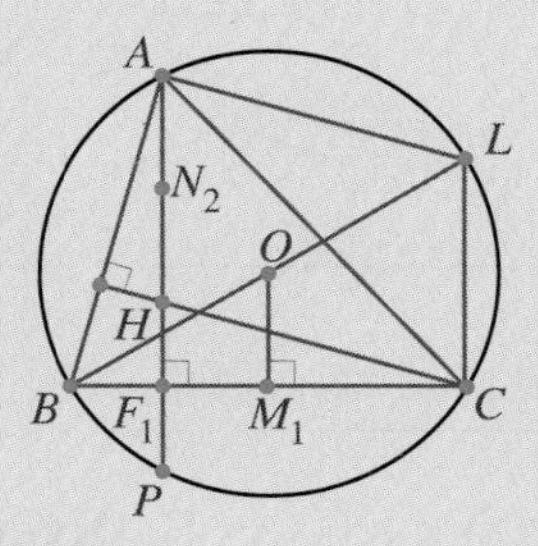

Figure 7.4

3. In Figure 7.5 (as in Figures 7.2 and 7.3), N_2 is the midpoint of $\overline{AH}$. Prove that $\overline{M_1N_2}$ and $\overline{OH}$ bisect each other, that $N_2Q = QM_1 = F_1Q$, and that $QF_1 = \frac{1}{2}OP$ ($\overline{QF_1}$ is a midsegment in $\triangle OPH$).

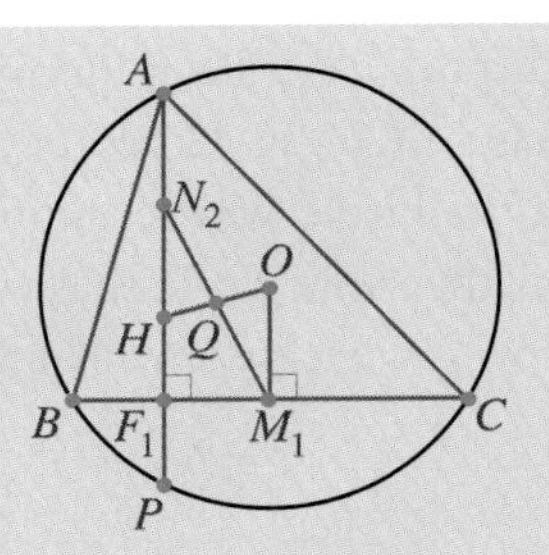

Figure 7.5

4. Prove that the circle centered at Q and with radius $\frac{1}{2}OP$ passes through F_1, M_1, and N_2.
5. Argue that the circle in question 4 passes through the other six points of the Nine-Point Circle.

Morley's Theorem

In (optional) Section 4.6, we proved Morley's Theorem using trigonometry. Here we will present another proof and state the theorem again for convenience. In 1899, Frank Morley, an English-born American mathematician, discovered an amazing property of triangles:

Theorem 7.2

The points of intersection of the adjacent angle trisectors of the angles of any triangle ($\triangle ABC$ in Figure 7.6*) are the vertices of an equilateral triangle ($\triangle PQS$).*

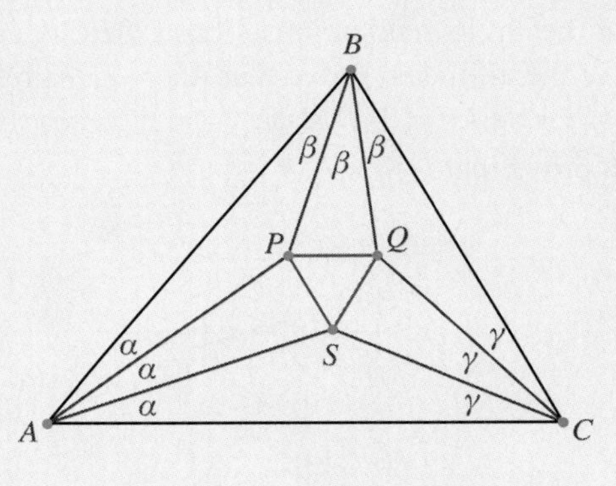

Figure 7.6

The following argument is based on a proof by H. D. Grossman (1943) and the article "Morley's Triangle" (M. E. Barnes, 2002). Our motivation for the proof, however, does not appear in either of these sources.

We begin our investigation by assuming that $\triangle PQS$ in Figure 7.7 is equilateral; we then search for properties that follow from this assumption. We extend the trisectors to intersect at H, K, and L and prove that $\triangle HPS$, $\triangle KSQ$, and $\triangle LQP$ are isosceles. To prove that $HP = HS$, notice that Q is the point in the interior of $\triangle HBC$ where the angle bisectors of $\angle HBC$ and $\angle BCH$ intersect. Because the three angle bisectors are concurrent, $\overrightarrow{HQ}$ must be the angle bisector of $\angle CHB$ and, therefore, Q is equidistant from the sides BH and HC; that is, $QT = QV$. Consequently, $HV = HT$ (notice that $\triangle HTQ \cong \triangle HVQ$). Because $HP = HT - PT$ and $HS = HV - SV$, we need to show only that $PT = SV$. This fact follows from the congruence of the shaded triangles PTQ and SVQ (H-L congruence condition). Similarly, we can show that $\triangle KSQ$ and $\triangle LQP$ are isosceles.

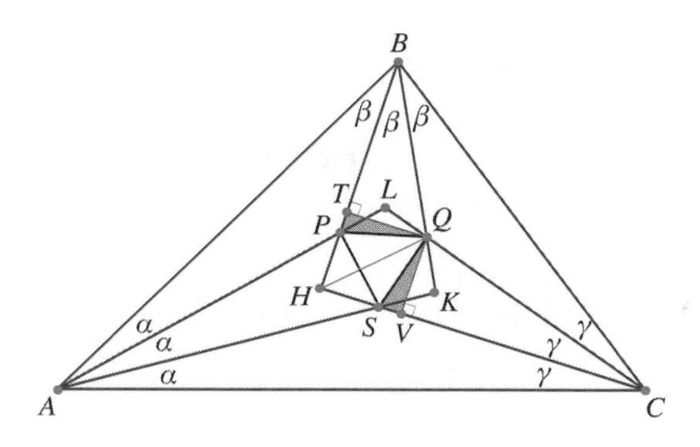

Figure 7.7

We will now find some of the angles in Figure 7.7 in terms of α, β, or γ. We find $\angle ASP$ as follows. First notice that

$$\angle ASP = \angle ASH + \angle HSP \tag{7.1}$$

Because $\angle ASH$[1] is an exterior angle in $\triangle ASC$,

$$\angle ASH = \alpha + \gamma \tag{7.2}$$

Because $\triangle HPS$ is isosceles, $\angle HSP = \dfrac{180° - \angle SHP}{2} = 90° - \frac{1}{2}\angle SHP$. But from $\triangle HBC$, we know that $\angle SHP = 180° - (2\beta + 2\gamma)$. Given that $3\alpha + 3\beta + 3\gamma = 180°$, $\alpha + \beta + \gamma = 60°$ and, therefore, $2\beta + 2\gamma = 120° - 2\alpha$. Consequently,

$$\angle SHP = 180° - (120° - 2\alpha) = 60° + 2\alpha \tag{7.3}$$

Thus $\angle HSP = 90° - \frac{1}{2}\angle SHP = 90° - \frac{1}{2}(60° + 2\alpha) = 60° - \alpha$. Given that $\alpha + \beta + \gamma = 60°$, $60° - \alpha = \beta + \gamma$ and

$$\angle HSP = \beta + \gamma \tag{7.4}$$

Substituting $60° - \alpha$ for $\angle HSP$ in Equation 7.1 and $\alpha + \gamma$ (from Equation 7.2) for $\angle ASH$, we get

$$\angle ASP = \alpha + \gamma + 60° - \alpha = 60° + \gamma \tag{7.5}$$

Similarly, we get that

$$\angle CSQ = 60° + \alpha \tag{7.6}$$

With the preceding results in mind, we now provide the actual proof. Let $\triangle ABC$ be an arbitrary triangle with angles 3α, 3β, and 3γ. We have $3\alpha + 3\beta + 3\gamma = 180°$ and therefore

$$\alpha + \beta + \gamma = 60° \tag{7.7}$$

Initially, we omit vertex B and construct only the trisectors of the other two angles as in Figure 7.8. (Thus P and Q will not be constructed using the trisectors of $\angle B$.)

---

1. In this chapter we follow the convention of not distinguishing between an angle and its measure.

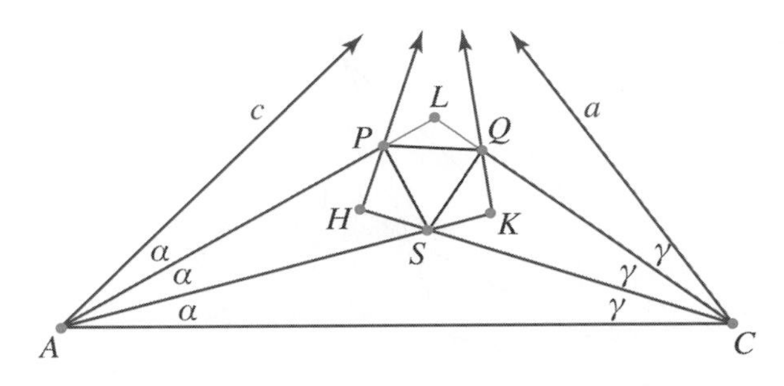

Figure 7.8

Next, we extend the trisector CS and construct the line through S that makes an angle equal to $\beta + \gamma$ with $\overrightarrow{CS}$. That line intersects the trisector of $\angle A$ closest to side AB at P. At P, we construct

$$\angle SPH = \beta + \gamma \quad \text{(See Equation 7.4 for the motivation)} \tag{7.8}$$

Thus we constructed the isosceles triangle SHP. Similarly, we construct $\triangle KSQ$ so that

$$\angle KSQ = \angle SQK = \alpha + \beta \tag{7.9}$$

Our next goal is to show that $\triangle PQS$ is equilateral. First we find that

$$\begin{aligned}
\angle PSQ &= 360° - (\angle PSH + \angle HSA + \angle KSC + \angle QSK + \angle ASC) \\
&= 360° - (\beta + \gamma + \alpha + \gamma + \alpha + \gamma + \alpha + \beta + 180° - (\alpha + \gamma)) \\
&= 180° - 2(\alpha + \beta + \gamma) \\
&= 180° - 120° = 60°
\end{aligned}$$

We proceed with the rest of the proof in Now Solve This 7.2.

Now Solve This 7.2

1. Show that $\angle APS = \angle CQS$ by showing that each equals $60° + \beta$.
2. Notice that S is the intersection of the angle bisectors of $\triangle ALC$. Show that $PS = SQ$ and hence that $\triangle PQS$ is equilateral.
3. Prove that $\angle APH = \alpha + \beta$. Show that the angle formed by line c (see Figure 7.8) and line KQ is 2β, and therefore that line HP bisects this angle. Similarly, show that line KQ bisects the angle formed by line a and line HP.
4. To complete the proof, show that lines c, a, HP, and KQ are concurrent.

Problem Set 7.1

In Problems 1, 2, and 3, prove the assertions.

- **1.** The center of the Nine-Point Circle is at the midpoint of the segment whose endpoints are the orthocenter (the intersection of the altitudes) and the circumcenter (the center of the circumscribing circle) of the triangle.
- **2.** The radius of the Nine-Point Circle is one-half of the radius of the circumcircle.

3. The centroid C of a triangle (the intersection of the medians) trisects the segment connecting the orthocenter and the circumcenter. (This fact was proved in 1765 by Leonhard Euler and the segment, called *Euler line*, is discussed in Section 5.4.)

★ **4.** Prove **Feurbach's Theorem** (proved by Karl Wilhelm Feurbach in 1822): The Nine-Point Circle is tangent to the incircle (inscribed circle) and the three excircles (an excircle is a circle tangent to the side of a triangle and an extension of another side).

You may want to "verify" the theorem experimentally using GSP (Geometer's Sketchpad) and search the Internet for a proof. Many proofs involve *inversion* in a circle—a transformation not discussed in this text.

5. Read and supply the details of the proof of Morley's Theorem given in 1909 by M. T. Naraniengar and found in Honsberger (1772, pp. 92–95). A good exposition is also given on the web at http://www.cut-the-knot.org/triangle/Morley/Naraniengar.shtml.

The following geometry facts have been discovered recently. The proofs are challenging.

6. (M. J. Zerger, *ΠME Journal*, II(7), Fall 2002, p. 392, problem 1051.) The two squares in the figures below are congruent. In the figure on the left, the octagon is formed by joining the midpoints of the sides of the square to vertices as shown. In the figure on the right, the trisection points of the sides are used instead.

a. Show that the octagons are similar and equilateral, but not equiangular.

b. Find the ratios of the areas.

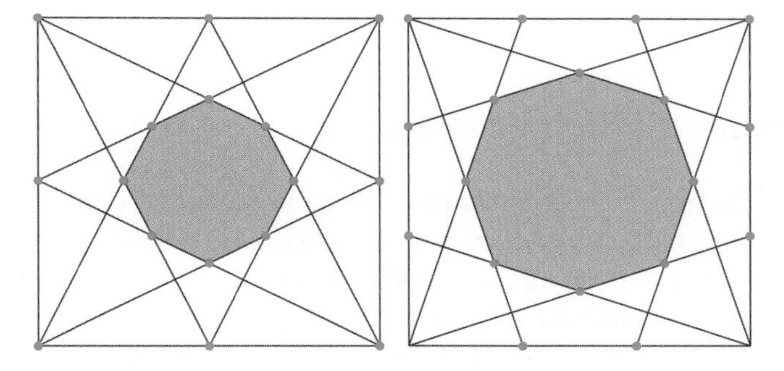

7. (Leon Bankoff, *ΠME Journal*, II(6), Spring 2002, p. 328, problem 1041.) The figure below shows a quarter circle with smaller circles inside.

a. Prove that the three larger circles have radii of equal length.

b. Prove that the remaining six smaller shaded circles have radii of equal length.

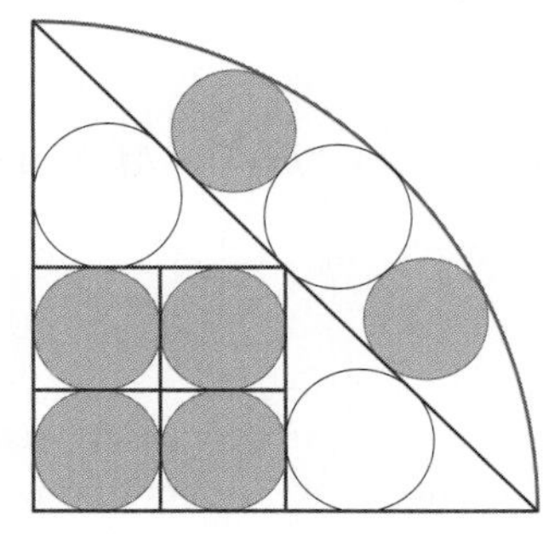

8. Prove **Marion Walter's Theorem** (1992): If the trisection points of the sides of any triangle are connected to the opposite vertices, the resulting hexagon has area one-tenth the area of the original triangle.

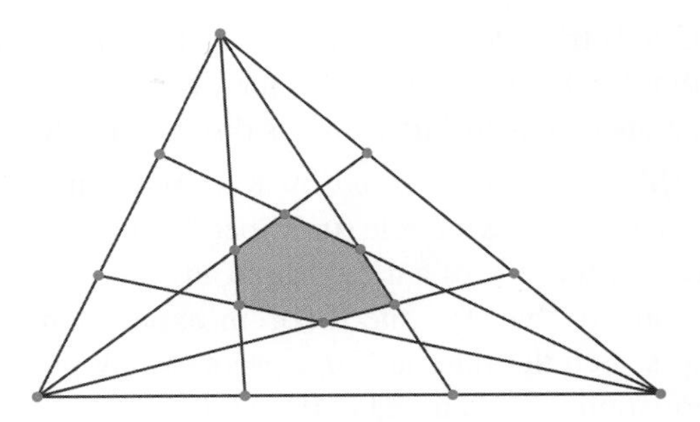

9. **A minimal inscribed quadrilateral** (G. Y. Sosnow, *AMM*, 1926, p. 161; and R. Honsberger, 1978, p. 138). *ABCD* is a cyclic quadrilateral whose diagonals meet at *X*. The points *P*, *Q*, *R*, and *S* are the feet of the perpendiculars from *X* to the sides of *ABCD*. Prove that, of all quadrilaterals having a point on each side of *ABCD*, *PQRS* has minimum perimeter.

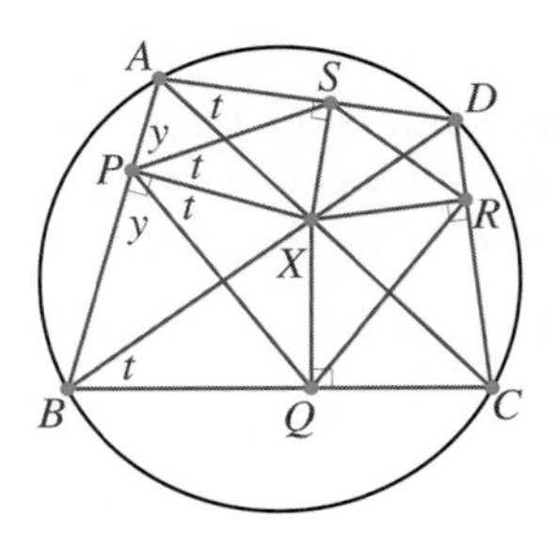

7.2 Complex Numbers and Geometry

If we extend the integers to rational numbers, then the equation $ax + b = 0$, where a and b are integers and $a \neq 0$, always has a solution. Similarly, we can extend the real numbers to include the solution of $x^2 + 1 = 0$. For that purpose, the symbol i is introduced and a **complex number** is defined as $a + bi$, where a and b are real numbers and $i^2 = -1$. Equality, addition, and multiplication are the same as for polynomials, where i^2 is replaced by -1. Complex numbers have numerous applications in mathematics, electrical engineering, and physics. In this section we focus on some of the applications of complex numbers to geometry.

Real numbers correspond to points on a line. In the x-y coordinate system, the real number a corresponds to the point $(a, 0)$ on the x-axis. If we view a as a complex number $a + 0 \cdot i$, we have the following correspondence:

$$a + 0 \cdot i \leftrightarrow (a, 0)$$

Thus it seems natural that the complex number $a + bi$ should correspond to the point (a, b) in the plane:

$$a + bi \leftrightarrow (a, b) \tag{7.10}$$

Consequently, we have a $1 - 1$ correspondence between the complex numbers and the points in the plane. For convenience, it is customary to omit the word "corresponds" and simply refer to the point that corresponds to a complex number as the complex number, and vice versa. Since $i = 0 + 1 \cdot i$, the point that corresponds to i is $(0, 1)$; however, we often say that *i is the point* $(0, 1)$.

We have seen that $(a, 0)$ on the x-axis correspond to $a + 0 \cdot i$—that is, to a real number. Similarly, the point $(0, b)$ on the y-axis corresponds to the complex number $0 + ib = ib$. For historical reasons (see the Historical Note later in this section), the complex number i is called the **imaginary number** and the y-axis is called the **imaginary axis**. The plane whose points correspond to complex numbers is referred to as the **complex plane**.

Complex numbers are commonly named by z, w, or the Greek letter ζ. If $z = a + ib$, where a and b are real, we say that z is in **standard form**. We define a to be the **real part** of z and b to be the **imaginary part** of z and write:

$$\begin{cases} a = \text{Re}(z) \\ b = \text{Im}(z) \end{cases} \tag{7.11}$$

We define the **conjugate** $\bar{z}$ of $z = a + ib$ as

$$\bar{z} = a - ib \tag{7.12}$$

If z corresponds to point $P(a, b)$, then $\bar{z}$ corresponds to the point $P'(a, -b)$, where P' is the reflection of P in the x-axis, as shown in Figure 7.9a. We say that $\bar{z}$ is the reflection of z in the x-axis and draw a diagram like the one in Figure 7.9b. Of course, if z is above the x-axis, then $\bar{z}$ is below the x-axis.

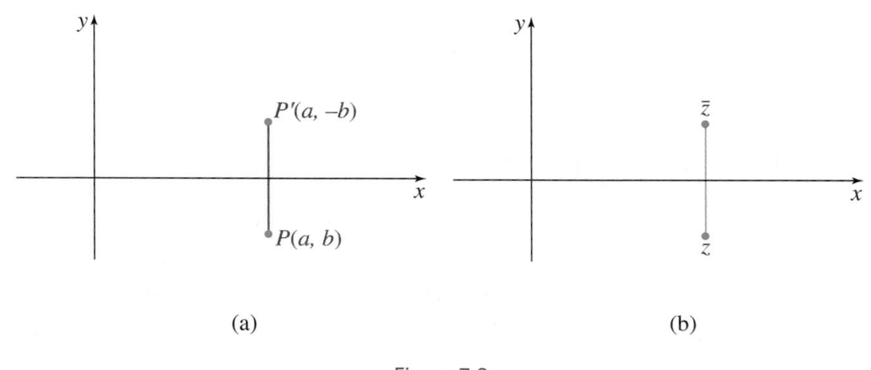

Figure 7.9

The **absolute value** $|z|$ of z is defined as the distance from the point that corresponds to z to the origin. That is, if $z = a + ib$, then

$$|z| = \sqrt{a^2 + b^2} \tag{7.13}$$

Notice that

$$|z|^2 = z \cdot \bar{z} \tag{7.14}$$

Now Solve This 7.3

1. Why in the complex plane is the x-axis often called the real axis and the y-axis called the imaginary axis? Is the complex plane different from the real plane?
2. Is the definition of the absolute value of a complex number valid if the complex number is real? Are Equations 7.13 and 7.14 valid if $b = 0$?

Historical Note: Complex Numbers

Like negative numbers, complex numbers were not easily accepted. The Italian mathematician Girolamo Cardano (1501–1576) considered several forms of quadratic equations to avoid using negative numbers. Cardano, who gained fame for finding a formula for solving any cubic equation, encountered square roots of negative numbers when applying his formula to a cubic equation that he knew had three real roots. His contemporary Rafael Bombelli (1526–1572), in his treatise *Algebra*, used complex numbers however reluctantly.

Reneé Descartes called a negative solution of an equation "false" and a solution of $x^2 = -1$ "imaginary." Leonhard Euler used complex numbers extensively, and introduced the symbol i for a solution of $x^2 = -1$. Nevertheless, he called complex numbers "impossible numbers."

The full geometric and vector representation of complex numbers was first given by the Norwegian mathematician and surveyor Caspar Wessel (1745–1818). Gauss obtained results about integers in *Number Theory* using complex numbers, "which gave a tremendous boost to the acceptance of complex numbers in the mathematical community" (Klein, p. 593).

William Rowan Hamilton (1805–1865), the famous Irish mathematician, used ordered pairs of real numbers and hence put complex numbers on a firm "real" foundation. Many mathematicians believe that complex numbers come "alive" only after studying abstract algebra.

It is also useful to see complex numbers as vectors. In Section 5.1, we viewed vectors as directed segments. In that section we introduced a translation τ_{AB} from A to B that was determined both by the directed segment AB and by infinitely many directed segments having the same length and direction.

In the plane, we define a vector as an ordered pair (a, b), where a and b are real numbers. If O is the origin and $P(a, b)$, then the vector $\overrightarrow{OP}$ is called the **position vector**; its **head** P gives the position of the point P. As shown in Figure 7.10, there are infinitely many vectors equivalent to $\overrightarrow{OP}$ but $\overrightarrow{OP}$ is the unique vector showing the position of the point P.

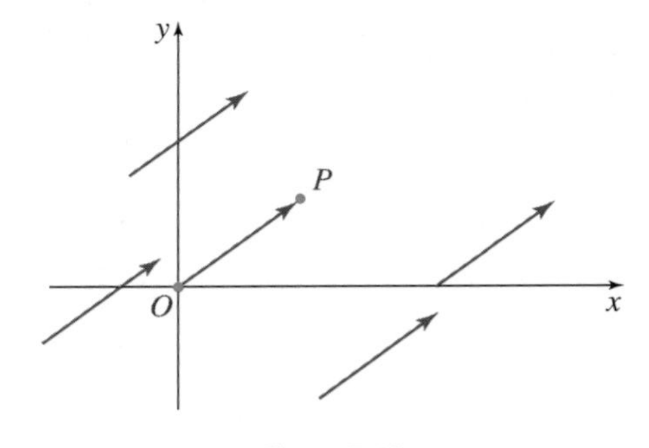

Figure 7.10

To each point $P(a, b)$, we correspond the position vector $\overrightarrow{OP}$. Because each complex number $a + ib$ corresponds to the unique point (a, b), each complex number corresponds to a unique position vector.

We define **addition of vectors** as follows:

$$(a, b) + (c, d) = (a + c, b + d) \tag{7.15}$$

This operation corresponds to the addition of complex numbers:

$$(a + ib) + (c + id) = (a + c) + i\,(b + d) \tag{7.16}$$

Notice that the results of the additions in Equations 7.15 and 7.16 correspond as well (in abstract algebra, we say that the set of vectors under addition is isomorphic to the set of complex numbers under addition).

The addition of vectors as defined in Equation 7.15 is shown visually in Figure 7.11a. The sum is the vector $\overrightarrow{OP}$, which is the diagonal of the parallelogram $OAPB$ starting at O. The vertex P of the parallelogram has coordinates $(a + c, b + d)$. Instead of drawing the parallelogram, we could draw, at the head A of $\overrightarrow{OA}$, the vector $\overrightarrow{AP}$ equal to $\overrightarrow{OB}$ and obtain $\overrightarrow{OP}$. Using the corresponding complex numbers in Figure 7.11b, the sum $z + w$ corresponds to the diagonal of the parallelogram starting at O. It can also be obtained by drawing the vector z (that is, the vector that corresponds to the complex number z) with the initial point at the origin and then drawing the equivalent vector w whose initial point is at the terminal point (head) of vector z.

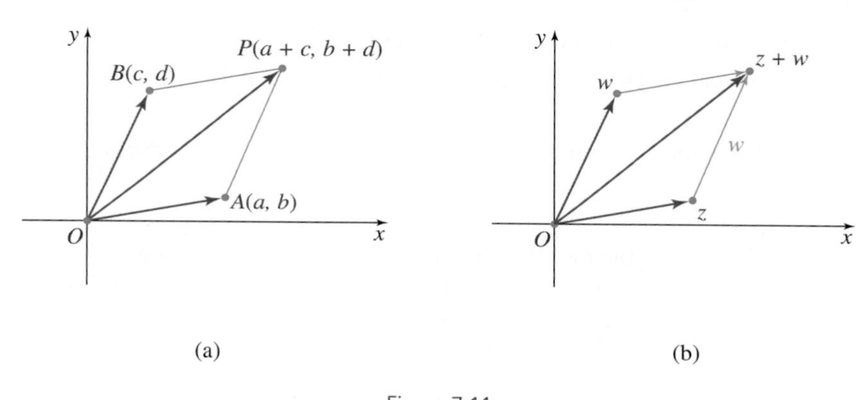

Figure 7.11

Multiplication of a vector (a, b) by a real number λ is defined as

$$\lambda(x, y) = (\lambda x, \lambda y)$$

The **additive inverse** $-(a, b)$ is defined as

$$-(a, b) = (-1)(a, b) = (-a, -b)$$

Subtraction of vectors is defined as

$$\begin{aligned}(a, b) - (c, d) &= (a, b) + -(c, d)\\ &= (a - c, b - d)\end{aligned}$$

We can visualize subtraction of complex numbers $z - w = z + (-w)$ as vectors (see Figure 7.12). (Notice the position of $-w$.) The subtraction can be performed by using addition or simply by connecting the points corresponding to z and w and obtaining the vector whose initial point is at w and whose terminal point is at z. (However, in this way, $z - w$ does not represent the corresponding position vector.) The position vector is the vector equivalent to $z - w$ whose initial point is at O. The validity of the vector for $z - w$ in Figure 7.12 can be confirmed by the addition of complex numbers (as well as the addition of vectors): $w + (z - w) = z$.

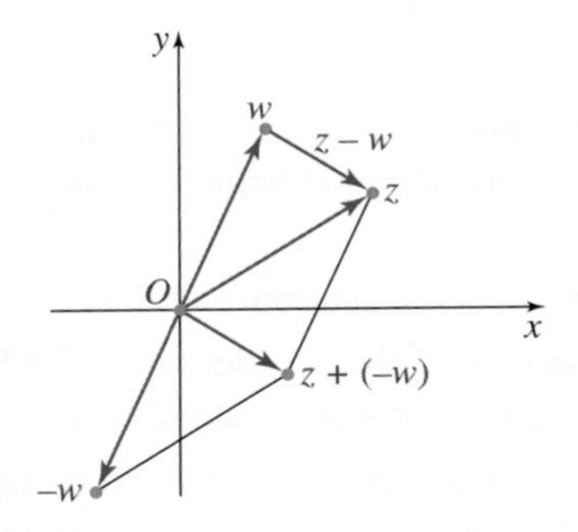

Figure 7.12

Example 7.1 If z_1 and z_2 are two points in the complex plane, find the complex number z that corresponds to

1. The midpoint of the segment connecting z_1 and z_2
2. The point between z_1 and z_2 that divides the segment connecting z_1 and z_2 into ratio $m{:}n$
3. The centroid (the intersection of the medians) of a triangle with vertices at z_1, z_2, and z_3

Solution

1. One approach is to use the fact that the diagonals of a parallelogram bisect each other. As shown in Figure 7.13, $z_1 + z_2$ corresponds to the diagonal with initial point at the origin and $\frac{z_1 + z_2}{2}$ corresponds to its midpoint M.

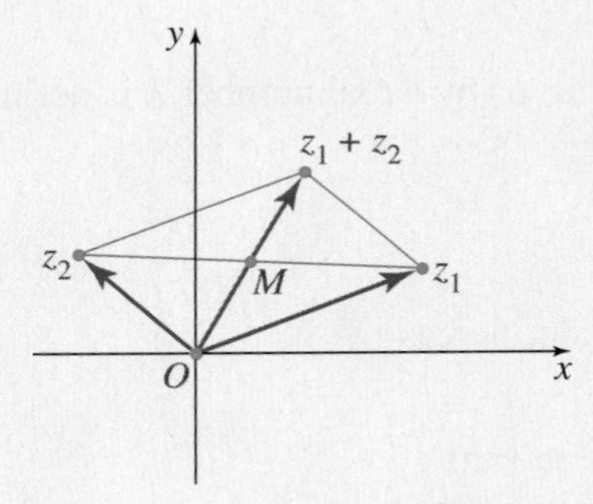

Figure 7.13

A second approach is to notice that the midpoint M of the segment connecting z_1 and z_2 corresponds to the vector $\overrightarrow{OM}$:

$$\begin{aligned}\overrightarrow{OM} &= \overrightarrow{Oz_1} + \overrightarrow{z_1M} \\ &= z_1 + \frac{1}{2}(z_2 - z_1) \\ &= \frac{z_1 + z_2}{2}\end{aligned}$$

2. In Figure 7.14, notice that $\frac{|z - z_1|}{|z_2 - z_1|} = \frac{m}{m + n}$. Consequently, $z - z_1 = \frac{m}{m + n}(z_2 - z_1)$, which implies

$$z = \frac{nz_1 + mz_2}{m + n} \tag{7.17}$$

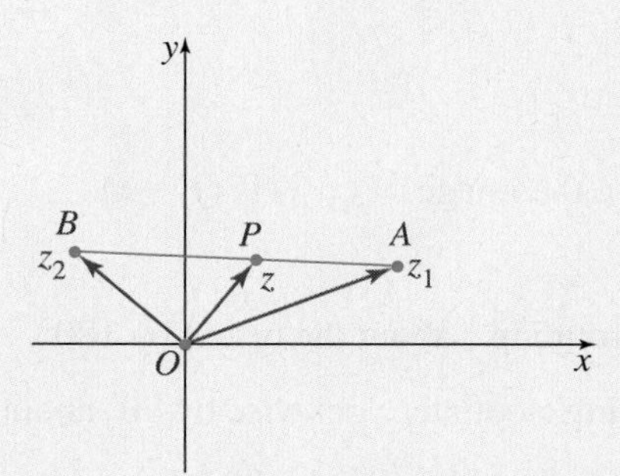

Figure 7.14

3. We can obtain the answer from Equation 7.17 or proceed independently as follows. Using the fact that the centroid G is on any of the medians, and in particular the median on z_1M (see Figure 7.15), two-thirds of the way from z_1, we have

$$\overrightarrow{z_1G} = \frac{2}{3}\left(\frac{z_2 + z_3}{2} - z_1\right)$$

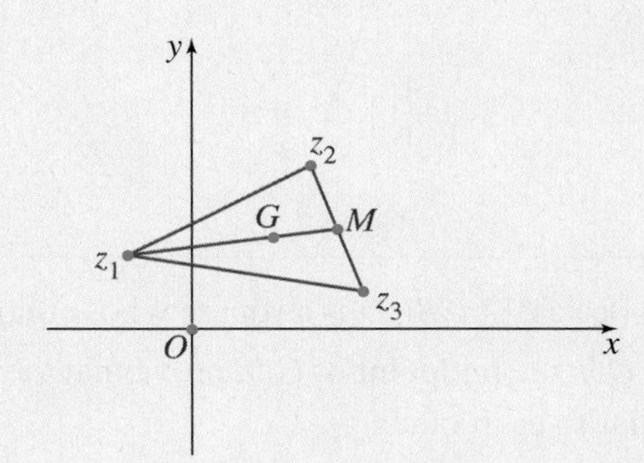

Figure 7.15

Notice that $\overrightarrow{z_1G}$ is not a position vector. We can obtain the position vector $\overrightarrow{OG}$ by adding z_1:

$$\begin{aligned}\overrightarrow{OG} &= z_1 + \frac{2}{3}\left(\frac{z_2 + z_3}{2} - z_1\right)\\ &= \frac{z_1 + z_2 + z_3}{3}\end{aligned}$$

Geometric Interpretation of Multiplication

Before addressing multiplication by an arbitrary complex number, let's investigate the special case of multiplying a complex number z by -1, i, or $-i$. Notice that

$$(-1)z = -z = -x - iy \leftrightarrow (-x, -y)$$

Hence for multiplication by -1, the image of (x, y) is $(-x, -y)$.

Also,

$$iz = i(x + iy) = ix - y = -y + ix \leftrightarrow (-y, x)$$

Hence for multiplication by i, the image of (x, y) is $(-y, x)$.

Similarly,

$$(-i)z = -(iz) = y + (-i)x$$

Hence for multiplication by $-i$, the image of (x, y) is $(y, -x)$.

Figure 7.16 shows that

$$z \rightarrow (-1)z \text{ amounts to rotating } z \text{ about the origin by } 180°, \tag{7.18}$$

$$z \rightarrow iz \text{ amounts to rotating } z \text{ counterclockwise by } 90° \text{ about the origin, and} \tag{7.19}$$

$$z \rightarrow -iz \text{ amounts to rotating } z \text{ clockwise by } 90° \text{ about the origin.} \tag{7.20}$$

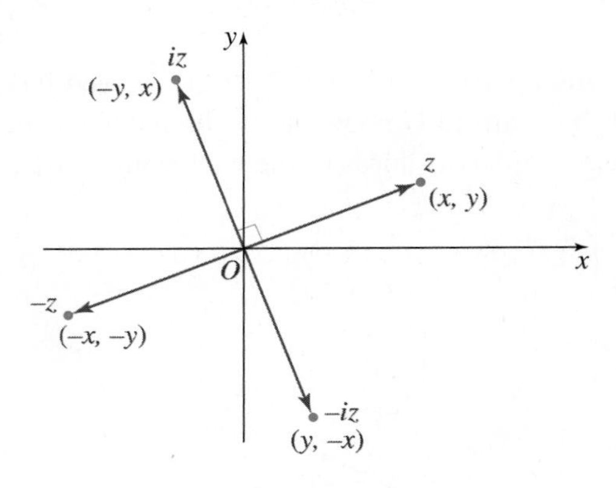

Figure 7.16

Example 7.2 In Figure 7.17, $ABCD$ is a square whose diagonals intersect at O. If P is the midpoint of $\overline{OB}$ and Q is the midpoint of $\overline{CD}$, prove that the segments AP and PQ are congruent and perpendicular to each other.

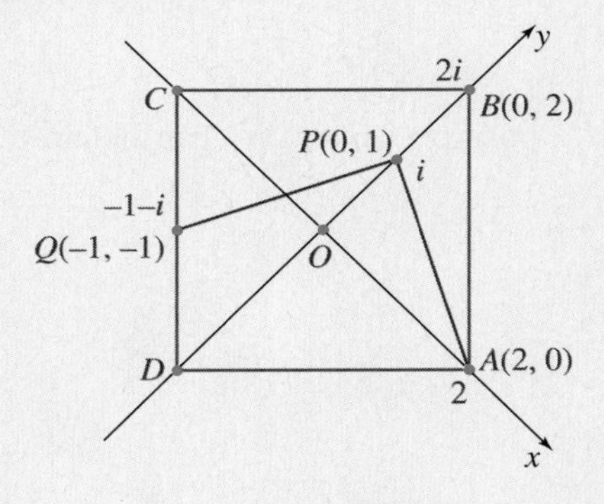

Figure 7.17

Solution

Because the diagonals of a square are congruent and are perpendicular bisectors of each other, we can set up a coordinate system such that O is at the origin, diagonal AC is the x-axis, and diagonal DB is the y-axis. Because any point on the x-axis can correspond to 1,

(the position at O is easy to follow). Alternatively, we can find the treasure by constructing the point (0, 1)—that is, by constructing the perpendicular bisector of $\overline{T_1T_2}$ and finding T (above $\overleftrightarrow{T_1T_2}$) such that $OT = \frac{1}{2}T_1T_2$.

Now Solve This 7.4

1. Prove that $|z_1z_2| = |z_1| \cdot |z_2|$.
2. On the sides of an arbitrary quadrilateral, squares have been constructed as shown in Figure 7.20. The centers of the squares are C_1, C_2, C_3, and C_4. Use complex numbers to prove that the segments C_1C_3 and C_2C_4 are congruent and perpendicular. (This is Problem 3 in Chapter 0 and Problem 6.3 in Chapter 6.)

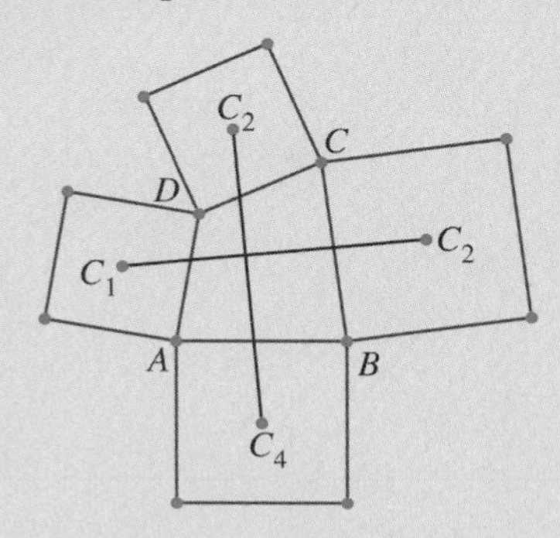

Figure 7.20

3. (For readers who know some linear algebra.) In this problem, we investigate the effect of multiplying any complex number $z = x + iy$ by a constant complex number $z_0 = a + ib$; that is, we consider the function $f: z \to z_0 z$ or, equivalently, $f(z) = z_0 z$. Notice that

$$\begin{aligned} z_0 z &= (a + ib) \cdot (x + iy) \\ &= ax - by + i(bx + ay) \end{aligned}$$

Therefore the image of (x, y) is $(ax - by, bx + ay)$. If we denote the image of (x, y) by (x_1, y_1), we have

$$\begin{cases} x_1 = ax - by \\ y_1 = bx + ay \end{cases} \quad (7.27)$$

Using matrix notation, we can write this image as follows:

$$\begin{pmatrix} x_1 \\ y_1 \end{pmatrix} = \begin{pmatrix} a & -b \\ b & a \end{pmatrix} \begin{pmatrix} x \\ y \end{pmatrix}$$

Suppose that $z_1 = z_0 z$ is transformed to $w_0 z_1$, where $w_0 = c + id$ (a constant). In other words, we have a new function $g : z \to w_0 z$ [that is, $g(z) = w_0 z$] and we want to find $g \circ f$, the composition of f and g.

(a) Show that $(g \circ f)(z) = (w_0 z_0)z$, where $w_0 z_0 = (a + ib)(c + id) = (ac - bd) + i(bc + ad)$.

(b) Write the matrix that corresponds to the transformation in part (a).

(c) Explain why the matrix you found in part (b) is equal to the product

$$\begin{pmatrix} a & -b \\ b & a \end{pmatrix}\begin{pmatrix} c & -d \\ d & c \end{pmatrix}$$

Polar (or Trigonometric) Representation of Complex Numbers

Figure 7.21 shows the polar coordinates (r, θ) of a point $P(x, y)$. Notice that $r = OP$ and that θ is the angle between $\overrightarrow{OP}$ and the positive x-axis. From the definition of trigonometric functions (see Section 4.6), we know that $\cos\theta = \dfrac{x}{r}$ and $\sin\theta = \dfrac{y}{r}$. Thus

$$\begin{aligned} x &= r\cos\theta \\ y &= r\sin\theta \end{aligned} \tag{7.29}$$

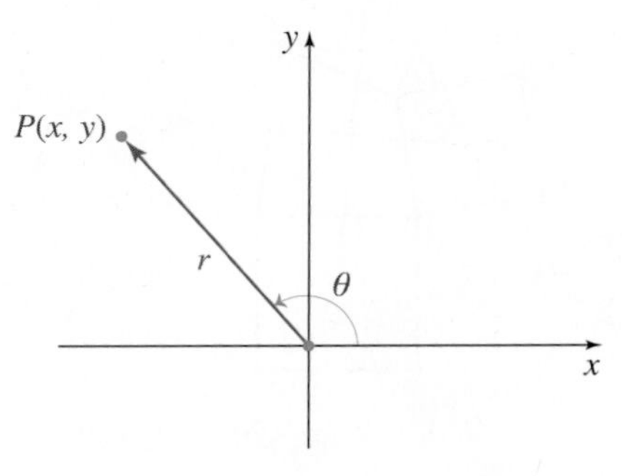

Figure 7.21

Consequently, we may write any complex number $z = x + iy$ as follows:

$$z = x + iy = r(\cos\theta + i\sin\theta) \tag{7.30}$$

Equation 7.30 is called the **polar** or **trigonometric representation** of z. Notice that $r = \sqrt{x^2 + y^2} = |z|$. It is common practice to refer to θ as the **argument** of z and to write

$$\theta = \arg z \tag{7.31}$$

The polar representation of z is very convenient when we are multiplying complex numbers, because $\arg(z_1z_2) = \arg z_1 + \arg z_2$. To see why, consider Figure 7.22, where we illustrate what happens when z is multiplied by z_0. Let $\alpha = \arg z$ and $\beta = \arg z_0$. We will show that $\arg z_0z = \alpha + \beta$.

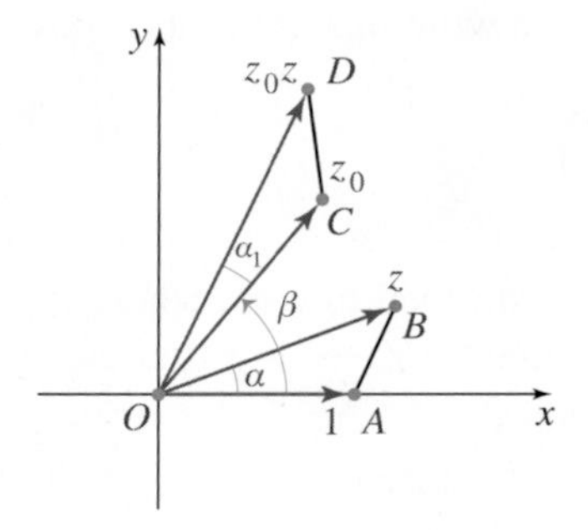

Figure 7.22

Notice that

$$\arg(z_0 z) = \angle AOD$$
$$= \alpha_1 + \beta$$

To show that $\alpha = \alpha_1$ we need to prove only that $\triangle COD \sim \triangle AOB$. The triangles will be similar if

$$\frac{CO}{AO} = \frac{OD}{OB} = \frac{DC}{BA} \tag{7.32}$$

Referring to Figure 7.22, we see that Equation 7.32 is equivalent to

$$\frac{|z_0|}{1} = \frac{|zz_0|}{|z|} = \frac{|z_0 z - z_0|}{|z - 1|} \tag{7.33}$$

Using the fact that $|z_1 z_2| = |z_1| \cdot |z_2|$ (see Now Solve This 7.3), we can immediately see that the relationship in Equation 7.33 holds true.

Since $|z_1 z_2| = |z_1| \cdot |z_2|$, we have proved the following theorem:

Theorem 7.3

If $z_1 = r_1(\cos\theta_1 + i\sin\theta_1)$ *and* $z_2 = r_2(\cos\theta_2 + i\sin\theta_2)$, *then*

$$z_1 z_2 = r_1 r_2(\cos(\theta_1 + \theta_2) + i\sin(\theta_1 + \theta_2)) \tag{7.34}$$

This result leads to our next theorem:

Theorem 7.4

De Moivre's Theorem

For all positive integers n,

$$(\cos\theta + i\sin\theta)^n = \cos n\theta + i\sin n\theta \tag{7.35}$$

Remark 1 Abraham De Moivre (1667–1754) was a French–British mathematician who established analytical trigonometry and the foundations of probability.

Remark 2 Equation 7.35 also holds true for negative integers.

Using the Taylor expansion for $\cos x$, $\sin x$, and e^x, it can be shown that

$$\cos x + i\sin x = e^{ix} \tag{7.36}$$

Substituting $x = \pi$, we have the famous Euler's formula

$$-1 = e^{i\pi} \quad \text{or} \quad e^{i\pi} + 1 = 0 \tag{7.37}$$

Now Solve This 7.5

1. Use Equation 7.34 to prove the following fundamental trigonometric formulas:
   (a) $\cos(\alpha + \beta) = \cos\alpha \cos\beta - \sin\alpha \sin\beta$
   (b) $\sin(\alpha+\beta) = \sin\alpha \cos\beta + \sin\beta \cos\alpha$
2. Using part 3 of Now Solve This 7.3 (but without using Equation 7.34 or part 1 of Now Solve This 7.5), show that the matrix that corresponds to multiplying a complex number by $\cos\theta + i\sin\theta$ is

$$\begin{pmatrix} \cos\theta & -\sin\theta \\ \sin\theta & \cos\theta \end{pmatrix}$$

   Then show that

$$\begin{pmatrix} \cos\alpha & -\sin\alpha \\ \sin\alpha & \cos\alpha \end{pmatrix}\begin{pmatrix} \cos\beta & -\sin\beta \\ \sin\beta & \cos\beta \end{pmatrix} = \begin{pmatrix} \cos(\alpha+\beta) & -\sin(\alpha+\beta) \\ \sin(\alpha+\beta) & \cos(\alpha+\beta) \end{pmatrix}$$

   Finally, use the product of the matrices to derive formulas (a) and (b) in part 1.
3. Prove De Moivre's Theorem
   (a) Using the Taylor series.
   (b) Using the following semi-legitimate approach: Let $z = \cos x + i\sin x$. Then

$$\begin{aligned} \frac{dz}{dx} &= -\sin x + i\cos x \\ &= i(\cos x + i\sin x) \\ &= iz \end{aligned}$$

   Hence

$$\frac{dz}{z} = i\,dx$$

   Complete the "proof" by integrating both parts of the above equation (don't forget a constant).
4. Why did we call the approach in question 3, part (b), "semi-legitimate"?
5. Why is the preferred measure of angles in De Moivre's Theorem radians and not degrees?
6. Prove De Moivre's Theorem for negative integers.
7. Show that $\arg\left(\frac{z_1}{z_2}\right) = \arg z_1 - \arg z_2$.

Example 7.5 (Problem 4 in Problem Set 4.6.) Without using a calculator, prove that if $AE = ED = DC = BC$ in Figure 7.23, then $\alpha + \beta + \gamma = 90°$.

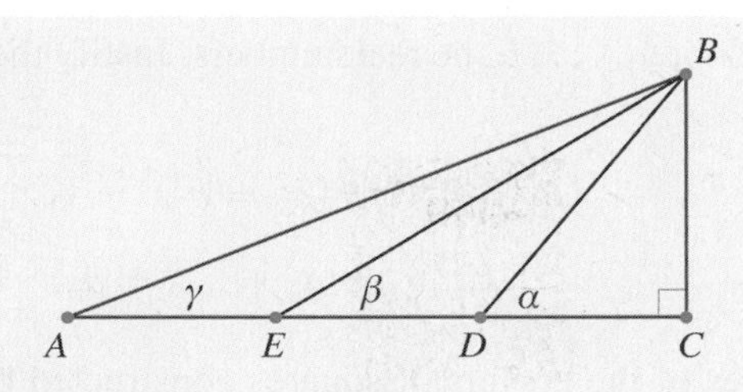

Figure 7.23

Solution

We create a coordinate system with C as the origin, $\overleftrightarrow{CB}$ as the y-axis, and $\overleftrightarrow{CA}$ as the x-axis. We choose $DC = 1$. If $B(0, 1)$, $D(-1, 0)$, $E(-2, 0)$, and $A(-3, 0)$, then the complex numbers representing these points are i, -1, -2, and -3, respectively. Notice that α, β, and γ are the arguments of the complex numbers represented by the vectors $\overrightarrow{DB}$, $\overrightarrow{EB}$, and $\overrightarrow{AB}$, respectively. These complex numbers are $i - (-1)$, $i - (-2)$, and $i - (-3)$, or $i + 1$, $i + 2$, and $i + 3$, respectively. The sum $\alpha + \beta + \gamma$ is the argument of the product $(i + 1)(i + 2)(i + 3)$. The argument of this product will be 90° (as anticipated) if the product is ki for some real number i (then ki will correspond to a point on the y-axis). Indeed,

$$\begin{aligned}(i + 1)(i + 2)(i + 3) &= (i + 1)(5 + 5i) \\ &= 10i\end{aligned}$$

Now Solve This 7.6

Use an approach similar to the one in Example 7.5 to show that if α and β are angles as shown in Figure 7.24, then $\alpha + 2\beta = 45°$.

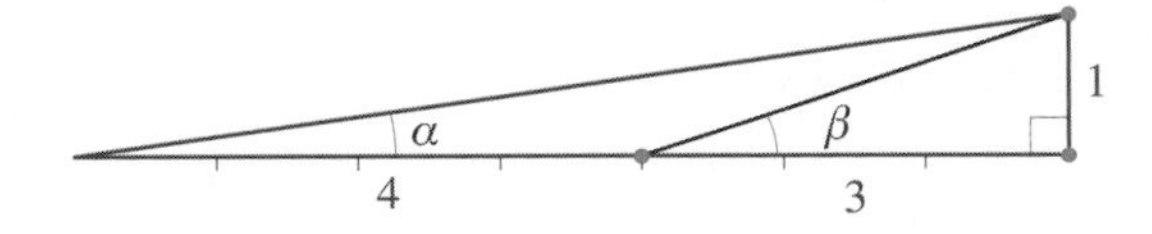

Figure 7.24

Problem Set 7.2

All proofs should be completed using complex numbers.

1. Let z_1, z_2, z_3, and z_4 be the complex numbers corresponding to the vertices of a quadrilateral. Find in terms of z_1, z_2, z_3, and z_4 each of the following:
   a. The complex number corresponding to the intersection of the two segments connecting the midpoints of the opposite sides of the quadrilateral.
   b. The complex number corresponding to the midpoint of the segment joining the midpoints of the diagonals.
   c. State the theorem based on your answers to parts (a) and (b).

•2. Let z_1, z_2, and z_3 be any three complex numbers. Show geometrically that

$$|z_1 + z_2| \leq |z_1| + |z_2|$$

•**3.** Let $a_1, a_2, \ldots, a_n$ and $b_1, b_2, \ldots, b_n$ be real numbers. Justify the following inequality geometrically:

$$(a_1^2 + b_1^2)^{\frac{1}{2}} + (a_2^2 + b_2^2)^{\frac{1}{2}} + \ldots + (a_n^2 + b_n^2)^{\frac{1}{2}}$$
$$\geq [(a_1 + a_2 + \ldots + a_n)^2 + (b_1 + b_2 + \ldots + b_n)^2]^{\frac{1}{2}}$$

4. Prove that the vertices of the centers of squares constructed on the sides of a parallelogram form a square.

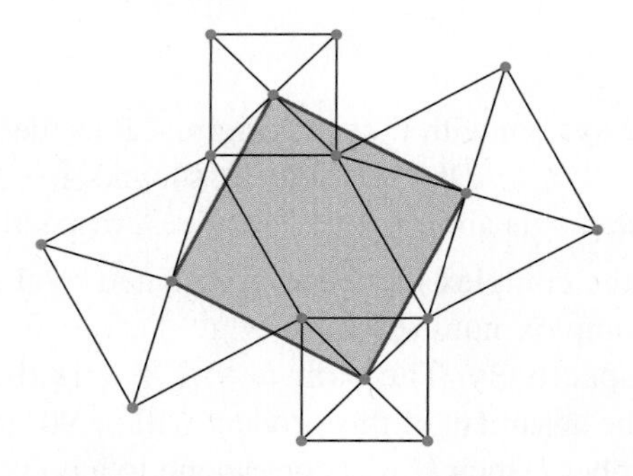

5. Prove Napoleon's Theorem: If Q, P, and S are the centroids (the points where medians intersect) of equilateral triangles constructed (as shown in the figure) on the sides of an arbitrary $\triangle ABC$, then $\triangle QPS$ is equilateral.

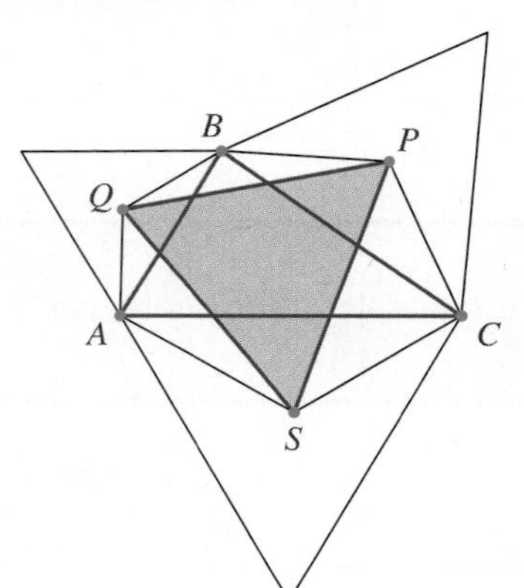

Napoleon's Theorem:
ΔPQS is equilateral.

•**6.** The accompanying figure shows a right triangle ABC and squares constructed on its sides. The quadrilaterals $JKDC$ and $BEFG$ are parallelograms.

a. Prove that $\triangle AKF$ is isosceles.

b. Is $\triangle AKF$ also a right triangle? Justify your answer.

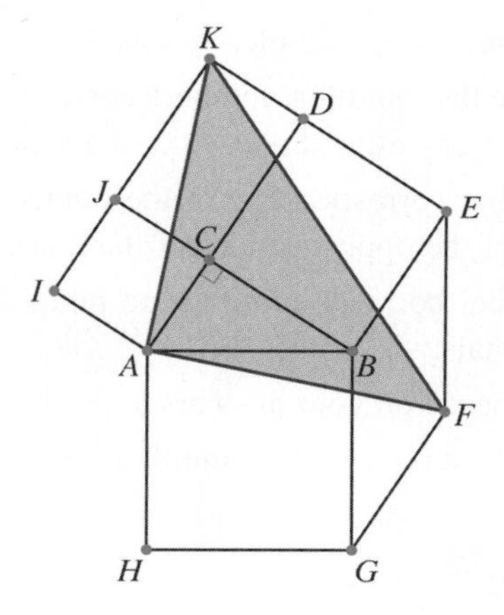

7. Suppose $ABCD$ is a square, and $A_1B_1C_1D_1$ is another square in the interior of $\triangle ABCD$ as shown in the figure. The points P, Q, R, and S are the midpoints of $\overline{AA_1}$, $\overline{BB_1}$, $\overline{CC_1}$, and $\overline{DD_1}$, respectively.
   a. Prove that $PQRS$ is a square (regardless of the location of the smaller square in the interior of the larger one).
   b. Will $PQRS$ still be a square even if $A_1B_1C_1D_1$ is not in the interior of $ABCD$? Justify your answer.

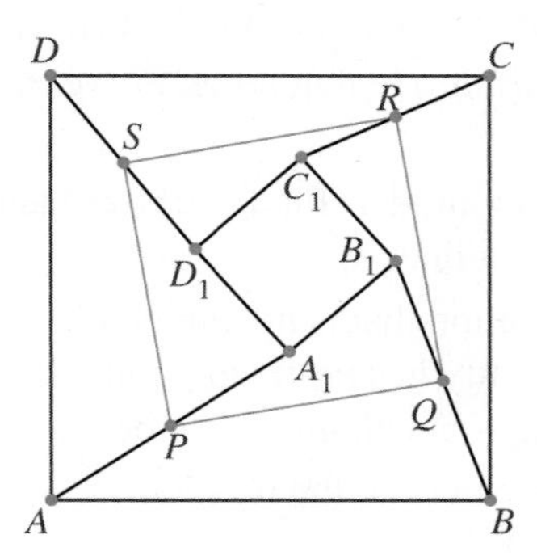

•8. Prove that a triangle inscribed in a unit circle with vertices at z_1, z_2, and z_3 is equilateral if and only if $|z_1| = |z_2| = |z_3| = 1$ and $z_1 + z_2 + z_3 = 0$.

The following problems are about isometries and similarity transformations expressed via complex numbers. All functions are from complex numbers to complex numbers.

9. Prove that $f(z) = az + b$ represents an isometry if and only if $|a| = 1$.
10. Justify the following:
   a. Every translation is given by the function $T(z) = z + z_0$, where z_0 is a complex number.
   b. Every rotation about the origin is given by $R_\theta(z) = az$, where $a = e^{i\theta}$.
   c. A rotation about z_0 is given by $R(z) = a(z - z_0) + z_0$, where $|a| = 1$. (The diagram should be helpful.)

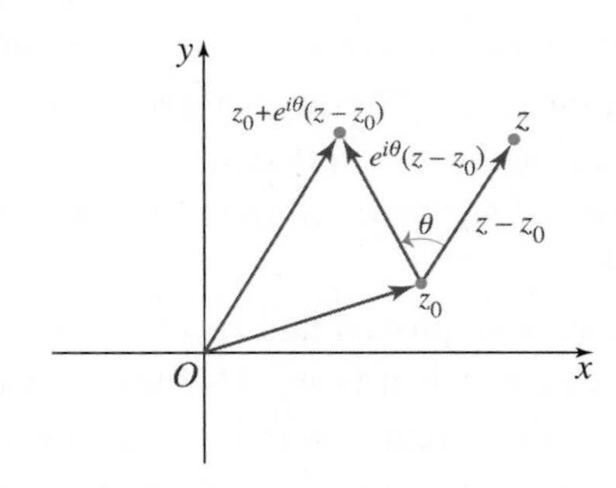

   d. Obtain the result in part (c) by showing first that the rotation about z_0 by θ equals $T_{z_0} \circ R_{0,\theta} \circ T_{-z_0}$, where T_{z_0} is the translation by the vector z_0, T_{-z_0}, is the translation by the vector $-z_0$ and $R_{0,\theta}$ is the rotation about the origin by θ.
   e. A reflection in line ℓ through the origin making an angle θ with the positive x-axis can be achieved by the composition $R_\theta \circ M_x \circ R_{-\theta}$. Thus the reflection in line ℓ is given by

$$z \to ze^{-i\theta} \to \overline{ze^{-i\theta}} \to (e^{i\theta}\,\overline{z})e^{i\theta} = e^{2i\theta}\,\overline{z}$$

   or $M_\ell(z) = e^{2i\theta}\,\overline{z}$.

f. A reflection in line k intersecting the x-axis at $(h, 0)$ and parallel to line ℓ of part (e) is given by

$$M_k = (\bar{z} - h)e^{2i\theta} + h$$

11. Use part (e) of Problem 10 to find the matrix that corresponds to the reflection in a line through the origin that makes angle θ with the positive x-axis.

•**12.** Show that the functions in Problem 10, parts (e) and (f), are reflections by proving that the composition of each with itself is an identity.

13. Let $f(z) = \bar{z} + ik$, where k is a real number. Prove that $f(z)$ is a reflection, and find the line of reflection.

•**14.** Let $g(z) = a\bar{z} + b$, where $|a| = 1$. Find necessary and sufficient conditions on a and b for $g(z)$ to be a reflection in a line.

•**15.** Using the fact that a rotation that is not an identity must have one fixed point and a translation that is not an identity has no fixed points, find necessary and sufficient conditions on a and b for $f(z) = az + b$ with $|a| = 1$ to be

a. A rotation. (Find the center of the rotation.)

b. A translation.

16. Complete the following part of a proof that an isometry with at least two fixed points is either the identity or a reflection in a line:

Let f be an isometry for which $f(A) = A$ and $f(B) = B$, with $A \neq B$. If f is not the identity, then there exists a point P such that $f(P) = P'$ and $P' \neq P$.

a. Let P be any point for which $P' \neq P$. Prove that $AP' = AP$ and $BP' = BP$. Conclude that A and B must be in the perpendicular of $\overline{PP'}$, which is the line AB.

b. Conclude that any isometry with at least two fixed points is either the identity or a reflection in a line.

17. Prove that every plane isometry is represented by either $f(z) = az + b$ or $a\bar{z} + b$, where $|a| = 1$, by following the steps below:

a. Let $g(z)$ be an isometry, and let $f(z) = az + b$. Notice that $f(0) = b$. Choose b such that $b = g(0)$. Notice that $f(1) = a + b$. Choose a such that $a = g(1) - b = g(1) - g(0)$. Show that for these choices of a and b, $g(0) = f(0)$ and $g(1) = f(1)$.

b. Consider the function $f^{-1} \circ g$. Show that it has two fixed points at 0 and 1 and, by part (a) that it is either the identity or reflection in the line through 0 and 1 (i.e., the x-axis). Thus $f^{-1} \circ g = I$ or $f^{-1} \circ g = M_x$. Show that in the first case $g(z) = f(z)$ and in the second case $g(z) = f(z)$.

18. In Problem 15 you may have proved that $f(z) = az + b$ is a rotation if and only if $|a| = 1$ but $a \neq 1$ (if not, prove this fact now). Use this fact to prove that the composition of two rotations about two different points is either a rotation or a translation. [*Hint:* Let $f_1(z) = e^{i\theta_1}z + b_1$ and $f_2(z) = e^{i\theta_2}z + b_2$.] Finally, generalize the Treasure Island Problem by considering turning at the trees by angles α and β and requiring a condition on $\alpha + \beta$.

19. Prove that every similarity transformation can be achieved by $f(z) = az + b$ or $f(z) = a\bar{z} + b$, where $a \neq 0$. (A similarity of the plane is a transformation that multiplies all distances by a constant; see Section 5.4.)

The following problems involve the concept of a *group*. The variables are complex numbers, a and b vary over all complex numbers with stated exceptions if any.

•**20.** **a.** Let $f(z) = az + b$, where $a \neq 1$. Find $f^{-1}(z)$ in terms of a and b.

b. In part (a), if $|a| = 1$, show that $f^{-1}(z) = \bar{a}z - \bar{a}b$.

21. **a.** Prove that the set of functions $f(z) = az + b$, where $a \neq 1$, forms a group with respect to composition of functions.

b. Prove that the set of functions $f(z) = az + b$, where $|a| = 1$, is a subgroup of the set in part (a).

c. Prove that the set of functions $T(z) = z + b$ is a subgroup of the set in part (b).

d. Prove that the set of functions $H(z) = -z + b$ is a subgroup of the group in part (b).

e. Prove that the set of functions $\mathcal{F}$ such that $\mathcal{F} = \{f \mid f(z) = az + b \text{ or } f(z) = a\bar{z} + b, \text{ where } a \neq 0\}$ is a group with respect to composition of functions and that each of the sets in parts (a) through (d) are subgroups of $\mathcal{F}$.

22. Consider the set of functions

$$G = \{f \mid f(z) = -z + b \text{ or } f(z) = z + b\}$$

Prove that G is a group.

23. Let $H = \{f \mid f(z) = z + b \text{ or } \bar{z} + b\}$. Is H a group? Justify your answer.

Appendix: Basic Notions

A

Introduction

In this appendix, we set out the basic definitions and axioms that form the foundation for the work we do in the rest of the book. Many of the notions presented here are intuitive, and we merely formalize them by stating them as axioms or theorems. We begin by taking the terms *point*, *line*, *plane*, and *space* as undefined (not in the sense that we don't know what they are, but in the sense that they have not been given formal definitions). We also take as undefined the concepts of *set*, *belongs to*, or *is an element of* a set. In the formal axiomatic approach that we will be taking, we use only the properties of the undefined terms that we state in the axioms—we are not allowed to make conclusions based on drawings. Our first axiom deals with lines, planes, and space as sets of points.

Axiom A.1 *Lines, planes, and space are sets of points. Space contains all points.*

In geometry, it is common to use the terms *on*, *in*, *passes through*, and *lies on*. We say that a point is *on* a line rather than *belongs* to a line. Synonymously, we say that a line *passes through* a given point. We also say that a point is *on* a plane or *in* a plane. A line *lies* in a plane if it is a subset of the plane—that is, if every point on the line is also in the plane. Points that are on the same line are called **collinear**. Points that are on the same plane are **coplanar**.

The following axioms describe the fundamental relationships among points, lines, and planes. The accompanying figures are merely models for the relationships; they do not represent the only possible configuration.

Axiom A.2 *Any two distinct points are on exactly one line. Every line contains at least two points.*

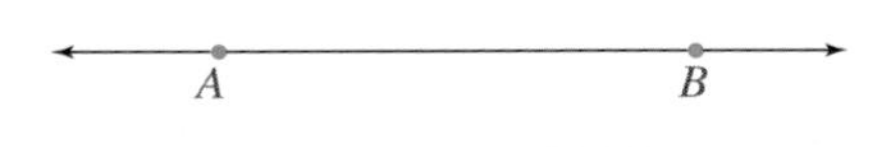

Figure A.1

Remark It may seem that Axiom A.2 implies that all lines are straight; otherwise, more than one line could be drawn through two points. First, notice that "straightness" has not yet been defined. Second, it is possible to show an example of geometry in which lines are objects that satisfy Axiom A.2 and other axioms in this section but are not "straight."

Axiom A.3 *Any three noncollinear points are on exactly one plane. Each plane contains at least three noncollinear points.*

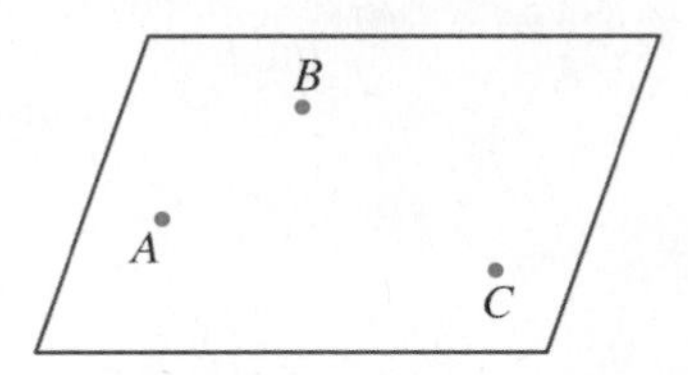

Figure A.2

Axiom A.4 *If two points of a line are in a plane, then the entire line is in the plane.*

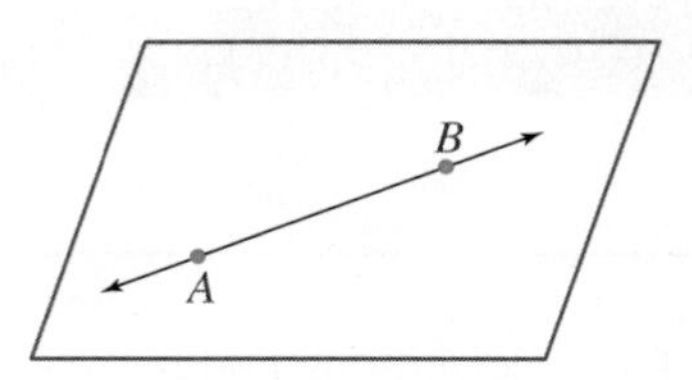

Figure A.3

Axiom A.5 *In space, if two planes have a point in common, then the planes have an entire line in common.*

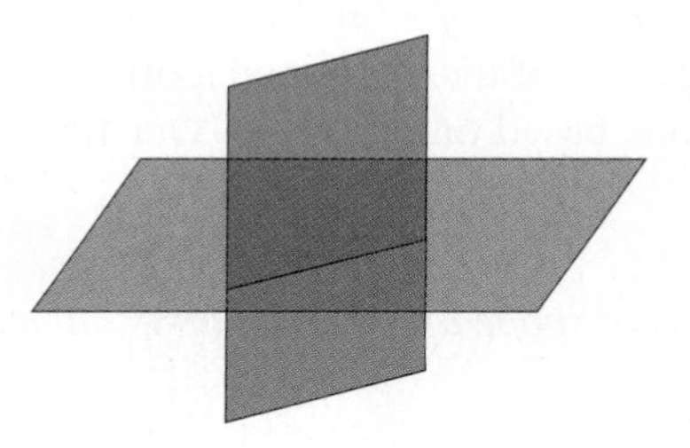

Figure A.4

Axiom A.6 *In space, there exist at least four points that are noncoplanar.*

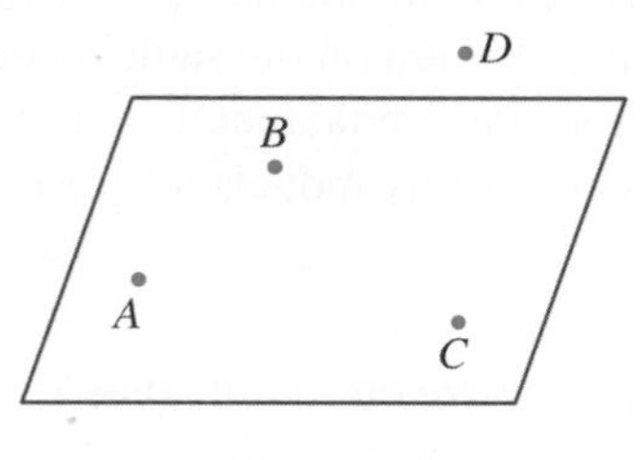

Figure A.5

Remarks

- Axiom A.2 is often encountered in an equivalent form: "Two points determine a unique line" or "There is one and only one line passing through two distinct points."

- Similarly, an equivalent form of Axiom A.3 is "Three noncollinear points determine a unique plane."
- Notice the second sentence in Axioms A.1 and A.2. We know intuitively that lines and planes contain infinitely many points, but this fact does not follow from the preceding axioms. Additional axioms will be needed to assure infinitude of points on a line.

A.1 Notation for Points, Lines, and Planes

It is customary to designate points by capital letters of the Latin alphabet. Axiom A.2 assures that a line can be named by any two points on it. The line containing points A and B, as in Figure A.1, will be denoted by $\overleftrightarrow{AB}$ or line AB. Whenever convenient, we also may name a line by a single letter. In this text, only lowercase letters from the Latin alphabet are used to name lines.

Axiom A.3 assures that a plane can be named by any three noncollinear points on the plane. For example, the plane containing the upper face $ABCD$ of the box in Figure A.6 can be named in each of the following ways: plane ABC, plane BCD, plane ADC, and plane ABD. (Of course, the plane can also be named by any three other noncollinear points not labeled in the figure.) Whenever convenient, we may also name a plane by a single letter. In this text, only lowercase Greek letters will be used to name planes.

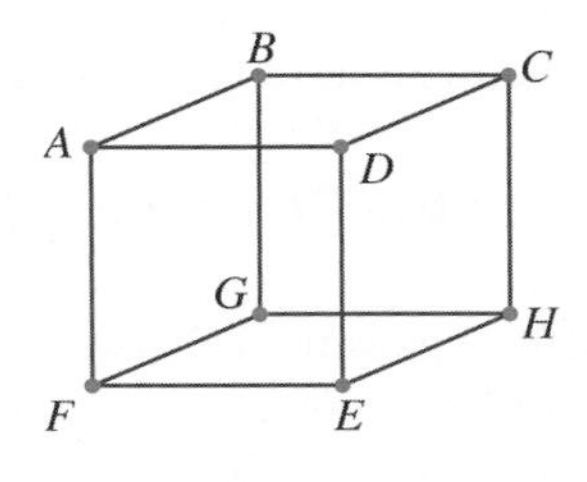

Figure A.6

Recall that the intersection of two sets is the set of all elements that are common to both sets. We know intuitively that two distinct lines either do not intersect (have no points in common) or intersect in exactly one point. This understanding leads to our first theorem.

Theorem A.1

If two distinct lines intersect, they intersect in exactly one point.

Proof

Let the lines be k and ℓ. It is given that the lines intersect. Therefore, there exists a point P that is on both lines. We want to show that k and ℓ have no other points in common. For that purpose, we use an indirect proof. Suppose there is another point Q on k and ℓ, as in Figure A.7.

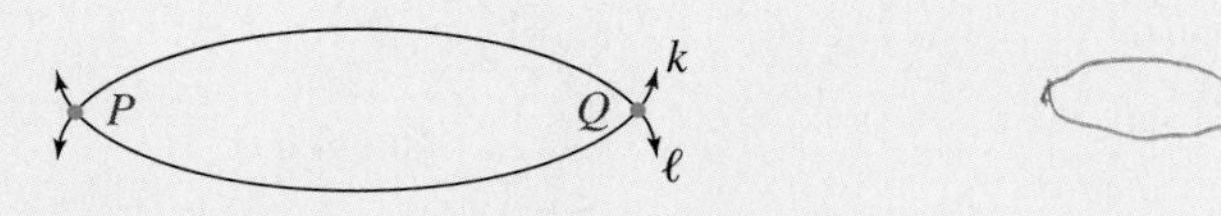

Figure A.7

By Axiom A.2, there is a unique line through P and Q. Hence, $k = \ell$, which contradicts the hypothesis that the lines are different. Consequently, the existence of another intersection point Q must be rejected. □

Axiom A.3 assured us that three noncollinear points determine a plane. The thoughtful reader might set his or her mind to work to see if he or she could come up with another way of uniquely determining a plane. In fact, you may have come up with another way already—namely, "two parallel lines uniquely determine a plane." Before the concept has any meaning in our system, however, we must formally define parallel lines.

Definition of Parallel and Skew Lines Two lines are parallel if they lie in the same plane and do not intersect. Lines that do not intersect and are not contained in any single plane are called skew lines.

If ℓ and m are parallel, we write $\ell \parallel m$. In Figure A.6, for example, $\overleftrightarrow{AB} \parallel \overleftrightarrow{DC}$, $\overleftrightarrow{AB} \parallel \overleftrightarrow{EH}$, and $\overleftrightarrow{AC} \parallel \overleftrightarrow{FH}$, but $\overleftrightarrow{AB}$ and $\overleftrightarrow{DE}$ are skew lines.

Now Solve This A.1

1. What is the maximum number of intersection points determined by n lines in the same plane?
2. What is the maximum number of lines determined by n points? Does it matter if the points are in the same plane?

A.2 Intuitive Background for the Coordinate System and Distance

Following G. D. Birkhoff's (1884–1944) axioms, we now introduce the concept of distance, assuming the existence and properties of real numbers. We start with an intuitive background that will motivate the axioms and definitions that follow.

Historical Note: George David Birkhoff

George David Birkhoff was one of the most distinguished leaders in American mathematics and the preeminent U.S. mathematician of his time. He taught at Harvard from 1912 until his death in 1944. Birkhoff proposed an axiom system for Euclidean geometry in his 1941 text *Basic Geometry*.

Given any line, a **coordinate system** on the line can be created by choosing an arbitrary point O on the line and having that point correspond to 0. This point O is called the **origin.** Next to the right of point O, another point P is chosen (see Figure A.8) to which corresponds the number 1. The segment $\overline{OP}$ is called a **unit** segment.

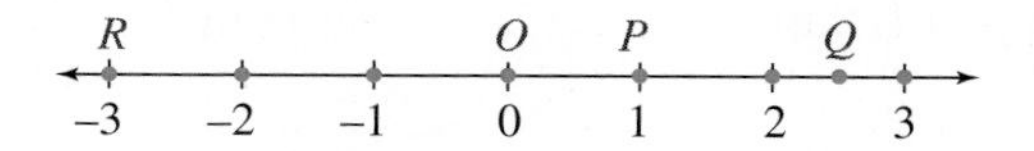

Figure A.8

By marking off segments equal to the length of OP repeatedly to the left and right of O, we find points corresponding to the integers. By dividing segments into an appropriate number of equal parts, we find points that correspond to all rational numbers. You most likely know that any real number (and not only rational numbers) corresponds to some point on the line and, conversely, that every point on the line corresponds to some real number. (In Chapter 3, we have shown how to find points that correspond to real numbers such as $\sqrt{2}$ and $\sqrt{5}$.) Thus there is one-to-one correspondence between the points on a line and the real numbers. Such a correspondence is called a **coordinate system** for a line. The number corresponding to a given point P is called the **coordinate** of P. Thus the coordinate of Q in Figure A.8 is 2.5.

If we denote the line in Figure A.8 by x, we write the coordinate of Q as x_Q. Thus $x_Q = 2.5$ and $x_R = -3$. We can find the distance between two points by using the coordinates of the points. For example, in Figure A.8 we have $PQ = OQ - OP = x_Q - x_P = 2.5 - 1 = 1.5$. We can also find RP by finding the difference between the coordinates of the points: $RP = x_P - x_R = 1 - (-3) = 4$. Because distance is a non-negative number and it is cumbersome to indicate which point has the greater coordinate, we use the absolute value function. Thus $AB = |x_A - x_B|$.

Based on this discussion we introduce the following axiom and definitions.

Axiom A.7 The Ruler Postulate *The points on a line can be put in one-to-one correspondence with the real numbers.*

Notice that this axiom implies that every line has an infinite number of points.

Definition of a Coordinate System for a Line The correspondence in Axiom A.7 is called a coordinate system for a line. A line with a coordinate system is called a number line.

Definition of a Distance Between Two Points The distance between points A and B, denoted by AB, is the real number $|x_A - x_B|$, where x_A and x_B are the coordinates of A and B, respectively, in a coordinate system for $\overleftrightarrow{AB}$.

You may have already observed that the distance between two points depends on the unit chosen for the coordinate system. If Q and R stay in the same place but we change the position of the point that corresponds to the number 1, then QR will change as well. Also, because the distance from a point A to the origin is the real number $|x_A|$, and there exist real numbers as large as we might wish, we can conclude that there are points on a line as far from the origin as we wish. Therefore, we can say that a line is infinite in length.

We can use the concept of distance to define what we mean when we say that a point is between two other points.

Definition of Betweenness B is between A and C if and only if A, B, and C are collinear and $AB + BC = AC$ (see Figure A.9). In this case we write A–B–C.

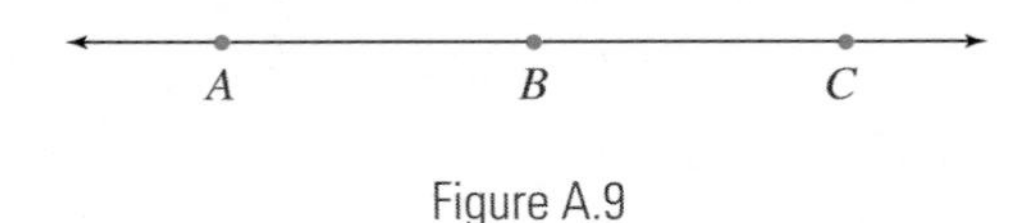

Figure A.9

Using the concept of betweenness for points, it is possible to define various geometric figures.

Definition of a Segment The segment $\overline{AB}$ consists of the points A and B and all the points between A and B.

The length of segment $\overline{AB}$ is the distance between A and B, denoted by AB. A point M between A and B such that $AM = MB$ is a **midpoint** of $\overline{AB}$. Our intuition tells us that every segment has exactly one midpoint. We can prove this fact by using Axiom A.7 and our definitions. You will be guided in the process of finding a proof in the problem set at the end of this appendix. Most of us probably find it more satisfying to prove statements that are not intuitively obvious. Nevertheless, you may find the challenge of proving a statement rewarding in itself even if the statement seems obvious. We also want you to realize that even intuitively obvious statements can be logically deduced from the axioms and definitions.

Example A.1 Given two points A and B on a number line with coordinates x_A and x_B, respectively, find the coordinate x_M of M, the midpoint of $\overline{AB}$. (See Figure A.10.)

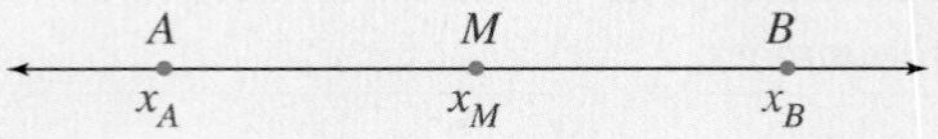

Figure A.10

Solution

Assume $x_B > x_A$. The definition of the midpoint implies that $AM = MB$. This equation implies

$$x_M - x_A = x_B - x_M$$

$$2x_M = x_A + x_B$$

$$x_M = \frac{(x_A + x_B)}{2}$$

Now Solve This A.2

A student approaches the solution of Example A.1 as follows: Because $AB = x_B - x_A$ and the midpoint of AB is halfway between A and B, the coordinate of the midpoint should be $\frac{1}{2}(x_B - x_A)$. The student realizes the answer is wrong but would like to know why and how to use her approach to obtain the correct answer. How would you respond?

Definition of a Ray The ray $\overrightarrow{AB}$ (shown in Figure A.11a) is the union of $\overline{AB}$ and the set of all points C such that B is between A and C. The point A is called the endpoint of the ray. The rays having a common endpoint and whose union is a straight line are opposite rays. In Figure A.11b, $\overrightarrow{AC}$ and $\overrightarrow{AB}$ are opposite rays.

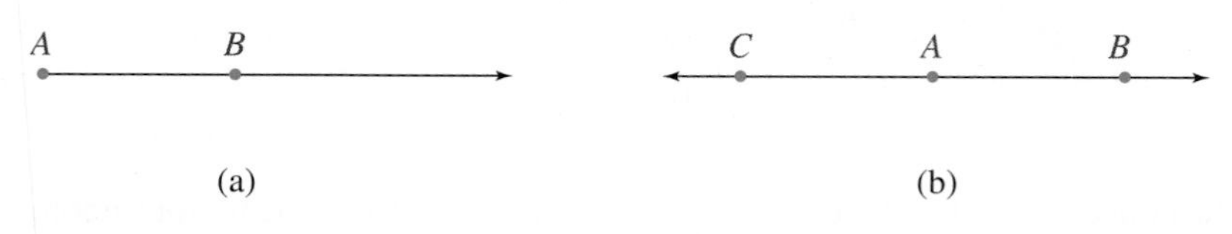

Figure A.11

Definition of an Angle An angle is a union of two rays with a common endpoint. The common endpoint is the vertex of the angle, and the two rays are called the sides of the angle.

In Figure A.12a, the angle shown is the union of $\overrightarrow{AB}$ and $\overrightarrow{AC}$ and its vertex is A. (Using set notation, the angle is $\overrightarrow{AB} \cup \overrightarrow{AC}$.) The angle in Figure A.12a is designated by $\angle BAC$ or $\angle CAB$. When there is no danger of ambiguity, it is common practice to name an angle by its vertex. Thus the angle in Figure A.12a can also be denoted by $\angle A$. If the rays $\overrightarrow{AB}$ and $\overrightarrow{AC}$ are on the same line—that is, if A, B, and C are collinear, as in Figure A.12b—then $\angle ABC$ is called a **straight angle**.

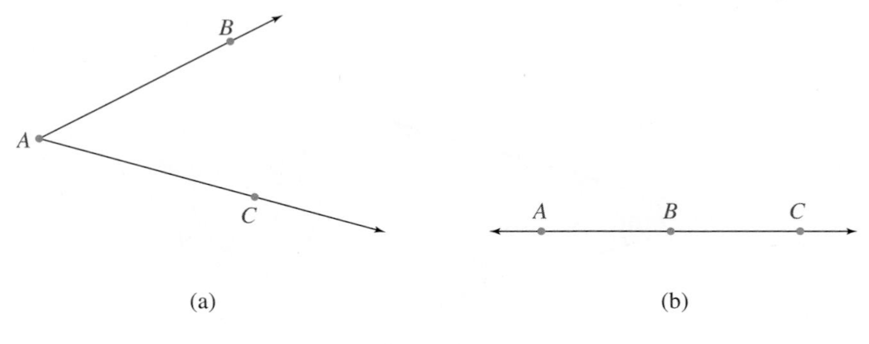

Figure A.12

If A, B, and C are three noncollinear points, then the union of the three segments $\overline{AB}$, $\overline{BC}$, and $\overline{AC}$ is called a **triangle** and is denoted by $\triangle ABC$ (shown in Figure A.13). The three segments are called the **sides** of the triangle, and the three points are called the **vertices**.

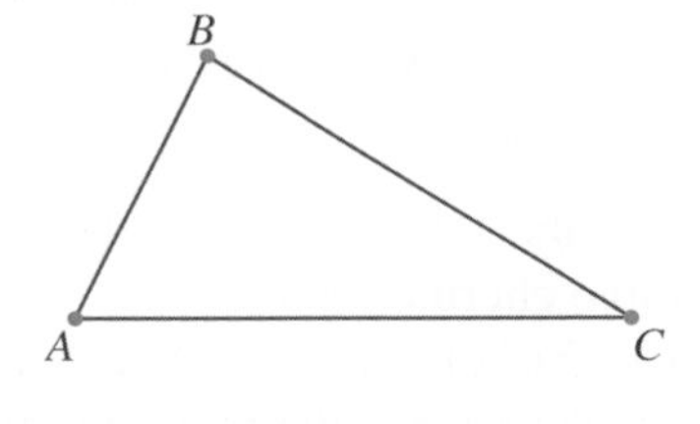

Figure A.13

At this juncture, we could prove several statements that are intuitively obvious. For example, if A–B–C (B is between A and C), then C–B–A (B is between C and A). Also, $AB = BA$ and $\triangle ABC = \triangle CBA$. By "equal," we mean "exactly" equal in the set theory sense: Two sets are equal if they contain the same elements. To give a rigorous treatment of the next topics we will discuss, we need the concept and properties of half planes. Intuitively, we know that any line divides the plane into two parts separated by the line and that each part is referred to as the **half plane**. This fact will be introduced in Axiom A.8, the Plane-Separation Axiom. Before we get to this axiom, however, it will be useful to define what we mean by a convex set.

Definition of a Convex Set A set is convex if for every two points P and Q belonging to the set, the entire segment $\overline{PQ}$ is in the set.

Notice that the interiors of the triangle as well as the circle in Figure A.14 are convex sets. (We are making this statement on an intuitive basis; the interior of a triangle has not been defined yet.) Also, the segment $\overline{AB}$ is a convex set. However, the interiors of the figures in Figure A.15 are not convex, as the segment $\overline{PQ}$ in each figure does not entirely belong to the figure.

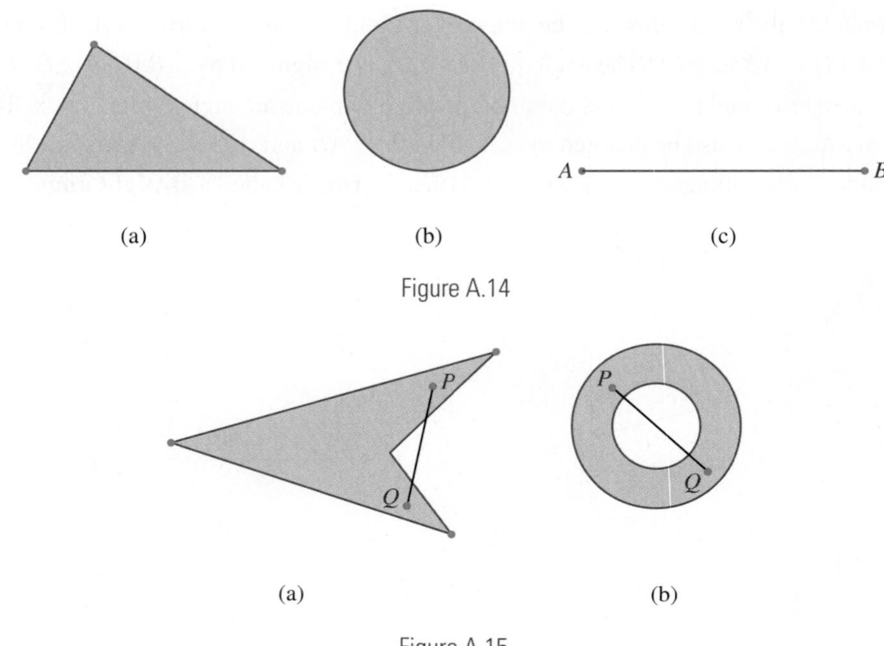

Figure A.14

Figure A.15

Problems Solved and Unsolved

Properties of convex sets have been extensively investigated, and the concept of convex sets has important applications in mathematics. We briefly describe here two properties of convex sets, one that has been proved and one that has not. We use terminology that should be intuitively clear but has not yet been precisely defined.

Chords intersecting at 60°: In the interior of any closed convex curve (see Figure A.16), there exists a point P and three chords through P (a chord of a region is a segment connecting any two boundary points of the region) with the following property: The six angles formed at P are 60° each and P is the midpoint of each of the chords. (The proof of this statement requires more extensive study of a convex sets. Figure A.16 illustrates the statement for a circle.)

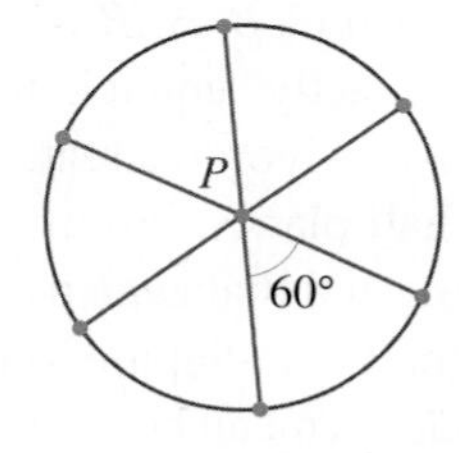

Figure A.16

Equichordal points: A point P in the interior of a region is an equichordal point if all chords through P are of the same length. For example, the center of a circle is an equichordal point of the circle. However, there exist noncircular regions that have an equichordal point. In 1916, Fujiwara raised the question of whether there exists a plane convex region that has two equichordal points. No one has been able to give a complete answer to this question. In 1984, Spaltenstein described a construction of a convex region on a sphere that has two equichordal points. (This does not answer the question posed in 1916, as the region is on a sphere and hence is not planar.)

Axiom A.8 The Plane-Separation Axiom *Each line in a plane separates all the points of the plane that are not on the line into two nonempty sets, called the half planes, with the following properties:*

1. *The half planes are disjoint (have no points in common) convex sets.*
2. *If P is in one half plane and Q is in the other half plane, the segment $\overline{PQ}$ intersects the line that separates the plane.*

Notice that neither of the half planes in Axiom A.8 includes the line. Thus a line divides the plane into three mutually disjoint subsets: the two half planes and the line. Also, it follows from Axiom A.8 that a half plane is determined by a line and a point not on the line. Thus we can refer to the two half planes in Figure A.17 as "the half plane of ℓ containing P" and "the half plane of ℓ containing Q."

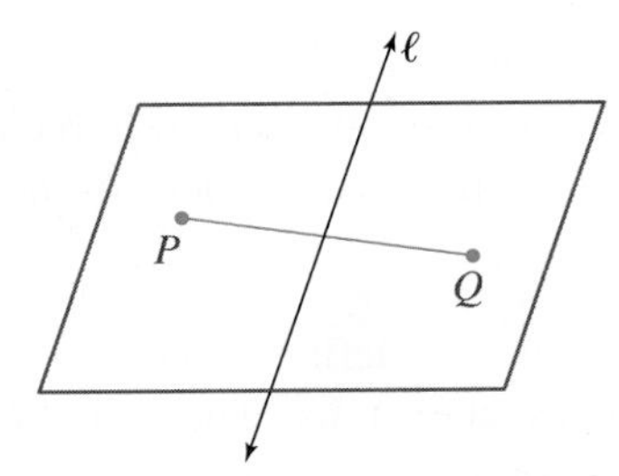

Figure A.17

Using Axiom A.8, it is possible to prove a theorem named after the German mathematician Moritz Pasch (1843–1930), which states that if a line intersects one side of a triangle and does not go through any of its vertices, it must also intersect another side of the triangle (see Figure A.18). In 1882, Pasch published one of the first rigorous treatises on geometry where he stated this theorem as an axiom (he did not use Axiom A.8 as an axiom). Pasch realized that Euclid often relied on assumptions made visually from diagrams and contributed to filling the gaps in Euclid's reasoning.

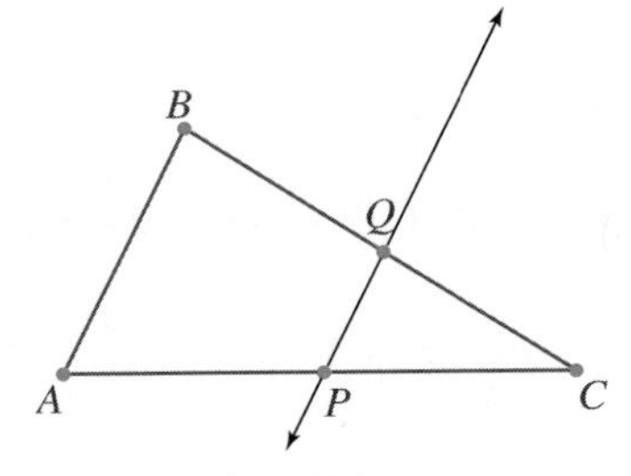

Figure A.18

We state Pasch's Axiom as a theorem and leave its proof for you to explore in the problem set at the end of this appendix.

Theorem A.2

Pasch's Axiom

If a line intersects a side of a triangle and does not intersect any of the vertices, it also intersects another side of the triangle.

Using Axiom A.8, it becomes possible to precisely define the interior of an angle and hence the interior of a triangle.

Definition of the Interior of an Angle If A, B, and C are not collinear, then the interior of $\angle BAC$ is the intersection of the half plane of $\overleftrightarrow{AB}$ containing C with the half plane of $\overleftrightarrow{AC}$ containing B. (See Figure A.19.)

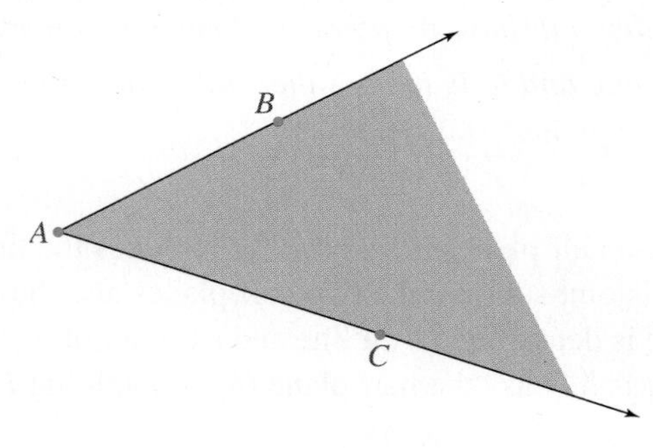

Figure A.19

Remark This definition is valid (or, as it is commonly referred to in mathematics, **well defined**) if it is independent of the choices of B and C on the sides of the angle. It can be proved that this indeed is the case.

Using the preceding definition, we can define the interior of a triangle. Coming up with an appropriate definition is left to you as an exercise. The definition of the interior of an angle can also be used to define **betweenness** for rays.

Definition of Betweenness of Rays $\overrightarrow{AD}$ is between $\overrightarrow{AB}$ and $\overrightarrow{AC}$ and if and only if $\overrightarrow{AB}$ and $\overrightarrow{AC}$ are not opposite rays and D is in the interior of $\angle BAC$. (See Figure A.20.)

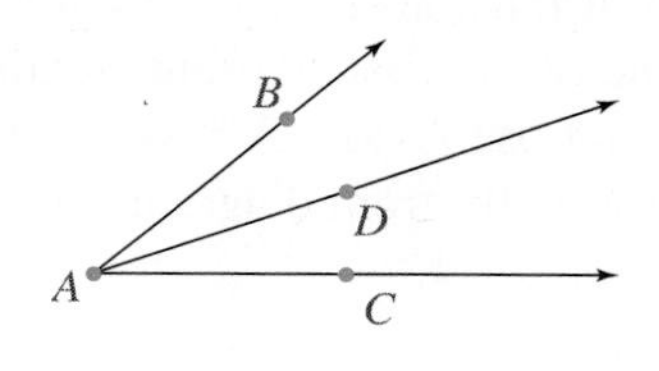

Figure A.20

We can use this definition to prove the following visually obvious theorem. (Its proof is also left as an exercise for you.)

Theorem A.3

$\overrightarrow{CD}$ intersects side $\overline{AB}$ of $\triangle ABC$ between A and B if and only if $\overrightarrow{CD}$ is between $\overrightarrow{CA}$ and $\overrightarrow{CB}$. (See Figure A.21.*)*

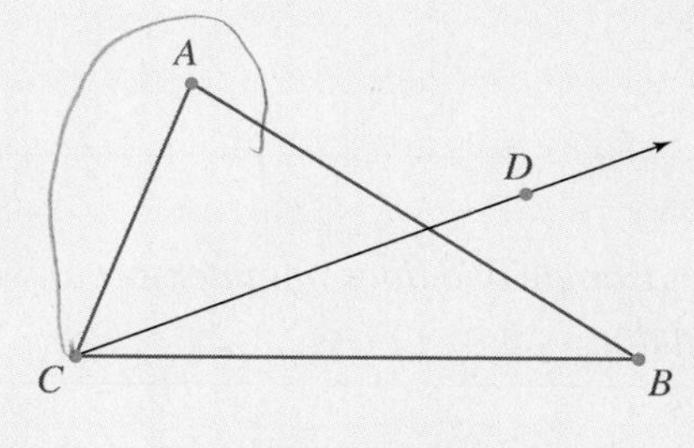

Figure A.21

A.3 Angle Measurement

Angles are commonly measured in degrees with a protractor. (Another unit of measurement for angles is the **radian**.) To measure $\angle EAB$, we place the protractor as shown in Figure A.22 and read off the measure of the angle as 140°.

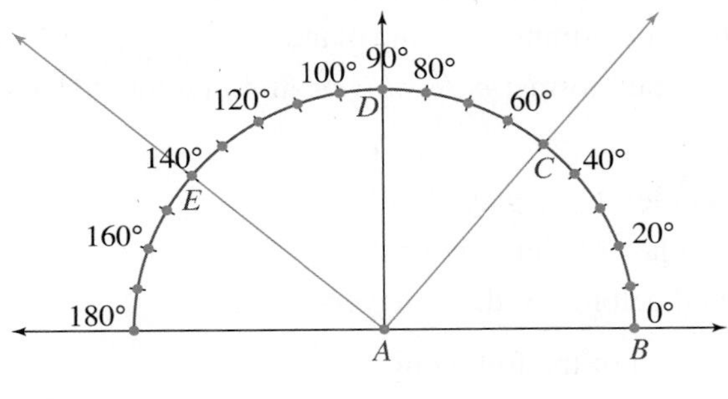

Figure A.22

The following axioms and definitions formalize our intuitive knowledge of the protractor.

Axiom A.9 The Angle Measurement Axiom *There is a real number between* 0 *and* 180 *that corresponds to every angle* $\angle BAC$. *The number* 180 *corresponds to the straight line.*

The number in Axiom A.9 is called the **degree measure of the angle** and is written $m(\angle BAC)$. In Figure A.22, $m(\angle BAC) = 50°$ and we say that $\angle BAC$ is a 50-degree angle, written as 50°.

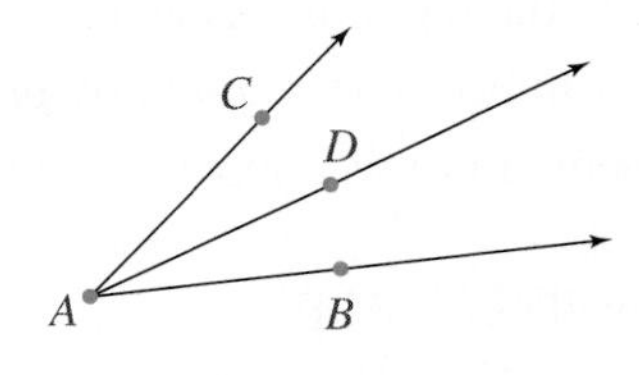

Figure A.23

Axiom A.10 The Angle Construction Postulate *Let* $\overrightarrow{AB}$ *be a ray on the edge of a half plane. For every real number r,* $0 < r < 180$, *there is exactly one ray, with C in the half plane, such that* $m(\angle CAB) = r$.

Axiom A.11 The Angle Addition Postulate *If D is a point in the interior of* $\angle BAC$ *(see* Figure A.23*), then* $m(\angle BAD) + m(\angle DAC) = m(\angle BAC)$.

Notice that Axiom A.11 implies that measures of angles can be computed by subtraction. For example, $m(\angle BAD) = m(\angle BAC) - m(\angle DAC)$.

Problem Set

In the following problems, use the axioms, definitions, and theorems presented in this appendix to prove what is required.

•**1.** Prove that

a. If two lines intersect, they lie in exactly one plane.

b. Two parallel lines determine a unique plane.

2. Consider the undefined terms: *ball*, *player*, and *belongs to*. Also consider the following axioms:

Axiom 1 There is at least one player.

Axiom 2 To every player belong two balls.

Axiom 3 Every ball belongs to three players.

Prove or disprove each of the following:

a. There are at least two balls.

b. There are at least three players.

c. There is always an even number of balls.

d. The number of players is always odd.

(*Hint:* You may want to model the balls by a point and players by segments.)

3. Consider the undefined terms: *line*, *point*, and *belongs to*. Also consider the following axioms:

Axiom 1 Any two lines have exactly one point in common.

Axiom 2 Every point is on (belongs to) exactly two lines.

Axiom 3 There are exactly four lines.

Answer each of the following:

a. How many points are there? Prove your answer.

b. Prove that there are exactly three points on each line.

(*Hint:* You may prefer to substitute the term *point* by *person* and *line* by *committee*.)

4. Using the terminology presented in this appendix, come up with a precise definition of the interior of a triangle.

5. Is the *betweenness* of *rays* well defined? Explain.

•**6.** Which of the following is true *always*, *sometimes* but not *always*, or *never?* Justify your answers.

•**a.** The intersection of two convex sets is convex.

b. The intersection of two nonconvex sets is a nonconvex set.

•**c.** The union of two convex sets is convex.

d. The union of two nonconvex sets is nonconvex.

7. Prove Pasch's Axiom.

Selected Hints and Answers

Chapter 0 (Prologue)

Problem Set 0 (pages 6–7)

1. **a.** Conjecture: No matter where the gallows is chosen, the treasure is in the same location.

   b. Yes, if the directions of turning right and left are strictly observed.

   c. The treasure is at the midpoint of the hypotenuse of the isosceles right triangle, with vertices at the trees and the spike corresponding to the banana tree, and right angle at the banana tree. (See figure.)

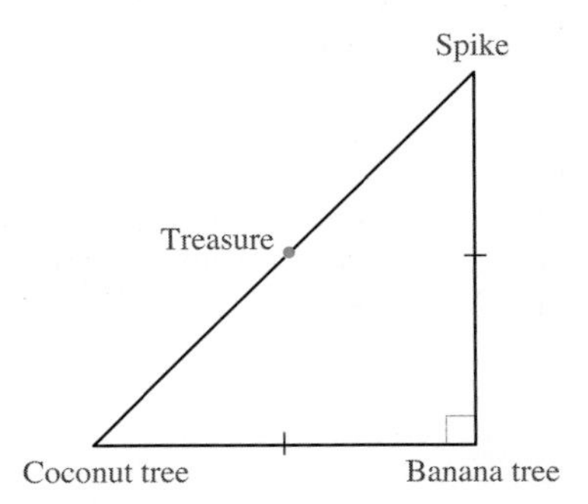

   d. The treasure is on the perpendicular bisector of the segment connecting the trees at the point T "above" the trees, which is half the distance between the trees from the line connecting the trees. Notice that it is simpler to describe the treasure using the concept of rotation: If Γ is at the midpoint of the segment connecting the trees, then the treasure T can be obtained by rotating the point corresponding to the coconut tree $90°$ clockwise about Γ. (See the figure on the next page.)

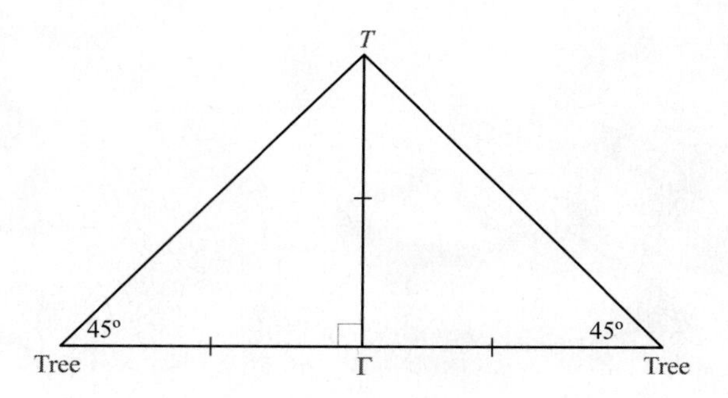

e. Answers vary.

2. **a.** Fold the circle onto itself. Each folding will determine a crease which will contain the center. Folding the circle twice (and unfolding after the first folding) will determine two creases whose intersection is the center.

b. Pick two points on the arc and fold the arc so that one point will fall on top of the other. The folding will determine a crease. Repeat the folding for two different points (it is sufficient that one point will be different). The intersection of the two creases determines the center. Notice that the folding may be easier and more accurate if the chosen points are first connected by a segment or a line. Then we need only fold the segment (or line) onto itself so that one of the points will fall on top of the other.

3. The segments are perpendicular to each other and congruent.

In Problems **4** and **5**, GSP is very useful.

6. The white right triangles with sides a, b, and c are all congruent. The shaded quadrilateral in the third drawing of Figure 0.8 is a square (This can be easily proved by showing that the measures of the two angles with a common vertex in every pair of neighboring triangles add up to 90°). The area of the two squares in the second figure, which is $a^2 + b^2$, equals the area of the square with side $a + b$ minus the area of the four right triangles. However, from the third figure we see that the area of the shaded square, which is c^2, also equals the area of the square with side $a + b$ minus the area of the four right triangles.

Chapter 1

Problem Set 1.2 (pages 37–40)

16. Use the fact that the diagonals are perpendicular bisectors of each other.

21. Prove that if $x > 0$ and $y > 0$, then $\sqrt{x} + \sqrt{y} > \sqrt{x+y}$.

Problem Set 1.3 (pages 56–63)

7. $m(\angle \mathrm{B}) = 120°$, $m(\angle \mathrm{D}) = 60°$

8. $m(\angle \mathrm{D}) = 90° + \dfrac{\alpha}{2}$

10. **c.** Bisect each other and are congruent.

18. That its diagonals are congruent and perpendicular. (They don't need to bisect each other.)

24. **a.** Construct the circumscribing circle.

32. The sum of the distances from any point on an equilateral triangle or in its interior is a constant equal to the height of the triangle.

31. Draw more squares and fit $x + y$ in an isosceles right triangle.

Problem Set 1.4 (pages 69–72)

6. Bisect $\angle A$.
9. Make the segment a median in a triangle.
12. In $\triangle ABC$, let P be the point of intersection of the medians. Construct $\triangle APC$
16. Show that the sides of the square must be parallel to the diagonals of the rhombus.
17. In your investigation, place point X on $\overline{PQ}$ so that $PX = AP$. Then show that $QX = QC$.
20. **a.** From G draw the perpendicular to $\overline{CD}$ and from F the perpendicular to $\overline{AD}$. Prove that the triangles created are congruent.

Chapter 2

Problem Set 2.1 (pages 87–90)

7. (a) $x = 120^\circ$, (b) $x = 80^\circ$, (c) $x = 80^\circ$, but z can take any value such that $0^\circ < z < 90^\circ$. For each z, $y = 180^\circ - z$. Reason: draw a diameter and construct an inscribed angle z $(0 < z < 90^\circ)$. There are infinitely many such values of z. For each construct a 100° inscribed angle whose vertex is the point on the circle where the side of z which is not the diameter intersects the circle.
8. $ABED$ is cyclic. Show that $m(\angle D) + m(\angle B) = 180^\circ$ by showing that $\angle B$ is congruent to an angle formed by the tangent m and chord $\overline{AC}$.

Problem Set 2.2 (pages 98–104)

1. Write $m(\angle ADE)$ as a sum of two angles and show that $m(\angle DAE)$ equals that sum.
4. Notice that the distance from O to each chord is the radius of the smaller circle.
6. Let O be the center of the circle. Show that $m(\angle AOB) = 180^\circ - \alpha$.
8. Show that $\overline{AQ}$ is a diameter.
11. **a.** There are several ways to prove this, one is to show that the distance from M to $\overleftrightarrow{AB}$ is the same as the distance from M to $\overleftrightarrow{CD}$. To justify this, construct a perpendicular from M to the bases of the trapezoid and prove that the two right triangles created are congruent. Hence each distance is half the height of the trapezoid. That height equals the diameter of the circle. Thus the distance from M to each base is the radius of the circle. Because the distance from O to $\overline{AB}$ also equals the radius, O must be on $\overleftrightarrow{MN}$.
b. Prove that $\angle BON \cong \angle OBN$.

Problem Set 2.3 (pages 112–113)

4. **d.** Let ABC be the required triangle and O its incenter. First show that $m(\angle BOC) = 90 + \frac{\alpha}{2}$. Then use Construction 2.2 to construct $\triangle ABC$.
6. Find α by constructing the circle that circumscribes the triangle.
7. Construct the circle through Q, C, and S and use the fact that $\angle ACS$ measures 45°. Thus locate the point where the diagonal $\overline{AC}$ intersects the circle.

Chapter 3

Problem Set 3.1 (pages 125–130)

4. **a.** $\frac{1}{2}\, d_1 d_2$

6. $\dfrac{a^2 - b^2}{4}$

10. **a.** Each equals half the area of the parallelogram.

14. Extend the sides of square B that contain the center of square A.

16. Try an approach similar to that used in Example 3.1 on page 123.

Problem Set 3.2 (pages 140–145)

4. **a.** Let a, b, c, and d be the distances from O to the four sides of the rectangle. Express $x^2 + z^2$ as well as $y^2 + w^2$ in terms of these distances.

7. **a.** **(i)** Write 19 as $4^2 + (\sqrt{3})^2$ or $(3\sqrt{2})^2 + 1^2$ or $10^2 - 9^2$.

(ii) Use $\sqrt{\dfrac{2}{3}} = \dfrac{\sqrt{6}}{3}$ and $6 = 2^2 + (\sqrt{2})^2$.

14. **a.** Show first that $2h^2 = a^2 + b^2 - (a_1^2 + b_1^2)$.

15. **a.** $\dfrac{a(\sqrt{3} - 1)}{4}$

16. **a.** Let the legs of the right triangles be c and b where $c > b$. Prove that the diameter of the incircle is $c + b - a$. Also prove that the side of the smaller square is $c - b$.

Problem Set 3.3 (pages 152–154)

4. One approach is to show that if $m_1 m_2 = -1$, then referring to the figure in Problem 3, $(AB)^2 = (OA)^2 + (OB)^2$. Another approach is to consider lines ℓ_1 and ℓ_2 whose respective slopes are m_1 and m_2. Given that $m_1 m_2 = -1$, the lines cannot be parallel. Why? Let P be their point of intersection and let ℓ_2^* be the line through P perpendicular to ℓ_1. Then from Problem 3, $m_2^* = \dfrac{-1}{m_1}$. Since $m_1 m_2 = -1$, $m_2 = \dfrac{-1}{m_1}$. Hence $\ell_2^* \parallel \ell_1$. Because ℓ_1 and ℓ_2 go through the same point P, $\ell_2^* = \ell_1$.

8. Perhaps the simplest approach is to compare the given equation to the equation of a circle. A necessary condition for the given equation to be an equation of a circle is that $A = B \neq 0$. Next, divide both sides of the given equation by A (or by B) and complete the squares. Answer: $C^2 + D^2 > 4AE$

12. $x_M = \dfrac{x_1 + y_1}{2}$, $y_M = \dfrac{y_1 + y_2}{2}$

13. Use Problem 12 and the fact that the diagonals of a parallelogram bisect each other. There are three possible locations for the 4th vertex.

Chapter 4

Problem Set 4.1 (pages 171–176)

1. $x = \dfrac{8}{3}$, y cannot be uniquely determined.

5. **a.** Show that $\dfrac{PS}{BT} = \dfrac{AS}{AT}$ and $\dfrac{AS}{AT} = \dfrac{SQ}{TC}$

b. $PS = SQ$

6. Use Example 4.2 (page 167) and Problem 5 above.

9. **a.** Use the fact that the diagonals of a rectangle bisect each other.

b. $FE = 2\sqrt{ab}$

11. If $AC = b$ and the height to $\overline{AC}$ is h then the length of the side of the square is $\dfrac{bh}{b + h}$.

14. **a.** First prove that each side of a triangle (starting from the second triangle on) is half the side of the previous triangle.

b. If the side of $\triangle ABC$ is a then the length of the path is also a.

19. **a.** The slope of $\overline{AB}$ is $\dfrac{y_B - y_A}{x_B - x_A}$. Show that a point $P(x, y)$ is on the line AB if the slope of $\overline{PA}$ equals the slope of $\overline{AB}$.

Problem Set 4.2 (pages 185–191)

1. **a.** $\dfrac{6}{13}$

b. $\dfrac{2m(c-a)}{3(2a+c)}$

4. **a.** $\dfrac{a+c}{b}$

8. **a.** $a\sqrt{2}$

11. $\dfrac{2a^2}{\sqrt{a^2+b^2}}$

16. Assume that $\overleftrightarrow{AC}$ intersects one circle at B_1 and the other at B_2. Then $(CD)^2 = (AC)(CB_1)$ and $(CE)^2 = (AC)(CB_2)$.

19. **a.** $\dfrac{d}{5}$

Problem Set 4.3 (pages 197–198)

2. $\dfrac{2\sqrt{S_2(S_1+S_2)}}{\sqrt[4]{4S_1^2 - S_2^2}}$

5. **a.** $\left(\dfrac{3}{4}\right)^5$

7. Use the Pythagorean Theorem repeatedly in several right triangles including triangles with the diagonals as hypotenuses.

Problem Set 4.4 (pages 203–206)

5. $a(\dfrac{\sqrt{5}+1}{2})$

7. Use the fact that we know how to construct arcs (or central angles) that are $\dfrac{1}{3}$ of a circle as well as $\dfrac{1}{5}$ of a circle and that $\dfrac{1}{3} - \dfrac{1}{5} = \dfrac{2}{15}$.

9. **b.** $\dfrac{1}{\phi^4}$ or $\dfrac{7-3\sqrt{5}}{2}$

11. **c.** The sequence approaches ϕ.

d. Divide both sides of the equation by F_{n-1}, and take a limit of each side.

13. The ratio is not fixed; it is a function of $\angle A$. It equals $\dfrac{1}{2\cos A}$.

Problem Set 4.5 (pages 212–215)

1. **b.** $2\pi r$

c. $2\pi r$

3. **b.** Approximately 21.5%

9. $(4.5\sqrt{3} - 2\pi)a^2$

11. Assume that the three circles are concentric, that the original circle has radius R, the second largest circle has radius R_1, and the smallest has radius R^2. First show that $R_1 = \sqrt{\frac{2}{3}}R$ and $R_2 = \frac{\sqrt{3}}{3}R$.

Problem Set 4.7 (pages 234–236)

2. **a.** **(i)** αr **(ii)** $\frac{\pi\alpha}{180}r$

4. **a.** First prove that $\tan(\alpha+\beta) = \frac{\tan\alpha + \tan\beta}{1-\tan\alpha\tan\beta}$.

5. **c.** Use the hint to 4(a) and show that $\tan(\pi-\gamma) = -\tan\gamma$.

9. **a.** $\frac{\sqrt{6}-\sqrt{2}}{4}$

11. Let a be the length of a side of the square, show that $AE = \frac{a}{2\cos 15°}$ and then use the Law of Cosines in $\triangle ADE$ to show that $DE = a$.

13. Use part(a) to show that the equation of the line ℓ in the figure is $x\cos\theta + y\sin\theta - d = 0$. The equation $Ax+By+C=0$ will be in the above form if and only if $\frac{cos\theta}{A} = \frac{sin\theta}{B} = \frac{-d}{C}$. Hence conclude that $d = \frac{|C|}{\sqrt{A^2+B^2}}$.

Chapter 5

Problem Set 5.1 (pages 248–255)

4. **c.** n lines of symmetry. Distinguish two cases: n even and n odd.

6. **a.** r' is parallel to r.

7. **e.** Consider first the case when one of the sides of the triangle is parallel to ℓ.

g. $x' = \frac{2y}{m} - x$ and $y' = y$, where $m = \tan\alpha$.

i. Use Law of Sines and Law of Cosines.

10. **c.** P. The composition is the identity.

d. τ_{MN}.

11. **a.** $\tau_{AB} \circ \tau_{BC} = \tau_{AC}$.

13. **e.** $\ell' \parallel \ell$

16. **a.** Domain: $\mathbb{R}^2$ i.e., the entire plane. Range: $\{(x,y) \mid y \geqslant 1 \text{ or } y < 1\}$, i.e., the plane excluding the strip $-1 \leqslant y < 1$.

20. **a.** **(i)** $y = a(x-h)^2 + k$

(ii) $y = -ax^2$

(iii) $y = ax^2$

(iv) $ay^2 = x$

21. **e.** Reflections in the lines $y=0, x=0, y=x, y=-x$, rotation about the origin by $\frac{\pi}{2}$, π, $\frac{3\pi}{2}$ and 2π (the identity).

g. Translation by any integer multiple of 2π and glide-reflection consisting of translation by the vector OP where O is the origin and $P(\pi,0)$ followed by a reflection in the x-axis.

Problem Set 5.2 (pages 261–265)

1. **b.** Show that the perimeter of $PXYZ$ equals the length of a path connecting P' and P'' and consisting of four segments.

d. Use the answer to Problem 6(a) of Problem Set 5.1 (page 250).

4. Look for those points on the circle whose image under half turn about P is on ℓ. For that purpose find the image of all the points on the circle.
7. Look for those points X with image under τ_{PQ} on street A.

Problem Set 5.3 (pages 273–275)

3. **c.** The length of the path is twice the perimeter of the pedal triangle.
5. Use the triangle inequality twice.
7. $a(\sqrt{3}+1) < 2a\sqrt{2}$

Problem Set 5.4 (pages 280–281)

2. **b.** $\frac{S}{5}$

9. $\frac{a}{b} = \frac{a_1}{b_1}$

Chapter 6

Problem Set 6.1 (pages 299–303)

4. See the proof of Theorem 6.1.
5. **a.** At most 2 reflections.
b. At most 1 reflections.
8. **a.** Line x is unique; it is the image of ℓ under rotation about O by θ.
10. Let k, l and m be three parallel lines then

$$M_m \circ M_\ell \circ M_k = M_m \circ (M_\ell \circ M_k)$$
$$= M_m \circ (M_m \circ M_x) = I \circ M_x = M_x,$$

where line x is the image of line m under the translation $M_k \circ M_\ell$ (which is the inverse of $M_\ell \circ M_k$).
13. $P(1,1)$, and the equation of ℓ is $y = x$.
18. Identify line n such that the following holds:

$$(R_{-\theta} \circ M_k) \circ R_\theta = R_{-\theta} \circ M_k \circ (M_k \circ M_n)$$
$$= R_{-\theta} \circ M_n$$
$$= (M_\ell \circ M_n) \circ M_n = M_\ell$$

Problem Set 6.2 (pages 312–316)

3. $ABCO$ is a parallelogram regardless of the location of P. Since O is the image of C under τ_{BA}, O is uniquely determined and hence the same for all P.
4. Think about this problem as the limiting case of Problem 6.4 (page 308), i.e., think about points A, D, and C being "almost" colinear. Let $T_1(P) = P'$, $T_2(P) = P''$, and $T_3(P) = P'''$. Prove that each side of the equation equals P'''.
12. Only the set in (c) is not a group.
14. **a.** Consider a composition of two symmetries. The image of a figure S under the first symmetry is S. The image of S under the second symmetry is also S, so the composition of two symmetries is a symmetry. Similarly argue that an inverse of a symmetry maps the figure onto itself, and the identity transformation is a symmetry. For associativity, use Problem 11.

Chapter 7

Problem Set 7.1 (pages 322–324)

1. Notice that $\overline{N_2M_1}$ is a diameter of the Nine-Point Circle (see Figure 7.1), also $\overline{N_2M_1}$ and $\overline{HO}$ bisect each other at Q (see Figure 7.5).

2. In Figure 7.5, the radius of the Nine-Point Circle is $\frac{1}{2}N_2M_1$ and $\overline{N_2Q}$ is a midsegment in $\triangle AOH$.

Problem Set 7.2 (pages 337–341)

2. Use triangle inequality and notice that $|z_1 + z_2|$ is the distance between points that correspond to z_1 and $-z_2$.

3. Consider a polygon $OA_1A_2...A_n$ where O is the origin and the vectors OA_1, A_1A_2, A_2A_3, ..., $A_{n-1}A_n$ correspond to $z_1 = a_1 + ib_1$, $z_2 = a_2 + ib_2$, ..., $z_n = a_n + ib_n$ respectively. Then notice that the vector OA_n corresponds to $z_1 + z_2 + ... + z_n$.

6. In a complex plane, let C be at the origin, $B(a, 0)$, and $C(0, 1)$. Show that $\overrightarrow{AK} = -1 - i(a + 1)$, $\overrightarrow{AF} = 1 + a - i$ and that $i\overrightarrow{AK} = \overrightarrow{AF}$.

8. One approach is as follows: if $|z_1| = |z_2| = |z_3| = 1$, the points are on a unit circle. If in addition $z_1 + z_2 + z_3 = 0$ then $z_2 + z_3 = -z_1$. Use this to show that the quadrilateral with vertices at $0, z_2, -z_1$, and z_3 must be a parallelogram with all sides equal to the radius of the circle, i.e., 1. Conversely, if z_1, z_2 and z_3 are the vertices of an equilateral triangle inscribed in the unit circle, then clearly $|z_1| = |z_2| = |z_3| = 1$. In addition, show that the triangles with vertices at $0, z_2$ and $-z_1$ as well as $0, -z_1$ and $-z_3$ are equilateral. Hence conclude that $z_2 + z_3 = -z_1$.

12. If $M_\ell(z) = e^{2i\theta}\overline{z}$ then

$$\begin{aligned}(M_\ell \circ M_\ell)(z) &= M_\ell(M_\ell(z)) \\ &= M_e(\ell^{2i\theta}\overline{z}) \\ &= e^{2i\theta} \cdot \overline{e^{2i\theta}\overline{z}} \\ &= e^{2i\theta} \cdot e^{-2i\theta} z \\ &= z\end{aligned}$$

14. $a\overline{b} + b = 0$

15. **a.** $a \neq 1$. The center is $\dfrac{b}{1-a}$

b. $a = 1$

20. **a.** $f^{-1}(z) = \dfrac{1}{a}z - \dfrac{b}{a}$

b. $|a| = 1$ if and only if $a \cdot \overline{a} = 1$

Appendix (page 322)

1. **a.** Let O be the point where the lines intersect. Choose point A on a line and B on the other. These three points determine a unique plane (Axiom A.3) and the intersecting lines lie in this plane. Any other plane that contains the intersecting lines will contain the points O, A and B and hence be the same plane.

b. By definition two parallel lines lie in a plane. Choose a point on one line and two other points on the other line and proceed as in part (a).

6. **a.** Always true.

c. Sometimes but not always true.

References

M. E. Barnes. Morley's Triangle. *Pi Mu Epsilon Journal* 2002;II(6):293–298

George David Birkhoff and Ralph Beatley. *Basic Geometry,* third edition. Providence, RI: AMS Chelsea Publishing, American Mathematical Society, 1940.

H. S. M. Coxeter. *Introduction to Geometry*. New York: Wiley, 1961.

H. S. M. Coxeter and S. L. Greitzer. *Geometry Revisited.* Washington, DC: New Mathematical Library, Mathematical Association of America, 1967.

William Dunham. *Journey Through Genius: The Great Theorems of Mathematics.* New York: Penguin Books, 1991.

George Gamow. *One Two Three ... Infinity: Facts and Speculations of Science.* New York: Dover, 1946.

H. D. Grossman. The Morley Triangle: A New Proof. *American Mathematical Monthly* 1943;50(9):552.

Thomas L. Heath. *Euclid: The Thirteen Books of the Elements.* New York: Dover, 1956.

D. Hilbert and S. Cohn-Vossen. *Geometry and the Imagination.* New York: Chelsea, 1952.

David C. Lay. *College Geometry: A Discovery Approach,* second edition. Boston: Addison-Wesley, 2001.

Edwin E. Moise. *Elementary Geometry from an Advanced Standpoint,* third edition. Reading, MA: Addison-Wesley, 1990.

Alfred S. Posamentier. *Advanced Euclidean Geometry.* Emeryville, CA: Key College Publishing, 2002.

Alfred S. Posamentier and C. T. Salking. *Challenging Problems in Geometry.* New York: Dover, 1996.

Index

Q

R

S

T

V

W

Y